Edison & the Electric Chair

Edison & the Electric Chair

A Story of Light and Death

Mark Essig

First published in the United States of America in 2003 by
Walker Publishing Company, Inc.
McClelland & Stewart edition published in Canada in 2004

National Library of Canada Cataloguing in Publication

Essig, Mark Regan, 1969-
Edison & the electric chair : a story of light and death / Mark Essig.

Includes bibliographical references and index.
ISBN 0-7710-3078-9

1. Edison, Thomas A. (Thomas Alva), 1847-1931. 2. Electrocution – United States – History. I. Title. II. Title: Edison and the electric chair.

HV8696.E88 2004 364.66 C2003-906775-0

We acknowledge the financial support of the Government of Canada through the Book Publishing Industry Development Program and that of the Government of Ontario through the Ontario Media Development Corporation's Ontario Book Initiative. We further acknowledge the support of the Canada Council for the Arts and the Ontario Arts Council for our publishing program.

Book design by Ralph L. Fowler
Printed and bound in Canada

This book is printed on acid-free paper that is 100% recycled, ancient-forest friendly (100% post-consumer recycled).

McClelland & Stewart Ltd.
The Canadian Publishers
481 University Avenue
Toronto, Ontario
M5G 2E9
www.mcclelland.com

1 2 3 4 5 08 07 06 05 04

TO MY PARENTS,
DOROTHY AND JOHN ESSIG

CONTENTS

PROLOGUE
Edison on the Witness Stand 1

CHAPTER 1
Early Sparks 5

CHAPTER 2
The Inventor 16

CHAPTER 3
Light 26

CHAPTER 4
Electricity and Life 40

CHAPTER 5
"Down to the Last Penny" 49

CHAPTER 6
Wiring New York 62

CHAPTER 7
The Hanging Ritual 74

CHAPTER 8
The Death Penalty Commission 85

CHAPTER 9
George Westinghouse and the Rise of Alternating Current 100

CHAPTER 10
The Electrical Execution Law 118

CHAPTER 11
"A Desperate Fight" 134

CHAPTER 12
"Criminal Economy" 148

CHAPTER 13
Condemned 163

CHAPTER 14
Showdown 174

CHAPTER 15
The Unmasking of Harold Brown 190

CHAPTER 16
Pride and Reputation 200

CHAPTER 17
The Electric Wire Panic 212

CHAPTER 18
Designing the Electric Chair 224

CHAPTER 19
The Conversion of William Kemmler 234

CHAPTER 20
The First Experiment 245

CHAPTER 21
After Kemmler 254

CHAPTER 22
The End of the Battle of the Currents 265

CHAPTER 23
The Age of the Electric Chair 277

EPILOGUE
The New Spectacle of Death 286

ACKNOWLEDGMENTS 295

NOTES 297

ART CREDITS 341

INDEX 343

Edison & the Electric Chair

PROLOGUE

Edison on the Witness Stand

"MR. EDISON, what is your calling—your profession?"

"Inventor."

"Have you devoted a good deal of attention to the subject of electricity?"

The hearing room erupted in laughter. It was a standard lawyer's question, intended to establish the qualifications of an expert witness, but it was hardly necessary in this instance. The men who packed the room—lawyers, electricians, doctors, and assorted gawkers—knew very well the qualifications of the man on the stand. In 1879 Thomas Edison had invented the first practical incandescent lamp—the lightbulb—and over the next decade he carried his light into homes and offices around the world. As the world's most celebrated electrical authority, Edison clearly had "devoted a good deal of attention to the subject of electricity."

The inventor took the question in stride. "Yes, sir," he replied.

The date was July 23, 1889, and the lawyer asking the questions was

William Poste, deputy attorney general of New York State. Edison was forty-two years old, his dark hair streaked with white, face smooth-shaven, gray eyes sparkling. In his black suit and white tie, Edison had the aspect, one reporter remarked, of "a benignant clergyman of middle age." No stranger to the American legal system, he sat thoroughly at ease in the witness chair. Inventing was a cutthroat business, and Edison spent a great deal of time dealing with lawyers–suing other companies for stealing his patents, getting sued in turn for stealing theirs. Invention, he might have said, was 1 percent inspiration and 99 percent litigation.[1]

The hearing on this day in July, though, was not concerned with patents. It had to do with a murder case.

"What amount of electrical energy," Poste continued, "do you think would be sufficient to produce instant, painless, death in all cases?"

"One thousand volts," Edison said.

"What experiments have you observed in your laboratory bearing upon that question?"

"Only of horses and dogs."

Edison referred to a series of tests that had taken place over the previous year at his New Jersey laboratory. The inventor's assistants attached electrodes to dogs–about two dozen in all, bought at a quarter a head from neighborhood boys–and killed them with powerful jolts of electricity. Six calves and two horses also died in the experiments.

"Now, Mr. Edison," the attorney said, "in your opinion, can an electrical current be applied to the human body by artificial means in such a manner as to produce death in every case?"

"Yes, sir."

"Instant death?"

"Yes, sir."

"Painless?"

"Yes, sir."[2]

A year and a half earlier New York State had abolished hanging and decreed that condemned criminals would be executed with electricity. The first murderer condemned under the new law filed an appeal, claiming that electrical execution was a cruel and unusual punishment and therefore unconstitutional. A judge ordered hearings to collect expert testimony on the matter, and Edison agreed to testify in support of the new method.

Electricity had long been considered a mysterious, miraculous force—telegraph operators sent it zipping along slender copper wires, showmen amused fairgoers by giving them mild electric shocks, and doctors claimed that the current could cure illness and revive victims of drowning or suffocation. It had even been used to kill: In the 1750s Benjamin Franklin slaughtered chickens and turkeys with static electricity.

But using electricity to execute criminals was unprecedented. One newspaper declared that state officials had been swept up in an electrical craze and were "merely endeavoring to show that there was no end to the wonders of electricity."[3]

Although in 1887 Edison had said he would "join heartily in an effort to totally abolish capital punishment," a year later he became the most powerful advocate of this new method of scientific killing. Like other defenders of the electrical execution law, he claimed that a powerful current would be far more humane than hanging.[4]

Edison's critics, however, believed there was more to the story. In the hallways outside the hearing room and in the pages of newspapers and electrical journals, insiders alleged that Edison's support for electrocution was motivated by a devious scheme to gain control of the electrical industry; that an Edison competitor was spending tens of thousands of dollars to defeat the electrocution law and foil Edison's plans; that the convicted murderer whose life was on the line had become a pawn in a bitter industrial struggle.[5]

If any of these rumors were true, Edison did not let on. Killing with electricity was simply "a good idea," he said. "It will be so lightning-like quick that the criminal can't suffer much."[6]

CHAPTER 1

Early Sparks

THE ANCIENT GREEKS were the first to record the observation that amber, after being rubbed, attracted bits of straw or cloth. Around 1600 the Englishman William Gilbert noted that materials such as diamond and glass shared amber's attractive qualities. He coined a new word, *electric,* based on *elektron*, Greek for amber. An *electric* was a substance that, when rubbed, drew light objects to itself; *electricity* was the property shared by these substances.[1]

After Gilbert the study of electricity languished for a century or so until it was taken up by members of London's Royal Society, a new association devoted to the study of the natural world. Using hollow glass tubes thirty inches long and one inch in diameter, Royal Society members produced the strongest electrical effects ever witnessed. In 1729 Stephen Gray, an experimenter with the society, corked the ends of his tube to keep dust from being sucked inside. After rubbing the glass, he noticed to his surprise that feathers were attracted to the cork as well as to the glass. The "attractive Virtue," as he put it, had been "communicated" from the glass to the cork. Curious to see how far this communication would extend, Gray attached ordinary thread to the cork, tied a shilling to the string, and found that the coin attracted feathers. He extended the string and tied on more objects–a piece of tin, an iron poker, a copper teakettle, various vegetables–and found that all became electrified. Gray attached thirty-two feet of thread to the corked end of the glass tube, tied a billiard ball to the other end of the string, and dangled it out a window. When he rubbed the glass, he found that the billiard ball still proved attractive.

Abandoning a plan to drop a string from the cupola of St. Paul's Cathedral, Gray decided to proceed horizontally. He snaked a long piece of iron wire along the ceiling of his workroom, suspending it from the beams with pieces of string. When he touched the wire with the rubbed glass wand, however, the attractive virtue did not communicate to the far end. Gray thought the string suspenders might be too thick, so he tried silk, which worked beautifully. Equally thin brass, however, failed, leading Gray to conclude that success depended upon the supports "being Silk, and not upon their being small." The differences between silk and brass wire raised the question of which objects could be *supports* and which *receivers*. (Before long another experimenter started calling these two classes *conductors* and *insulators*.) To test the electrical properties of the human body, Gray persuaded an orphan boy to allow himself to be suspended horizontally from the ceiling, supported at his chest and thighs by stout loops of silk. Gray rubbed his glass tube, touched it to the lad's feet, and found that he attracted feathers to his fingers.[2]

Philosophers at the time believed that electricity–as well as light, heat, and magnetism–consisted of exquisitely fine "fluids" that passed through ordinary matter. The electrical phenomena of attraction and repulsion were thought to be caused by jets of subtle fluids blowing into and out of tiny pores in larger objects. The public, however, was less concerned with theories of electricity than with the thrilling effects it produced. Members of polite society in the eighteenth century flocked to scientific lecture-demonstrations, where they learned about planetary motion, the shape of the Earth, and the size of the solar system. Newtonian physics could be a bit dull, but a suspended human body attracting objects to its fingers–that was

In the electrical craze of the 1740s, "human conductors" were sometimes suspended from the ceiling by silk cords and charged with electricity.

magic. Electrical displays swept Europe in the 1740s, and a French entrepreneur sold electrical kits that included a glass wand for rubbing, light objects for attracting, and thick silk cords for hanging human conductors. In darkened rooms lecturers drew sparks—"electrical fire"—from the noses of suspended men.

Experimenters in Germany produced more flamboyant effects. They replaced the glass wand with a spinning globe and used a "rubber" of leather or paper to excite it. They also suspended *prime conductors*—usually a sword or gun barrel—near the globe to collect the charge. Experimenters were soon killing flies with shocks from their fingers and showcasing the "Venus electrificata," a woman whose kisses threw sparks. When a glass of brandy was lifted toward the lips of a charged man, the spark from his nose set the liquor aflame.[3]

Human conductors began to complain that these shocks were unpleasant, but they did not know true pain until they experienced another new device. In 1746 Pieter van Musschenbroek of the University of Leyden attempted to produce electricity with a glass globe and then store it in a jar of water. He attached a wire to the gun barrel that served as his prime conductor and placed the wire's end in a water-filled glass jar. While an assistant spun and rubbed the glass globe, Musschenbroek held the water jar in his hand and reached toward the gun barrel. The shock knocked him to the floor. Unwittingly, he had invented what became known as the *Leyden jar,* which could build up charges of remarkable strength. One experimenter used the jar to knock children off their feet, and another reported that his wife could not walk for a time after being shocked. The *discharge* (a new word coined to describe the Leyden jar shock) could be communicated through several people. In France, to amuse the king, a powerful Leyden jar was discharged first through a circle of 140 courtiers, then through 180 gendarmes. Two hundred Cistercians felt the jolt in their Paris monastery and leaped toward the heavens in unison. The experimenters found they could make the shocks even more powerful by linking several jars to form a *battery.* One man wrote, "Would it not be a fatal surprise to the first experimenter who found a way to intensify electricity to an artificial lightning, and fell a martyr to his curiosity?"[4]

ATMOSPHERIC LIGHTNING—the type that shot from the heavens—posed greater dangers and provoked nearly as much curiosity. According to prevailing theories, lightning resulted from colliding clouds or some unknown chemical reaction in the atmosphere, but no one knew for sure what it was. A few believed that it was composed of electrical fluid—the spark and crackle of electricity made the connection obvious—but this theory had not been proved. Inspired by an itinerant lecturer, the Philadelphia printer Benjamin Franklin began experimenting with electricity in 1745. A few years later he proposed an experiment to "determine the Question, Whether the Clouds that contain Lightning are electrified or not." He attached a silk handle to the end of a kite string and tied a key where silk and string met. Standing in a doorway to keep himself and the silk dry, he flew the kite into a "Thunder Gust." Electricity tingled down the wet string, and Franklin drew sparks from the key first with his knuckle, then with his tongue.[5]*

Many experimenters in Europe tried variations on Franklin's experiment. Most survived the dangerous test unscathed, either through dumb luck or because they carefully insulated themselves from the lightning. In 1753, however, Georg Wilhelm Richmann, a German working in St. Petersburg, drew a bolt directly through his body. He became the first man to sacrifice his life in the pursuit of electrical knowledge.[6]

Franklin himself knew something about death from electricity. Not

---

*Although Franklin was the first to propose this type of experiment, he was not the first to perform it. He described a lightning experiment in a letter published in England in 1751. In May 1752 French experimenters followed his instructions and confirmed that lightning was electrical in nature. A month or so later—probably before he had heard of the French success—Franklin flew his kite into the storm.

long after he proposed his famous lightning experiment, he informed a friend that the discharge from a battery of two Leyden jars was "sufficient to kill common Hens outright." The birds died so quickly, he said, that "compassionate persons" might adopt it as a method of killing. Butchers could build a battery of six Leyden jars, link the battery to a chain, wrap the chain around the thighs of a turkey, and lift the bird until its head touched the prime conductor. "The animal dies instantly," Franklin wrote. He warned experimenters to be cautious. While killing turkeys, he accidentally administered the shock to himself: "It seem'd an universal Blow from head to foot throughout the Body. . . . My Arms and Back of my Neck felt somewhat numb the remainder of the Evening, and my Breastbone was sore for a Week after, [as] if it had been bruiz'd. What the Consequence would be, if such a Shock were taken thro' the Head, I know not." But electrical slaughter, Franklin averred, was worth the danger: "I conceit that the Birds kill'd in this Manner eat uncommonly tender."[7]

FRANKLIN ALSO GAVE Leyden jar shocks to people in an attempt to cure them of paralysis. Like others caught up in the electrical mania of the mid-eighteenth century, he believed that the remarkable new force could be used as a medical therapy. John Wesley, the founder of Methodism, was one of England's strongest advocates of electrical cures. Some physicians sealed a drug inside a glass wand, electrified it, and applied sparks to patients, claiming that the essence of the medicine penetrated the body along with the subtle electrical fluid. Although physicians sold electricity as a panacea capable of curing everything from constipation to venereal diseases to hysteria, most, like Franklin, focused on paralysis. Victims of the Leyden jar reported that the shocks made their muscles contract, and doctors claimed that electricity could restore paralyzed limbs.[8]

Electricity's ability to contract muscles also caught the attention of physiologists. A popular theory at the time held that the brain produced subtle "animal spirits" that were carried by the nerves to move the muscles of the body. Once it was found that electricity caused muscle contraction, some proposed that electrical fluid and animal spirits were one and the same—that electricity was the natural substance coursing through the nerves of animals.

In the 1780s the Bolognese physiologist Luigi Galvani was testing the effects of electricity on muscles. When he ran brass hooks through frog legs and hung them on an iron railing, he was surprised to see that the legs contracted spontaneously, without any application of the spark. He found that he could induce contractions by touching the frog leg in two places with different metals. Galvani supposed that the frog leg was a miniature Leyden jar, which he was discharging by the touch of two pieces of metal. Since there was no external source of electricity, the jolt must have come from within the frog leg—it was "animal electricity," he said, created and stored in muscle tissue.[9]

Galvani's results, published in 1791, did not convince everyone. Alessandro Volta, a professor of physics, claimed that the electricity that contracted Galvani's frog leg arose not within the leg itself but from the contact of brass hook and iron railing. This statement itself was controversial. All known electricity was created by rubbing glass or other insulators; Volta claimed that he could create electricity simply by bringing two different metals into contact. Volta convinced few people of this new theory of electrical generation until he created a device to demonstrate his point. He stacked multiple pairs of silver and zinc disks, placing a piece of wet cardboard between each pair. This electrical column, or *pile*, multiplied the effects of the individual pairs of disks and, when touched at either end, produced a palpable shock. Volta built a pile of forty pairs and gave himself a jolt through the ears: "The disagreeable sensation, and which I apprehended might be dangerous, of the shock in the brain, prevented me from repeating this experiment."[10]

The voltaic pile, created to quash the notion of animal electricity, had effects Volta never imagined. The pile could be used to charge Leyden jars, which confirmed that this new electricity was similar to that produced by rubbing glass. But there were crucial differences. Previously, all electricity had been what is now called *static*—the buildup of a charge, followed by its transitory discharge. The pile created an electric *current* that flowed indefinitely and could be made stronger by adding more pairs of metal disks.

Volta's pile, described in a letter to London's Royal Society in 1800, set off a frenzy of experimentation. One man built a battery from two types of silverware, although it was more common to pair silver half crowns with zinc disks. By summer experimenters reported that when they attached two wires to a pile and ran the "galvanic current" through water, hydrogen bubbles formed on one electrode while oxygen formed a compound with the metal of the other electrode. The current, in other words, had decomposed water into its component parts, and the science of electrochemistry was born. Humphry Davy, a professor of chemistry at the Royal Institution in London, ran the current through two common substances—potash and soda—and produced tiny globules of previously unknown metals, which were named potassium and sodium.

Davy's prestige in London rested as much on his skills as a popular lecturer as on his scientific discoveries. Though important to science, electrochemistry offered little drama in the lecture hall, so Davy found ways to please his audience. In an 1809 demonstration he ran the current from a powerful battery across a small gap between two carbon rods. As the current jumped the gap, it created a brilliant, arc-shaped, blue-white light that flooded the lecture hall and astonished the crowd.[11]

DAVY HAD INVENTED what came to be known as the *arc light*. At the time it had few practical applications, since batteries powerful enough to produce the effect consumed large amounts of rare metals—silver, copper, zinc—and were therefore enormously expensive. Around 1830, however, scientists discovered a new way to produce electricity. Michael Faraday, who started his scientific career as Davy's assistant, became intrigued by a report that an electric current caused movement in a nearby compass needle. This suggested that electricity produced magnetism. Faraday wondered if the reverse was true—whether magnetism could produce electricity. In 1831 he showed that rotating a coil of conducting wire within the lines of force of a magnetic field caused a current to flow in the wire. Following Faraday's lead, instrument makers in France created the first *magneto-electric generators* (often shortened to *magnetos*), hand-cranked machines that spun coils of wire relative to magnetic fields, creating electrical current.

The coils of conducting wire in a generator were known as an *armature.* As figure 1 shows, when the armature was in the first half of its

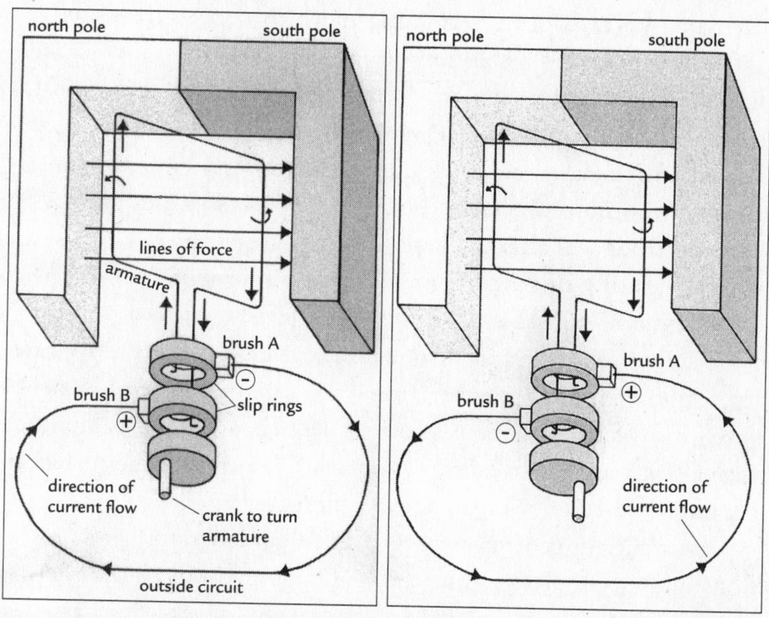

Figure 1: First half of rotation (left): When part of the armature cuts the magnetic lines of force near the magnet's *north* pole, current moves up the wire and produces a *positive* charge at the lower slip ring. The current is transferred from the slip rings through the brushes and flows through the outside circuit in a clockwise direction. Second half of rotation (right): The same part of the armature now cuts the lines of force near the *south* pole, causing current to move down the wire and producing a *negative* charge at the lower slip ring, reversing the current flow. The frequency of current reversal depends on the speed at which the coil rotates.

rotation, the current moved along the conductor in one direction, from point *A* to point *B*. But in the second half of the turn, the relationship between the coil and the north and south poles of the magnet was reversed, causing the current to flow from point *B* to point *A*. For every 360-degree turn of the coil, the current changed direction twice: from *A-B* to *B-A*, then back again. This became known as *intermittent*—or *alternating—current.*

Electricity so produced behaved differently from battery current, which flowed continuously in one direction. Electrochemistry—decomposing water or isolating sodium from soda, for instance—depended on one electrode remaining positive and the other negative. The same was true for electroplating, in which a brass object such as a spoon was placed in a solution of potassium cyanide in which gold had been dissolved. When an electric current was run through the solution, the spoon—which served as the negative electrode—became coated with a layer of gold. Electroplating and electrochemistry required continuous current, because the processes did not work if each electrode was alternately positive and negative, as was the case with alternating current. Magnetos created a form of electricity that appeared to be unusable.

To solve this problem, instrument makers developed a way to transform the alternating current from a generator into *continuous*—or *direct—current*, like that from a battery. This change was accomplished with a switching device called a *commutator*, which kept the current in the outside circuit flowing in one direction only.[12]

Direct-current generators proved useful for laboratory demonstrations and electroplating, but the one large mid-nineteenth-century industry that relied on electricity—telegraphy—stuck with batteries, which provided a steadier current. In the late 1830s electric telegraph systems were developed in England by W. F. Cooke and Charles Wheatstone and in the United States by Samuel F. B. Morse. In Morse's system, the transmitter consisted of a simple key that opened and closed a circuit, transmitting pulses of electricity that conveyed a message via a dot-dash code. At the receiving end, the electricity caused movement in a magnetic device attached to a pencil, which recorded dots and dashes on paper tape. These paper tape receivers soon were replaced by *sounders,* devices that translated the arriving pulses into clicking noises. Rather than decoding the message after the dots and dashes were printed on paper, operators listened to the coded clicks and transcribed on the fly.[13]

Morse built an experimental line from Baltimore to Washington, D.C., and on May 24, 1844, transmitted his telegraph's inaugural message—"What hath God wrought!" His next transmission—"Have you any news?"—proved prophetic, as within a few years a torrent of information gushed down the slender copper wires. Whereas all previous long-distance communication depended on transportation—horses, ships, or trains carrying words on paper—the telegraph carried messages at the blazing speed of electricity. Newspapers, ever eager to scoop their rivals, were quick to embrace the technology, as were railways. Trains dispatched according to timetables tended to get off schedule and collide with each other; telegraphs allowed railroad managers to coordinate traffic safely. Telegraph lines followed railroad rights-of-way, and the two technologies advanced in tandem, copper wires stretched out alongside iron rails.[14]

At the end of the Civil War the Western Union Telegraph Company, the industry leader, owned more than 44,000 miles of telegraph wire, more than the combined total of its two strongest rivals, American Telegraph and U.S. Telegraph. At that time long-distance transmission between cities remained the core of the industry, but new telegraph-based services began springing up rapidly—most notably, fire alarm call boxes on city streets that allowed citizens to report the location of fires, and stock and gold quotation systems that linked banks and brokerage houses with central exchanges. Competition was fierce in the young industry, and companies were eager to gain an edge through technical innovation. The situation created rich opportunities for ambitious young inventors.[15]

CHAPTER 2

The Inventor

Thomas Alva Edison began his working life in 1860, at the age of thirteen, when he took a job as a "news butch" selling newspapers and candy on the Grand Trunk Railway that ran between Detroit and his home in Port Huron, Michigan. The job did not pay well, but he found ways to supplement his income. On April 6, 1862, the *Detroit Free Press* was filled with news of the Civil War battle at Shiloh. Before the train started its return trip to Port Huron, Edison acquired 1,000 instead of his usual 100 papers and bribed a telegraph operator to wire news of the battle to stations along the line. At each stop the train was greeted by crowds of men eager for details of the battle, and Edison made a small fortune selling copies of the *Free Press* at five times the usual price.

Even as a boy, Edison displayed the skills that would serve him well for the rest of his life: an eye for the main chance, a knack for publicity, and a grasp of the possibilities of the latest technology.[1]

The young Edison—Alva to his mother, Al to his friends—never received much formal education. Born in February 1847, he spent the first seven years of his life in Milan, Ohio, before his father, a shingle maker, moved the family to Port Huron, where he attended school for less than a year. "Teachers told us to keep him in the streets, for he would never make a scholar," Edison's father reported. "Some folks thought he was a little addled." Edison's mother taught him to read at home.[2]

The Grand Trunk left Port Huron each morning at seven and arrived about four hours later in Detroit, where it stopped over until the return journey started in the evening. To fill the long afternoons, Edison joined the Detroit Young Men's Society and pored over the science books in its library. After reading a chemistry text, he bought chemicals, crucibles, and beakers, installed a laboratory in the train's baggage car, and spent many happy hours experimenting–until a broken bottle of phosphorus set the car on fire, and the enraged baggage master dumped the boy's laboratory onto the tracks. Edison later recalled that the baggage master "boxed my ears so severely that I got somewhat deaf thereafter." (Hearing problems would plague him for the rest of his life.)[3]

Seeking other outlets for his curiosity, Edison practiced on the equipment at railway telegraph offices. In 1862 he plucked a small boy from the path of a rolling freight car, and the boy's father, a telegraph operator, gave Edison formal telegraph lessons as a reward. His training complete, he left home at age sixteen to become a "tramp" telegrapher, moving from city to city in search of new jobs and new experiences. Between 1863 and 1868 he lived in

Edison at the age of fourteen, when he worked as a "news butch" on the Grand Trunk Railway in Michigan.

Ontario, Toledo, Indianapolis, Cincinnati, Memphis, Louisville, and New Orleans.[4]

Within the hard-drinking, hard-living fraternity of tramp telegraphers, Edison stood apart. Not content to listen and transcribe, he wanted to understand the principles underlying the devices he used each day. While other men avoided night shifts as an impediment to carousing, Edison embraced them because they left his days open for experimenting on equipment and reading technical journals and books (one of his favorites was Michael Faraday's *Researches in Electricity*). Even Edison's amusements were scientific. At one point he and a friend acquired a device called an *induction coil*—which transformed low voltages from an electric battery into painful shocks—and wired it to the metal sink in a railroad roundhouse. As unwitting victims touched the sink, they shouted and jumped into the air. "We enjoyed the sport immensely," Edison said.[5]

In 1868 Edison landed a job with Western Union in Boston and appeared for his first day of work dressed in a blue flannel shirt, jeans, a wrinkled jacket, and a hat with a torn brim. The finely dressed Boston operators, amused by his rough appearance, took to calling him "the Looney" and hatched a plan to haze him. On his first shift, he was asked to receive press copy from New York. The New York operator, who was in on the plot, began to transmit at the rapid clip of twenty-five words per minute, but Edison kept pace easily. When the speed increased to a dizzying thirty, then thirty-five words per minute, Edison still did not waver. Finally, when the New York operator began to skip and abbreviate words, Edison was forced to break. He tapped a message down the wire to New York: "You seem to be tired, suppose you send a while with your other foot." The Boston operators burst into laughter. The episode "saved me," Edison said. "After this, I was all right with the other operators."[6]

Despite his receiving skills, Edison was less interested in operating than in experimenting with the equipment. At the time, Boston was second only to New York as a center of telegraphy in the United States. Using contacts forged during his years as an operator, Edison found investors willing to fund his experiments on telegraph printers, fire alarm systems, and a vote recorder. The last device—for tallying votes in legislative houses—never found a market, but it was historic nonetheless: It became Edison's first patented invention, U.S. Patent 90,646, issued June 1, 1869. The lawyer who filed this patent application for Edison described the young inventor as "uncouth in manner, a chewer rather than a smoker of tobacco, but full of intelligence and ideas." With a couple of partners Edison started a business providing gold and stock quotations for banks and brokerage houses. Buoyed by initial success, he quit his job with Western Union and in January 1869 placed a notice in the journal *Telegrapher* announcing that he would now "devote his time to bringing out inventions."[7]

Within a few months of this announcement, he withdrew from the Boston enterprise and moved to New York, where he formed a partnership with Franklin Pope, a prominent expert in telegraphy, and James Ashley, an editor of *Telegrapher*. By the end of 1870, the three had developed a telegraph printer and sold the rights to it for $15,000 (roughly the equivalent of $250,000 today). Shortly after the sale, the partnership split apart on bitter terms. Edison claimed Ashley and Pope tried to cheat him out of money, while they accused Edison of violating the partnership agreement by striking his own deals with manufacturers.[8]

The rift did not slow Edison's career. After the quick success of the printer, his inventing skills were in high demand. He signed contracts to develop inventions for several telegraph firms and, with the money they provided, opened a laboratory and manufacturing operation in Newark that employed more than forty-five hands. (In a letter to his parents back in Michigan, he described himself as a "Bloated Eastern Manufacturer.") A telegraphic news service he started failed after a few months, but it proved important nonetheless: The twenty-four-year-old Edison courted one of the company's employees—Mary Stilwell, age sixteen—and married her on Christmas Day 1871.[9]

The main financial backer of Edison's Newark shops was the Gold & Stock Telegraph Company, for whom he developed the Universal stock printer, a device that became the industry standard, ticking out stock prices in brokers' offices all around the world. When Western Union bought control of Gold & Stock in 1871, Edison came under the wing of the industry giant. Western Union was particularly eager to finance Edison's research into duplex telegraphy, which allowed two messages to be sent simultaneously on the same wire, one in each direction. The company expected Edison simply to refine existing duplex technology, but Edison developed something revolutionary: the quadruplex telegraph, which allowed simultaneous transmission of two signals in each direction over one wire. By doubling the capacity of its wires, the quadruplex promised to save Western Union a great deal of money by limiting its biggest cash drain, the need to build and maintain wires. After Edison perfected his quadruplex and patented it, Western Union gave him a $5,000 advance payment and opened negotiations for purchase of full rights.[10]

Although Western Union had funded Edison's research, he had never signed a formal contract with the company. When the company was slow in coming to terms, the inventor therefore felt free to entertain an offer from Jay Gould. A notorious financier who had nearly cornered the gold market a few years earlier, Gould now wanted to challenge Western Union for control of the long-distance telegraph industry, and Edison's quadruplex was the key to his plan. When Western Union finally tendered a firm offer to Edison in January 1875, it was shocked to learn that the rights were no longer available; two weeks earlier Edison had sold out to Gould's company, Atlantic & Pacific, for $30,000. Western Union's president remarked that Edison had "a vacuum where his conscience ought to be."[11]*

WITH THE MONEY from the quadruplex and his other inventions, in 1876 the twenty-eight-year-old Edison built himself a new laboratory in the sleepy hamlet of Menlo Park, New Jersey, about twenty-five

*Western Union sued Atlantic & Pacific over rights to the quadruplex, and the dispute was resolved only by the merger of the two companies in 1877.

R. F. Outcault's illustration of Edison's laboratory complex in Menlo Park, New Jersey, as it appeared in 1881. The long building in the center is the main laboratory.

miles southwest of Manhattan. A few other men operated electrical and telegraphic laboratories at the time: The inventor Moses Farmer, for instance, worked on telegraphic and other electrical equipment from a small laboratory in Boston, and Elisha Gray conducted experiments at the shops of the Western Electric Manufacturing Company. But Edison's early triumphs allowed him to operate on a different scale. In a two-story frame building in Menlo Park, he built the best laboratory in the country and hired the most talented mechanics. He called it his "invention factory," and he had such faith in himself, his men, and his new lab that he predicted "a minor invention every ten days and a big thing every six months or so."[12]

The first major invention to emerge from Menlo Park was a refined version of the device Alexander Graham Bell had first unveiled in 1876: the telephone. In Bell's telephone transmitter, the sound waves of the human voice vibrated a metal diaphragm, which induced an electric current in an electromagnet. The current traveled over a wire and into a receiver, which essentially was a transmitter in reverse—an electromagnet vibrated a metal diaphragm, which (at least in theory) reproduced the original sound. Early users of Bell's telephone found, however, that voices emerging from the receiver were nearly unintelligible. The problem, Edison discovered, was the transmitter's electromagnet, which did a poor job of translating sound waves into electric current. Sensing an opportunity to win a patent on a crucial component of the telephone, Edison began experimenting on transmitters. He found that by replacing the electromagnet with buttons of compressed carbon, he could faithfully reproduce the modulations of the human voice. Edison's new carbon transmitter transformed the telephone from a novelty into a practical means of communication–and soon provided him with a fresh stream of income.

The telephone work led directly to another invention. At the time, when telephones were still largely in the experimental stage, no one had imagined a day when the instrument would be in every home and office, allowing people to speak directly to each other. Edison, like most other observers, expected that the telephone would function just as the telegraph system did, with an operator transcribing a voice message and delivering it to the recipient. The electrical pulses of a telegraph message could be preserved and replayed at a later time; Edison believed a telephone system should have a similar capability. The goal, as Edison described it in July 1877, was to "store up & reproduce automatically at any future time the human voice."[13]

Edison earlier had invented a device to record the electrical impulses of telegraph messages on waxed paper tape, and he also knew that the human voice created vibrations in the diaphragm of a telephone transmitter. He decided to see if these vibrations, like the telegraph messages, could be embossed and repeated. Late in 1877 Edison designed a lathelike machine consisting of a cylinder wrapped with tinfoil and attached to a hand crank. There were two diaphragms, each attached to a needle. One needle would emboss the sound waves into the foil; the other, in passing over the indentations, would reproduce the original sound.

One of the Menlo Park workers built the device to Edison's specifications, and it surprised everyone by working the first time it was tried. Edison called it the *phonograph*, or "sound writer."

Edward H. Johnson, a business associate of Edison's since the early 1870s and a master showman, took charge of promoting the phonograph in public exhibits up and down the East Coast. The concept was so novel that many people refused to believe it. A Yale professor insisted that the "idea of a talking machine is ridiculous" and advised Edison to disavow published accounts of the phonograph in order to protect his "good reputation as an inventor." Many were convinced that Edison was simply a ventriloquist, throwing his voice into the machine. A visiting minister rapidly shouted a tongue-twisting string of biblical names into the phonograph; only when the machine spit them back precisely did he believe that Edison had no tricks up his sleeve. On an April 1878 trip to Washington, D.C., Edison entertained President Rutherford B. Hayes and his wife with the phonograph until three in the morning.[14]

Before he invented the phonograph, Edison was well known to Wall Street and telegraph men. Afterward, he became one of the most famous men in the world. When reporters flocked to Menlo Park to interview the creator of this marvelous machine, they were surprised to encounter not a solemn man of science but a beaming, boyish inventor. Pants baggy and unpressed, vest flying open, coat stained with grease, hands discolored by acid, Edison "looked like nothing so much as a country store keeper hurrying to fill an order of prunes." Newspapers described a man who rarely slept and who appeared to subsist entirely on pie, coffee, chewing tobacco, and cigars. Because of his partial deafness, which had grown worse since boyhood, Edison's face took on an aspect of gravely serious concentration when he listened, but when he described his latest inventions in his high-pitched voice, his gray eyes flashed and his smooth-shaven face lit up with joy. In 1878 the *New York Daily Graphic* coined the nickname that would follow him the rest of his life: "the Wizard of Menlo Park." Edison seemed to be a distillation of America's self-image—unpolished and unpretentious yet gripped by an ambition to transform the world.[15]

Edison with his newly invented phonograph. The famed Civil War photographer Mathew Brady took this portrait during Edison's 1878 trip to Washington, D.C.

. . .

EDISON CRAVED the public's attention, but it also exhausted him. By the late spring of 1878, he was tired and ill. He had been working at a frantic pace for more than a year and had not had a vacation since his honeymoon nearly seven years before. When a friend invited him to join an expedition traveling to the Wyoming Territory to view a solar eclipse, he jumped at the chance.[16]

Edison spurned the comforts of the expedition's reserved railroad car and instead spent much of the journey perched precariously on the cowcatcher, the wedge at the front of the locomotive designed to pitch cows and other obstacles out of the train's path. Despite its dangers—

at one point, Edison later recalled, "the locomotive struck an animal about the size of a small cub bear, which I think was a badger," and he barely dodged out of the way—the spot allowed Edison to breathe the clear air of the West, untroubled by the black smoke billowing from the train's smokestack.[17]

During the western trip, Edison talked to other scientists about new discoveries in the field of electric lighting. Edison had toyed with lighting experiments before, but other projects intervened. Even before his train returned to Menlo Park, he had decided to take up the problem again. Much of the attraction was financial. In an interview in April 1878 Edison had said of the phonograph, "This is my baby, and I expect it to grow up to be a big feller and support me in my old age." He soon learned, though, that the phonograph was a solution without a problem: Everyone recognized its brilliance, but no one could figure out what to do with it. Edison imagined it as a tool for business dictation, but the machine was temperamental and slow to catch on, and the market for recorded music was a decade away. In 1878 the phonograph did not appear likely to turn a profit anytime soon.[18]

With electric light, on the other hand, the business plan was clear. People needed ways to dispel the darkness, and the existing technologies—candles, illuminating gas, kerosene lamps—were far from perfect. A good electric lamp might well support the inventor in his old age.

CHAPTER 3

Light

On August 26, 1878, Edison arrived back in Menlo Park and was reunited with his wife, Mary, and his daughter and son, five-year-old Marion and two-year-old Tom Jr.—nicknamed Dot and Dash by their father, ever the telegraph man. When reporters appeared to collect news of the trip, the inventor spoke rapturously about the West, complaining only about the springless stagecoaches at Yosemite: "If they had only fastened a good stout plank on the seat of a fellow's trousers, and employed an able-bodied mule to kick him uphill and over the canyons, it would have been a big improvement." The day after his return, Edison headed to his laboratory and started research on the electric light.[1]

The source of excitement among the scientists on the western trip was a new version of the electric arc lamp that had just been unveiled at the Paris Universal Exposition by the Russian engineer Paul Jablochkoff. The light's basic principle—running a strong electric current across a gap between two slender carbon rods—had been discovered by Humphry Davy seventy years before, but the technology had changed dramatically over the decades. Whereas Davy had used electricity created by a chemical battery, the Jablochkoff lamps used the latest design of electrical generator.

Because the understanding of electricity as a movement of electrons in a conductor would not emerge until around 1900, in the 1870s not even the greatest electricians could claim to know just what happened inside a copper electric wire. But this lack of theoretical understanding did not prevent scientists from becoming adept at manipulating electrical force. Faraday had discovered that moving a coil of conducting wire through the lines of force of a magnetic field caused current to flow in the conductor, and later experimenters learned that they could increase the strength of the current by using stronger magnets and multiplying the number of coils of conducting wire. The first generators employed permanent steel magnets, which were relatively weak. To skirt this difficulty, inventors in the late 1860s turned to the discovery that first inspired Faraday—the ability of electric current to produce a magnetic field—and built generators that replaced permanent magnets with far more potent electromagnets. At first the current for the electromagnets was supplied by batteries or smaller generators, but in the 1860s and 1870s inventors designed generators that produced the current for their own electromagnets. Because these machines excited their own magnetic fields, they were known as *dynamo-electric generators,* or *dynamos.* The new machines could produce a current that was powerful, steady, and inexpensive enough for arc lighting.[2]

In the Jablochkoff arc lamp system, the creation of light started with the burning of coal, which heated water in a boiler and produced steam. The steam drove the piston in an engine, and the piston moved a driveshaft, which was connected via a leather belt to the dynamo. The belt turned the dynamo's armature, an iron core wrapped with coils of copper wire. As it spun at hundreds of revolutions per minute, the armature repeatedly cut through the lines of force of an electromagnet. The movement of a conductor (the armature) through the magnetic field produced an electric current, which flowed through copper wire to the lamps, each of which contained two pencil-thin rods of carbon a fraction of an inch apart. The current leaped the gap between the carbon rods—producing a powerful light—then flowed on to the next lamp. The process moved from coal to steam to mechanical motion to electricity; it was a simple matter of the transformation of energy, from black coal to white light.

On September 8, 1878, two weeks after returning from his western trip, Edison visited the Connecticut factory of William Wallace, who in the previous few months had developed his own system of arc lighting. A newspaper reporter described the inventor's reaction to Wallace's factory: "Mr. Edison was enraptured. He fairly gloated over it. . . . He ran from the instruments to the lights, and from the lights back to the instruments. He sprawled over a table with the SIMPLICITY OF A CHILD, and made all kinds of calculations." Edison ordered a generator from Wallace, then returned to Menlo Park.[3]

When he began his lighting experiments, Edison chose not to focus on the arc lamp, because he noticed its limitations: It produced an intense glare and—because the carbon rods combusted—emitted choking fumes, making it suitable for use only outdoors or high overhead in factories. The arc lamp's blaze was measured in the thousands of candlepower, whereas home lighting required only a dozen or so. In most larger cities, homes and offices were lit with illuminating gas, which gave a soft, gentle light but also flickered and released gases (ammonia, sulfur, carbon dioxide) that poisoned the air, and soot that blackened the walls. Worse, the open flames of gaslight set buildings on fire with alarming regularity.[4]

Edison believed he could domesticate electric light. As he explained it, electric light "had never been made practically useful. The intense light had not been subdivided so that it could be brought into private houses."[5]

The principle behind Edison's "subdivided" light was known as *incandescence*—using electricity to heat a material until it glowed. The flow of electric current along a conductor depends on the relationship of voltage, resistance, and amperage. *Voltage* is the electrical pressure that causes current to flow. *Resistance* (measured in units called *ohms*) is the opposition that a conductor offers to current; when current encounters resistance, some of the electrical energy is converted into heat. *Amperage* is the rate of flow of electricity along the conductor, established as voltage overcomes resistance. Whereas the arc lamp relied on brute force–high-pressure electricity (500 to 1,000 volts) hurtling down a copper wire and leaping a gap between carbon rods–incandescence required a delicate touch. In the system Edison envisioned, a low pressure of 100 volts or so flowed smoothly from the generator through low-resistance copper conducting wires until it encountered the burner, which had a higher resistance and therefore impeded the flow, causing some of the electricity to be transformed into heat; the heat raised the temperature of the burner to incandescence, producing light.

The theory was simple, the practice excruciatingly difficult. When heated to temperatures high enough to incandesce, most materials either oxidized (burned) or fused (melted). The two most promising candidates were carbon, which had an extremely high melting point but tended to burn; and platinum, which did not burn but tended to melt. Inventors–including the Englishman Frederick De Moleyns and the Americans J. W. Starr and Moses Farmer–had experimented with these two substances as far back as the 1840s, but no one had created an incandescent lamp that glowed for more than a few seconds before disintegrating.[6]

AS SOON AS Edison returned from his visit to Wallace on September 8, he sketched a plan for a light in his laboratory notebook. Two days later he conducted his first experiments, and three days after that he wired to Wallace to inquire about the generator he had ordered: "Hurry up the machine. I have struck a big bonanza." Edison's laboratory notebooks reveal the nature of his alleged success. The inventor decided that carbon's tendency to burn rendered it useless, so he discarded it in favor of platinum. To skirt the problem of platinum's melting, he devised a thermal regulator: When the temperature of the platinum approached the melting point, a piece of metal would expand and break the current; when the regulator cooled, the current flowed again.[7]

"I have it now!" Edison proclaimed in the pages of the *New York Sun* on September 16. "When the brilliancy and cheapness of the lights are made known to the public," he said, "illumination by carburated hydrogen gas will be discarded."[8]

Upon the announcement of Edison's invention, stocks of illuminating gas companies plunged on the New York and London exchanges and investors scrambled to buy a stake in the new light. The Edison Electric Light Company was incorporated and capitalized at $300,000, with Edison receiving $50,000 to develop his invention. Investors included William H. Vanderbilt, the principal shareholder in Western Union; Norvin Green, the president of Western Union; and Egisto Fabbri, a partner at Drexel, Morgan & Company, the nation's leading investment banking firm. J. P. Morgan himself, normally a cautious investor who avoided risky new ventures, had no doubts about Edison—at his direction Drexel, Morgan snapped up British rights to Edison's light patents and became his agents for all of Europe.[9]

New York's financiers poured money into the creation of the Edison Electric Light Company because they were terrified of being left behind. Vanderbilt and other Western Union stockholders had seen firsthand how Edison's quadruplex and other inventions reshaped the telegraph industry, and they expected that he would do the same to the world of lighting. The investors expected nothing less than a technological and social revolution, a new service that no home or office could do without. The potential profits were immeasurable.

There was only one problem with Edison's announcement and the frenzy it produced: He had not, in fact, invented a working incandescent light.

Edison certainly thought he was closer to success than he was, but there may have been another motive behind his premature announcement. To invent the lightbulb, Edison needed a great deal of money, far more than investors would give him for early-stage experiments.

So he simply said he had already finished. By making the premature announcement in the *Sun*, he hoped to fire public enthusiasm and pry open the coffers of Wall Street. The ploy worked. With the investors' money in hand, Edison set to work on the invention he claimed to have already perfected.[10]

"I WAS ALWAYS AFRAID of things that worked the first time," Edison had said two years earlier, after his surprisingly quick success with the phonograph. He had nothing to fear from the electric light. As work proceeded in the fall of 1878, the thermal regulator remained balky and the platinum burners still melted. Edison began to understand that the task was much larger than he had imagined.[11]

Fortunately, he was well prepared. Edison's successes depended in part upon the work environment he created at Menlo Park. The location in rural New Jersey offered seclusion from but also proximity to the centers of capital in New York. Although he complained about the "damned capitalists," it was their money that built him the best laboratory in the world–complete with a new machine shop, a stockroom filled with every metal and chemical known to science, and an enormous library of scientific journals and books.[12]

The money also allowed Edison to hire assistants of extraordinary talent. Foremost among the Menlo Park staff was Charles Batchelor, an Englishman with a bushy black beard and considerable skill as a machinist and draftsman. Batchelor had started working for Edison in Newark in 1871 and immediately emerged as the inventor's chief assistant, his methodical work habits complementing Edison's cut-and-try enthusiasm. Another top associate, John Kruesi, trained as a clockmaker in the legendary shops of Switzerland before joining Edison's team, and those skills served him well when he was called upon to translate Edison's crude sketches into working models. Kruesi built the first phonograph in just six days, and it worked the first time it was tried. The newest arrival, Francis Upton, was a Princeton-trained mathematical physicist who had studied in Berlin with Hermann von

Helmholtz; Upton joined the staff in 1878 to help with the lighting experiments. A touch insecure about his own lack of formal training in mathematics, Edison liked to tease Upton about his fancy degrees. But he was venturing into territory where mathematical ability was essential, and part of his brilliance was in recognizing that Upton's mathematical talents balanced his own more intuitive grasp of technology.[13]

Newspapers reported that in the early evening, when most workers could expect to go home to their families, Edison's men were just hitting their stride. After assembling to review accomplishments and chart strategy, they dispersed to their individual tasks. Edison hustled from bench to bench, observing experiments and giving instructions. Then he would stop and become absorbed in a particular experiment. His thin hands floated above an instrument, darting in to make minute adjustments, while the rest of his body stood as rigid as stone.[14]

A little before twelve o'clock on many nights, two apprentices and a huge Newfoundland dog would set out for the local grocery. Menlo Park had no streetlights, electrical or otherwise, so the dog led the way with a lantern clamped in his teeth. After rousing the grocery keeper from bed, the party returned with baskets laden with soda crackers, cheese, butter, and ham. A boy fetched buckets of beer from Davis's Lighthouse, the local tavern, and the Menlo Park crew gathered for their midnight supper. After the meal Edison passed out cigars, and amid the smoke the men gossiped and told jokes. Some nights there was clog dancing, or an impromptu boxing match, or a sing-along to popular tunes. The German glassblower Ludwig Boehm might play the zither and yodel. Often "the old man"—as the workers called Edison—would sit down at the pump organ and pound out the few chords he knew. Then the boss would stand and hitch up his trousers—the signal to get back to work. Visiting reporters often got so caught up in the fun that they missed the last train back to New York and spent what was left of the night sleeping on the laboratory floor.[15]

Their host often chose similar accommodations, even though his wife and a warm bed awaited him just a short walk away. One of Edi-

Edison (seated in the middle with a scarf around his neck) with some of his assistants at the Menlo Park laboratory, February 1880.

son's favorite locations was a small storage closet under the laboratory's stairwell. He would crawl in, pull the door shut, and sleep for a few hours on the floor. (This space doubled as his hiding spot when unwanted visitors arrived.) He also liked to stretch out under one of the lab benches, using his coat as a pillow—but not before giving his men orders to wake him if anything important developed. According to one reporter, "Life in the Menlo Park laboratory partakes more of the character of a camp pitched near the battlefield than of anything else."[16]

EVEN AFTER FOUR MONTHS of unsuccessful experiments, Edison remained convinced that platinum was the best material for an incandescent burner. Previously, he had tested his platinum burners in the open air, but when he still could not keep them from melting, he decided to try a new technique. In late January 1879 he started placing the burner within a glass container evacuated of air—for the first time, he was working on a light *bulb.* Earlier inventors had tried coupling a vacuum with *carbon* burners in an attempt to avoid oxidation, but they had trouble creating a good vacuum. Edison at first believed that his decision to focus on platinum, which did not burn, had freed him from the need for a vacuum, but by late January he began to think otherwise. He discovered that bubbles of gas were being trapped within the platinum burners, causing them to melt more easily. If he heated platinum in a vacuum, Edison reasoned, he would release the occluded gases and raise the melting point of the platinum. The available vacuum pumps—complex contraptions of glass tubing and liquid mercury—did not work well enough, so Edison devised a new one that evacuated nearly all of the air from a glass globe.[17]

The vacuum pump was not the only new device at Menlo Park. Although initially impressed by William Wallace's electrical generator, Edison discovered that it, like all of the other generators on the market, could not produce a current efficient enough for economical incandescent lighting. Edison and his men began to experiment on designs of their own. By the spring of 1879 they had created what Upton called "the best generator of electricity ever made," one that converted mechanical energy to electrical with very little waste. The Edison dynamo featured two large, cylindrical magnets standing on end, an arrangement that, to the fertile imaginations of the men at Menlo Park, resembled a woman on her back with her legs in the air. They duly nicknamed the new machine the "long-legged Mary-Ann," although prudish newspaper editors confusingly revised that to "long-waisted."[18]

With the new vacuum pump and the new dynamo design, Edison believed he stood on the eve of triumph, so in March of 1879 he once again called in the newspaper reporters. A few minor problems remained to be cleared up, Edison said, but even now his light could be "put in practical operation everywhere, and electricity supplied at less than half the cost of gas."[19]

The announcement, as before, turned out to be premature—the platinum burners still did not work properly. When it became clear

The "long-legged Mary-Ann" dynamo.

Edison again could not make good on his claims of success, his investors became nervous, gas stocks rebounded, and critics sharpened their knives. "Day after day, week after week and month after month passes and Mr. Edison does not illumine Menlo Park with his electric light," the normally loyal *Daily Graphic* observed. "The belief has become rather general in this country and in England that for once the great inventor has miscalculated his inventive resources and has utterly failed."[20]

Franklin Pope, Edison's erstwhile friend and mentor, wrote a bitter anonymous letter to the *Telegraphic Journal*: "I know of no one here (whose opinion is worth anything) who has any confidence in the practical success of Edison's scheme. The way that the world stands agape waiting for the Edisonian mountain to bright forth its mouse is really absurd."[21]

As criticism mounted, Edison remained calm. "It has been just so in all of my inventions," he explained to a friend. "The first step is an intuition and comes with a burst– Then difficulties arise. This thing gives out then that. 'Bugs' as such little faults and difficulties are called, show themselves– Months of intense watching, study and labor are required before commercial success–or failure–is certainly reached." He neglected to mention that, back in September, he had already guaranteed commercial success.[22]

Although Edison's chosen material–platinum–still refused to work, Edison did hit upon a key insight into the *theory* of burners. All previous inventors who worked on the incandescent lamp employed a burner of fairly low resistance, one ohm or so, because they assumed that raising the resistance of the burner would require the use of more energy, thus boosting costs. Edison was the first to understand that energy consumption was proportional to the burner's radiating surface, not to its resistance. As Edison explained to a newspaper reporter, "The point is that the more resistance your lamp offers to the passage of the current, the more light you can obtain with a given current." Edison set out to create a burner with 100 times or more the resistance of those used by earlier inventors.[23]

Putting the theory of high resistance into practice proved more difficult. The resistance of a conductor was inversely proportional to its diameter–the thinner the wire, the higher the resistance. An appropriate platinum burner would have to be long and slender, and a long piece of wire would fit within a small glass globe only if it were wound into a tight spiral. This required the wire to be insulated, so that the turns of the spiral could touch each other without shorting out. Edison and his crew tried dozens of insulating substances–including barium nitrate, sodium tungstate, calcium acetate, and silk coated with magnesia–but none worked.[24]

The breakthrough finally came in October of 1879–a year after he first announced success–and, as with the phonograph, it resulted from his practice of working on several different projects at once. When Edison's carbon telephone transmitters entered the market, a crew was assigned to produce them. In a small shed beside the laboratory, kerosene lamps burned constantly, and workmen periodically scraped off the soot that collected on the lamp chimneys. The lampblack, a high-grade carbon, was used in the carbon buttons for the transmitter, and there was always a great deal of the material around the laboratory. A newspaper account described the eureka moment:

"Sitting one night in his laboratory reflecting on some of the unfinished details, Edison began abstractedly rolling between his fingers a piece of compressed lampblack mixed with tar for use in his telephone . . . until it had become a slender filament. Happening to glance at it the idea occurred to him that it might give good result [*sic*] as a burner if made incandescent."[25]

When he first started work on the lamp, Edison abandoned carbon because of its tendency to burn, and because all earlier inventors had used thick carbon rods of low resistance. In the fall of 1879, however, he realized that he could make carbon just as thin as platinum wire. With his new, powerful vacuum, the carbon would not burn—no oxygen, no oxidation. After experimenting with different types of carbon burners, Edison and Batchelor took a piece of cotton thread, .0013 of an inch in diameter, and carbonized it in an oven. The *filament*—as the slender burners were now called—was attached to platinum lead-in wires and sealed inside an evacuated glass bulb. The lab notebook entry tells the tale: "on from 1:30 AM till 3 pm[:] 13 1/2 hours and was then raised to 3 gas jets for one hour then cracked glass & busted." It was an understated entry for a historic event. Edison and his men had finally created a practical incandescent lamp—one that would burn for hours and use very little energy.[26]

"It is an immense success," Edison told a friend. "Say nothing." Although it went against his nature, he remained silent because he wanted to be absolutely sure of success before the press learned of it. Dissatisfied with the carbon thread, Edison and his men tested hundreds of different sources of carbon. Finally, at Bachelor's urging, they tried a horseshoe-shaped piece of cardboard boiled in sugar and alcohol and then carbonized. It worked even better than carbon thread. "I think the Almighty made carbon especially for the electric light," Edison told a reporter.

Now Edison was ready to exhibit his light.[27]

He invited the public to Menlo Park for New Year's Eve, 1879, and before nightfall the roads to the town were clogged with carriages,

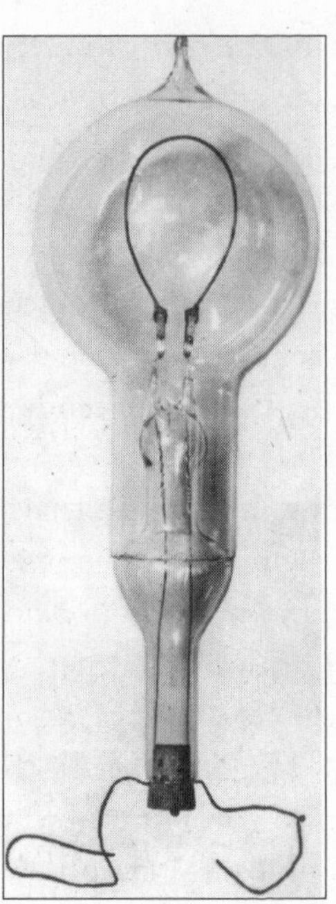

The Edison incandescent lamp as it appeared in 1880.

wagons, and pedestrians, and railroad companies ordered special trains to carry the crush. Thousands of spectators thronged the streets until past midnight. When Edison appeared, attired in a rough suit of work clothes, the crowd surged toward him. Some shouted questions, ranging from "How'd you get the red-hot hairpin into that bottle?" to more informed queries about the horsepower required to power each bulb. Edison had become an expert at working a crowd, playing the role of the modest genius, explaining complex science in simple terms.[28]

The system was powered by three long-legged Mary-Ann dynamos and controlled by a telegraph key in the machine shop. The visitors never tired of pressing the key, turning the lights off and on. When one of Edison's men plunged a lamp into a jar of water, the crowd was astonished to see that the water did not quench the flame. But the lights in open air were astonishing enough. Two lamps glowed softly at the entrance to the library, eight more atop wooden poles along the roadway, and a string of thirty lit up the laboratory building.

To modern eyes, it would have seemed a rather modest display. But those assembled were among the first people in the world to see the marvelous glow of incandescent light. No flame, no flicker, no soot, no fumes—just pure, steady light.[29]

CHAPTER 4

Electricity and Life

EDISON'S ELECTRIC LIGHT inventions marked another triumph in the great tradition of electrical innovation that included Volta's battery, Faraday's researches in electromagnetism, Morse's telegraph, and the first powerful electrical generators built in the 1870s. In the shadows of this march of progress, however, a very different electrical tradition survived. In the eighteenth century electricity had served primarily as a source of amusement and a form of medicine, and those uses persisted into the nineteenth century. The mysterious fluid that carried telegraph messages and produced light also was sent coursing through the bodies of animals and humans for the purpose of entertaining, healing, and killing.

Physicians in the 1740s had discovered that some people who appeared to be dead could be revived by forcing air into their lungs. Suddenly, the boundary between death and life became blurred, and doctors began to distrust their ability to diagnose death. In the 1760s these doubts inspired the creation of the first "humane societies," organizations dedicated not to the welfare of animals but to reviving the apparently dead. Resuscitation techniques included not only assisted breathing but also vigorous shakes and thumps that were intended to get the blood moving again. The revivalists did not trouble themselves with those who expired after long illness; they focused, rather, on those felled by the sudden misfortune of drowning, suffocation, or lightning strikes. Hoping to learn how to revive lightning's victims, the English experimenter and radical democrat Joseph Priestley used a large Leyden jar to kill a mouse, a rat, "a pretty large kitten," and a dog in the 1760s. He then tried to reanimate his victims by blowing into their lungs through a quill. The attempts failed, and he stopped the experiments, judging that "it is paying dear for philosophical discoveries, to purchase them at the expence of humanity."[1]

Others thought electricity might help bring back those who had died from some other means. One experimenter revived a suffocated dog with electricity in 1755, and twenty years later another claimed to have shocked a drowned man back to life. The invention of the chemical battery opened new avenues of experimentation. Giovanni Aldini, nephew of Luigi Galvani, staged experiments to determine the value

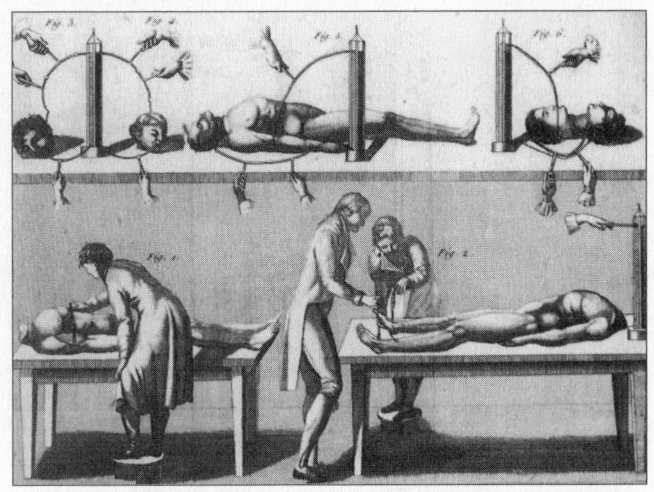

Giovanni Aldini, a nephew of Luigi Galvani, tested the effects of electricity on the corpses of executed criminals in 1803. The columns are voltaic piles.

of electricity as a means of resuscitation in cases of asphyxiation. A strong current sent through a dead ox produced such a flailing of limbs that "several of the spectators were much alarmed, and thought it prudent to retire to some distance." Before London's Royal Society in 1803, Aldini conducted experiments on the body of a freshly hanged criminal. When the poles were touched to the jaw and ear, the face quivered and the left eye opened, while a shock from ear to rectum produced a reaction so strong as "almost to give an appearance of re-animation." Aldini concluded that "Galvanism affords very powerful means of resuscitation."[2]

In 1818 a Glasgow chemist brought the body of a hanged man to his laboratory ten minutes after it was cut down. When the current from a battery was applied, "laborious breathing instantly commenced," but the man did not revive. At an 1827 hanging in Albany, New York, "eminent surgeons" stood ready "to try galvanic experiments upon the body, in order, if possible, to resuscitate it," but the authorities would not let them try. When John Skaggs was hanged in Bloomfield, Missouri, in 1870, the attending physicians pronounced him dead after ten minutes, then carried his corpse into the courthouse and tried to revive him with a hand-cranked magneto generator. The sheriff, who considered it odd to kill a man and then try to bring him back to life, suspected that the doctors cut Skaggs down prematurely to improve the odds of reviving him. "The intention of the law is to hang him till dead," the sheriff told the doctors. "It means dead in the strict sense of the word–enough to stay dead." The physicians nonetheless applied the current and provoked muscular action. "The right leg moves on the table like that of a clog-dancer," the *New York Times* noted. "Left arm swings around like a pugilist's." Skaggs reportedly developed a pulse and began breathing, but he died later that night.[3]

Electrical experiments had become such a popular fad in Germany that officials banned tests with the severed heads of executed criminals. Denied human subjects, a German named Karl August Weinhold took to killing kittens and replacing their brains and spinal columns with an amalgam of zinc and silver. One kitten so treated reportedly developed a pulse and heartbeat, opened its eyes, and hopped around.[4]

Weinhold's tests may have inspired Mary Shelley's *Frankenstein*, first published in 1818, the story of a doctor who, using body parts scavenged from the "dissecting room and the slaughter-house," cobbled together a creature and managed to "infuse a spark of being into the lifeless thing." Edgar Allan Poe played corpse revival for comic effect in "The Premature Burial," in which a man is buried alive but then exhumed before he expires. He "seemed to be in a fair way of ultimate recovery," Poe wrote, "but fell a victim to the quackeries of medical experiment. The galvanic battery was applied; and he suddenly expired in one of those ecstatic paroxysms which, occasionally, it superinduces."[5]

TO THE WITNESSES of resuscitation experiments, the contortions of dead creatures proved that there was a link between electricity and the spark of life. Although no one managed to revive the dead with electricity, there was a widespread belief that it could improve the health of those still living. "Electricity is life" became the mantra of those touting the medical uses of electricity. In the 1830s some hospitals created "electrifying rooms" for therapeutic shocks, and instrument makers in Boston sold small magnetos with electrode attachments that could be applied to, or inserted in, various parts of the body. During the Civil War, the U.S. surgeon general set aside wards for soldiers with nervous system illnesses and used electricity in attempts to cure them. Although elite physicians claimed to use electricity only for a few complaints such as nervous disorders and paralysis, those less interested in respectability treated electricity as a cure-all. One company promised that its electrical device would heal "rheumatism, paralysis, neuralgia, sciatica, asthma, dyspepsia, consumption, erysipelas, catarrh, piles, epilepsy, pains in the head, hips, back or limbs, diseases of spine, kidneys, liver and heart, falling, inflammation or ulceration." The Sears catalog offered the "Giant Power Heidelberg Electric Belt" as a cure for impotence.[6]

History has not been kind to the nineteenth century's medical therapies. In 1860 Oliver Wendell Holmes famously said that if most drugs then in use, such as mercury and arsenic, "could be sunk to the bottom of the sea, it would be all the better for mankind,—and all the worse for the fishes." Physiological theory held that health depended on maintaining the equilibrium between the body's intake (food, liquid, air) and outgo (bodily excretions). At a time when physicians lacked the instruments to see inside the body, their primary diagnostic tools were what came out of it. Their job was to manage a patient's delicate balance of forces and fluids, and they did so with drugs that caused sweating, urination, defecation, and vomiting. The therapies helped the body's systems regain balance. Just as important, the dramatic physiological reactions reassured patient and doctor that something was being done to cure the illness.[7]

Electrical medicine fit neatly into this scheme. The current from a battery or magneto was thought to preserve the healthful balance of a "fluid"—nerve force or animal electricity—that had become blocked or depleted. Although milder in its effects than many drugs, a medicinal electric shock produced tingles and shocks and sparks that served as clear evidence of therapeutic action. And it was impressive: By using electricity, physicians showed that they were masters of the arcane secrets of the era's most advanced technology. George Beard and A. D. Rockwell distilled the wisdom of decades of electrical medicine into *A Practical Treatise on the Medical and Surgical Uses of Electricity*, which was first published in 1871 and became the standard American text on the subject. The authors complained of "travelling charlatans" who sold electricity as a cure-all, but their own claims were nearly as sweeping. They advocated what they called "general faradization," in which the patient stood on a large copper electrode while the doctor passed the other electrode—contained in a moist sponge—all over her body, producing muscle contractions. The therapy, Beard and Rockwell claimed, invigorated the system and cured "all forms of pain and debility whatsoever."[8]

George Beard became famous for inventing an illness. In the 1860s many of his patients complained of vague ailments that included fatigue, anxiety, indecision, and sexual debility. Whereas earlier physicians claimed the problem was all in the head, Beard deemed it physical. Borrowing freely from Galvani's theory of animal electricity, he claimed the human body manufactured a "nervous force," electrical in nature, that carried messages between the brain and the body. But people possessed limited stores of this force, and nineteenth-century life—with its trains and telegraphs and bustling cities—easily exhausted it, producing what Beard called *neurasthenia*, or weakness of the nerves. He treated his patients with electricity, convinced that the fluid from a battery could recharge a depleted human system. Neurasthenia—or *American Nervousness*, as Beard titled his popular book—became the fashionable illness of America's upper classes. Like Sigmund Freud a few years later, Beard pioneered in the study of neuroses and the social causes of mental disease. For Beard, though, neurasthenia was not a mental illness to be talked through; it was the symptom of a disordered mechanism in need of a minor manipulation and a fresh infusion of energy.[9]

BEARD AND THOMAS EDISON became acquainted in 1874, when Edison branched out from telegraphy into medical machinery and Beard offered to endorse the inventor's new product: the *inductorium*. Edison noticed that instrument makers were collecting tidy profits selling medical induction coils, which were used to transform low voltages from a battery into more powerful shocks, so he decided to enter the market himself. "This instrument should be in every family as a specific cure for rheumatism," according to an Edison advertisement that ran in more than 300 newspapers. In three months he sold more than 100 inductoriums.[10]

Edison's induction coil had uses beyond the medicinal. He suggested creating a burglar alarm by connecting the battery's wires to a door or window and the electrodes to a cat: "When a window is raised or a door opened it will close battery ckt [circuit] & the handles being connected to a cat she will give an unearthly & diabolical yell & wake all up." This idea, contained in Edison's scribbled notes, never made it into print, but the newspaper advertisement for the inductorium does describe it as "an inexhaustible fount of amusement." Edison, who had played pranks with his induction coil in his days as a tramp telegrapher, thought administering shocks to unsuspecting victims was good fun. When he considered starting a "Scientific Toy Company," one of the devices he proposed was a "Magneto-elec-shocking Machine."[11]

The induction coil was a common toy even before Edison took hold of it. One electrical expert fondly recalled the "dreadful shock . . . given to our school-fellows when we became the proud possessors of our first electrical machine." The Ward B. Snyder catalog of sportsmen's goods advertised its electric battery as "an endless source of amusement for an evening party." In Salem, Massachusetts, an itinerant lecturer performed what a member of his audience described as "the old experiment of sending a sharp shock of electricity through the joined hands of some scores of people, each one of whom really believed he was the first one hit, so synchronous was the blow." At carnivals, fairgoers paid showmen for the pleasure of receiving shocks from an induction coil. Similar amusements took place at dime museums—those catchall institutions, brought to perfection by P. T. Barnum, where visitors might see Siamese twins, a wax statue of a famous murderer, a temperance play, and the latest scientific apparatus. One New York dime museum advertised "New and Wonderful Galvanic Batteries. Always ready, (free of charge) for the use of visitors." The kinds of use were not specified, but they certainly included the surprising pleasures of receiving a mild electric shock.[12]

Electricity was also administered for darker purposes. In 1878 the Ohio state penitentiary began using strong jolts from an induction coil to punish inmates, applying the electrodes "to the bare skin of the convict in various places." The coil "is so small that it looks like a toy," the *New York Sun* reported, "but it makes the subject of punishment yell sometimes, as though he was badly hurt."[13]

Killing experiments, first attempted by Benjamin Franklin and others in the 1750s, continued in the nineteenth century. The British physi-

EDISON'S INDUCTORIUM,

This is an exceedingly powerful induction coil, designed expressly for medical and family use. It is constructed upon a principle recently discovered by its inventor, whereby most extraordinary effects are produced, without a corresponding increase in the size of coil, or battery power, as has heretofore been necessary. Hence we are enabled to furnish the public with an apparatus at one-third the price asked by other makers for coils of equal power.

The workmanship is of a very superior style, and of a solid and substantial character.

The battery which accompanies the coil is the same as used by the telegraph companies.

An almost infinite number of experiments may be tried, by manipulating the electrodes. This instrument should be in every family as a specific cure for rheumatism, and as an inexhaustible fount of amusement.

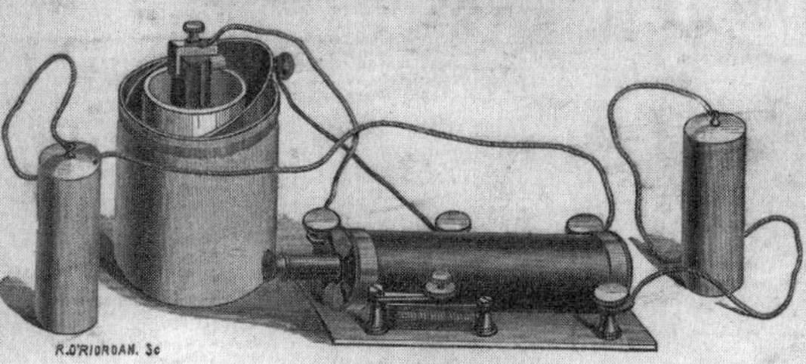

The second cut shows the manner of connecting the coil with the battery, and electrodes. The latter when applied to different parts of the body for medical purposes, should have moistened sponges placed in the hollow part of each. The handles, which are held by the persons applying the current, should be wrapped with dry paper to prevent its passing through his own body.

Directions for charging the battery accompany each box.

PRICE of each complete apparatus, SIX DOLLARS, sent C. O. D. by express to any part of the U. S. Liberal discount to Agents.

Address,

EDISON & MURRAY, 10 & 12 Ward St., Newark, N. J.

Manufacturers of Recording-Telegraph Instruments for learners and private use, Telegraph Supplies, Batteries, Mirror Galvanometers, Resistance Coils, Condensers, Philosophical and Experimental Apparatus, etc.

Edison's inductorium, a device for giving mild electric shocks, was sold as a cure for rheumatism and as "an inexhaustible fount of amusement."

cian B. C. Brodie killed a guinea pig with a Leyden jar in 1827 and concluded that in victims of lightning strikes "there is an instantaneous and complete destruction of the vital principle in every part of the animal machine." Fifty years later Benjamin Ward Richardson, a noted British physician, applied sparks from a big induction coil to various animals. A frog survived twenty-five shocks with no obvious ill effects, and a rabbit emerged from thirty jolts with only singed fur. When Richardson coupled the induction coil with a large Leyden jar, however, he managed to kill pigeons, reporting that after receiving such a shock a bird will appear "perfectly, livingly, natural, and yet it will be dead. No mark will be left on its body."[14]

Richardson, like Brodie, was investigating the physiological effects of lightning strikes, but others became interested in electrical killing for other reasons. In November and December of 1879, as Edison was preparing to unveil his incandescent lamp, the *New York Herald* published a series of articles on the possibility of executing condemned criminals with electricity. The *Herald* printed no comments from Edison on the matter, but the inventor's friend Dr. George Beard endorsed the idea. Beard claimed that sending a large Leyden jar shock from ear to ear would kill a man "in the small fraction of an instant."[15]

CHAPTER 5

"Down to the Last Penny"

SHORTLY AFTER Edison's demonstration of incandescent lighting on New Year's Eve 1879, an English expert published his view of the matter in the *Saturday Review*: "What a happy man Mr. Edison must be! Three times within the short space of eighteen months he has had the glory of finally and triumphantly solving a problem of world-wide interest. It is true that each time the problem has been the same, and that it comes up again after each solution, fresh, smiling, and unsolved, ready to receive its next death-blow. But . . . there is no reason why he should not for the next twenty years completely solve the problem of the electric light twice a year without in any way interfering with its interest or novelty." After crying wolf so many times, Edison deserved the mockery, but the doubts that greeted his announcement were quite genuine. Many engineers considered an inexpensive electric light to be a mathematical impossibility, something like a perpetual motion machine. One called Edison "a fraud, a willful deceiver of the public" who was interested only in booming his stock price. The humor magazine *Puck* offered a backhanded defense: "Edison is not a humbug. He is a man of a type common enough in this country–a smart, persevering, sanguine, ignorant, show-off American. He can do a great deal and he thinks he can do everything."[1]

The criticism infuriated Edison, but before reporters he maintained his genial public persona. Early in 1880 he announced plans to build his first commercial lighting station in New York, the city where he had made his fortune in telegraphy, where his investors had their offices, and where an aggressive and influential press would carry news of his achievements around the world.

Some experts believed that in the future every house would have its own small electrical generator, but Edison always imagined his system as similar to that of gas lighting, in which many homes and offices were served by one central plant. Lighting gas was produced by distilling coal, then trapping the resultant gas, purifying it, and storing it in vast reservoir tanks. Big main lines carried the gas from the reservoirs under the streets, with smaller pipes branching off into homes and businesses and terminating in lighting fixtures. By turning a key on the fixture, customers opened a valve and released gas, which they then lit with a match. Similarly, Edison imagined that large electrical generators in a central station would produce current, which would then be carried under the streets through copper wires and into homes and offices.[2]

When first introduced in American cities in the 1840s and 1850s, gaslight marked a big improvement over the reeking whale-oil lamps and tallow candles commonly in use. When the electric lamp came along, its advantages over gaslight seemed similarly obvious: Electric lamps produced less heat and no smoke, soot, or poisonous fumes. Gas lighting, though, had habit and tradition on its side, as well as the power of wealthy gas lighting companies. Cheap coal drove gas production costs down, while the gas companies–thanks to cozy relationships with city aldermen–enjoyed government-protected monopolies. By the early 1880s more than 70 percent of the consumer price of gas represented pure profit for the corporations, which meant that they would make money even if they slashed prices to compete with electric light. Gas company executives used political clout and bribery to try to keep Edison from getting a franchise for his system; they failed only because the Edison Electric investors were equally powerful. The biggest obstacle to incandescent lighting, however, was not political manipulation but economics. Gaslight was cheap, and electricity was not. Edison closely studied the costs of gas lighting and tried to eliminate all waste from his system. "Everything must be got down to the last penny," he said.[3]

Edison used a high-resistance filament for his bulb because it would give more light with less current, and he built the most efficient generator to date. In 1880 he made a breakthrough in circuit design that reduced the use of copper, one of the single biggest costs of an electric lighting system. Edison initially had planned to use what was known as a *tree circuit* to carry electricity from the central station to the lamps. All conductors of electricity offered some resistance to the current that flowed through them, which caused a portion of the electrical energy to be converted into heat and therefore wasted. Because of this lost energy, voltage dropped at the farther reaches of the circuit, and the lamps most distant from the generator were dimmer than those closest to it. To correct for this, the conductors in a tree circuit were thicker (and therefore had a lower resistance) near the generator, then tapered like tree branches as they got farther away. The initial thickness of the conductors prevented losses due to resistance, thereby keeping the lights shining with equal brightness throughout the circuit.

This worked well on a small scale, but a large system like the one planned for New York would have required the conductors to be as thick as tree trunks near the generators, and the copper costs would have been prohibitive. To skirt this problem, Edison designed a new circuit known as the *feeder-and-main* (see figure 2). Rather than having one thick, tapering conductor leading from the generator to serve all of the lamps, as in the tree circuit, a number of conductors (feeders) branched off from the generator and carried current to local circuits (mains) that served homes and offices. The mains, in other words, were

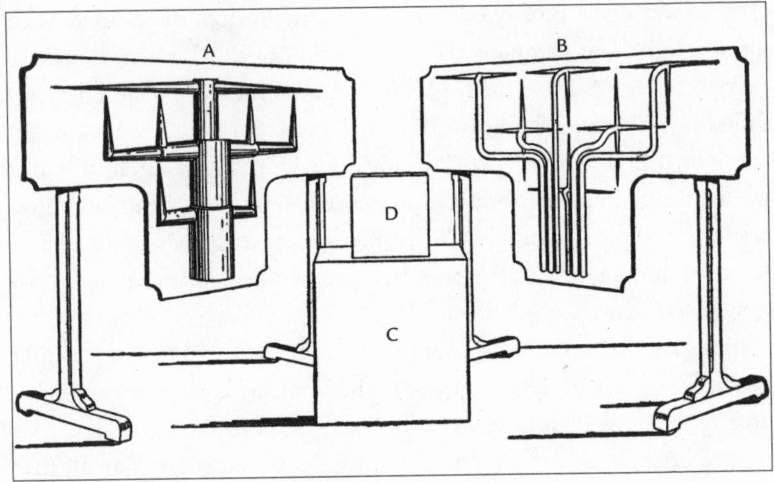

Figure 2: Compared to the older tree system (A), Edison's new feeder-and-main circuit design (B) required far less copper, one of the most expensive elements of an electric lighting system. The two cubes represent the amount of copper needed to light the same number of lamps using the tree system (C) and the feeder-and-main (D).

laid near the buildings to be illuminated, at some distance from the central station, while the feeders were longer, connecting directly to the station. The voltage drop (roughly from 120 volts to 110) was localized in the feeders, so a steady pressure of about 110 volts was maintained in the local circuits, where the lamps were located. The cost of copper for the feeder-and-main was about one-eighth that of the tree system. When asked why no one had thought of this design before, the great British physicist Lord Kelvin replied, "The only answer I can think of is that no one else is Edison."[4]

Even with the feeder-and-main circuit, the system required so much copper that the only way to make it pay was to build it in an area of high population density, so that many customers could be served with relatively few feet of feeders and mains. Lower Manhattan, the site of Edison's chosen district, fit the bill perfectly. Edison sent a small army of canvassers into the fifty-one-square-block area to ask residents how many gas jets they had, how much gas they used at various times of the year, what their gas bills were, and whether they would be willing to try electric light. The canvassers found about 1,500 customers using 18,043 gas jets. Promised free wiring and prices comparable to those of gas, almost all of the customers agreed to try electricity.[5]

As he worked toward the big dream of a Manhattan system, Edison was distracted by another project. Henry Villard, a major stockholder in the Edison Electric Light Company, also headed the Oregon Railway and Navigation Company. Impressed by Edison's New Year's demonstration, Villard decided that the incandescent light should be installed in the Oregon company's new steamship, the SS *Columbia*. Although reluctant to be distracted from his central station work, Edison decided that the *Columbia* offered a priceless opportunity to showcase his invention. The ship docked in Manhattan, and Edison's men set to work.

The *Columbia*'s lighting system—the first commercial installation of incandescent lamps—was rather crude. Rough cotton dipped in paraffin served as insulation for the wires, which were tacked to the woodwork with iron staples; when wires crossed over metal, a sleeve of soft rubber was slipped over them. Edison's men developed the first lamp sockets, wooden devices with two metal strips inside that pressed against two strips at the base of the bulb. (In the laboratory, the lamp's lead-in wires had been wound around terminal screws, a method too complex for general use.) No one had yet conceived of the screw socket, so the bulbs sat rather loosely in their sockets. The *Columbia* project inspired the invention of the first "safety catch," or fuse, a strip of soft metal that melted and broke the circuit if the line overheated, thus preventing fires and allowing the insurance underwriters to sleep easier. The last items to be installed were the bulbs themselves. The filaments were so fragile that Edison's men wrapped them in cotton batting, placed them in baskets, and carried them by hand through the streets of New York, careful to avoid jostling by passersby.

If a passenger in a stateroom wanted the lamps turned on or off, he called a steward, who unlocked a box outside the room to turn the switch. Few passengers complained about this inconvenience, though, because electric lights were such an improvement over oil lamps, which quickly fouled the air of the small staterooms. The *Columbia* reached San Francisco on July 26, 1880, and Californians flocked to see the spectacle of the new incandescent lamps. They were a bit disappointed, because about half the bulbs had expired en route, their brittle filaments fractured by the ship pitching in rough seas. A shipment of improved bulbs–cushioned in paper and packed in barrels–was rushed overland to fill the *Columbia*'s empty sockets.[6]

Edison and his men had been fiddling with filaments to lengthen bulb life even before the steamship project arose. Manufacturing the lamps seemed to be a hit-or-miss affair: A few lamps lasted as long as 500 hours, while apparently identical ones expired after just a few minutes. Starting in January 1880, dozens of materials–horsehair, fish line, teak, vulcanized rubber, cork, celluloid, grass fibers, linen twine, tar paper, wrapping paper, parchment, rattan, California redwood, corn silk, flour paste, leather, sassafras, cinnamon bark, eucalyptus, turnip, ginger, and macaroni–made their way into the carbonizing furnace and thence into evacuated bulbs. A carbonized thread of spider silk displayed a beautiful light pink phosphorescence but expired quickly.

By late spring of 1880 Edison and his men had a fairly good idea of what they were looking for: raw plant materials with dense, long, uniform fibers, which tended to hold their shape after carbonization. Rejecting jute and hemp, Edison finally settled on bamboo. Not quite satisfied with the quality of available specimens, Edison ransacked New York for new varieties of bamboo, then dispatched explorers to Japan and the Amazon. Another explorer, John Segredor, slogged through the Florida swamps. "What renders this job interesting is the strong probability of getting bitten by a snake," Segredor wrote to Edison. He traveled on to Cuba, where Edison sent him $150 for expenses. Two weeks later the explorer was dead of yellow fever. Sent news of the death, Edison drafted a response: "Bury him my expense with funds his possession." Reconsidering, he crossed that out and wired simply, "Bury him my expense." Edison located a source of high-quality bamboo in Japan.[7]

When the lamp experiments started in 1878, Edison employed about two dozen men, but by 1880 his staff had ballooned to sixty-five. With many different projects proceeding simultaneously—lamps, generators, insulators, conductors—Edison's role became that of director of research, but he took the lead role in every project. Although the staff had more than doubled in size, the atmosphere stayed largely the same. The boss still worked superhuman hours, and his men were expected to work whenever the boss did, day or night. Few seemed to mind. "The strangest thing to me is the $12.00 I get each Saturday," Francis Upton told his father, "for my labor does not seem like work, but like study and I enjoy it."[8]

The black sheep of Menlo Park's happy family was the glassblower Ludwig Boehm, whose arrogance made him the target of pranks. To escape harassment, Boehm moved out of his Menlo Park boardinghouse and into the attic room of the glassblowing shed, but it proved an unsafe retreat. Late at night the boys used one of Menlo Park's lesser-known inventions, an enormous ratcheting rattle mounted in a soapbox and turned by a crankshaft. It was known as "the corpse-reviver." They pressed it against the wall outside Boehm's bedroom, turned the crank, and produced a sound like a dynamite explosion, knocking Boehm clear out of bed.[9]

Boehm complained to Edison, who never bothered to solve the problem, though later he probably wished he had. Boehm quit in October 1880 and found work with a new Edison rival, the Manhattan-based United States Electric Lighting Company.

U.S. Electric was founded in 1878 to exploit the lighting patents of Hiram Maxim. Late in 1880 U.S. Electric installed 150 incandescent lamps at the Mercantile Safe Deposit Company in Manhattan. The lamps were round, not pear shaped like Edison's, and the filament was shaped not like a horseshoe but like a capital *M*. They nonetheless looked familiar, and there was a good reason for this. Earlier in 1880, Maxim had appeared in Menlo Park. In a display of courteousness that he later regretted, Edison spent an entire day showing him around and explaining the lamp-making process. Maxim then went back to his own lab at U.S. Electric, copied Edison's design, and went into business. The hiring of Ludwig Boehm helped U.S. Electric perfect bulb production.[10]

Newspaper reporters, uninterested in the niceties of patent law, offered high praise to Maxim's work in New York. It had been nearly a year since Edison had unveiled his lamp and promised to install lights in New York, yet he remained in Menlo Park, tinkering with his system. People were growing impatient. Maxim's "name will be remembered long after that of his boastful rival is forgotten," *Illustrated Scientific News* claimed, and others agreed. "The present stir over the Maxim incandescent light has a better basis than the Edison excitement a year ago," a newspaper reported. Edison "is no more called the 'magician of Menlo Park.'"[11]

Edison professed not to worry about what he was called or what his electric light rivals were up to. "I put the lights on the *Columbia*, but what good did it do? It put me back six weeks in the work I am doing here," Edison explained. "I could have done six months ago what Maxim has done, had I desired to make a show." In Menlo Park and on the *Columbia*, Edison had already installed lighting plants similar to Maxim's in New York. He was working on something big—not a handful of small dynamos running a few hundred lamps but a major system able to power entire city blocks.[12]

In preparation for the Manhattan system, Edison built a model in Menlo Park. He plotted a grid of imaginary streets in the fields around the laboratory and had his men put up white pine lampposts at fifty-foot intervals. At the top of each pole was a lamp, covered with a fishbowl-shaped globe to protect it from the elements. The posts leaned crookedly, but Edison gave them little thought, because the real work was going on underground. Laborers attacked the earth with plow and shovels, digging trenches and then laying long, shallow wooden boxes into them. Workers laid copper wires into the boxes, then poured tar over the wires. The job was completed by July 1880, and the conductors seemed to work well enough. When it rained, however, the wet ground became a better conductor, and electricity leaked everywhere. The lamps on the poles barely flickered.

Edison sent one of his men into the library to read up on insulation. The state of the art was not advanced, so the Edison workers made it up as they went along. For the next two months the men boiled up batches of noxious compounds, driving everyone else out of the laboratory. Finally, they settled on a mixture of paraffin, beeswax, linseed oil, and Trinidad asphalt. The trenches were dug up and the conductors laid again with the new insulation. The work was delayed by box turtles that, investigating the new smell in the neighborhood, got stuck in the tar and had to be rescued. The system was finally ready for another full test on Tuesday, November 2, which also happened to be the day of a presidential election. Edison, a staunch Republican, gave orders that the lights should be lit only if James Garfield won. He did, and the long rows of lamps glittered atop their poles until after midnight.[13]

With this successful test, Edison was ready to tackle New York. On December 17, some of the same investors who two years before had formed the Edison Electric Light Company–most notably, the investment bankers of Drexel, Morgan & Company–incorporated a new firm, the Edison Illuminating Company of New York, to finance a Manhattan central station.

Republican though he was, Edison knew that building his system in New York meant fraternizing with New York's Tammany Hall Democrats, so he invited New York's Board of Aldermen out to Menlo Park. On December 21, the politicians chartered a private car on the Pennsylvania Railroad and arrived at Menlo Park about half past five. As the aldermen approached the station, they saw the model light system, parallel strings of stars against the backdrop of a moonless night. After detraining, the aldermen strolled up a brightly lit plank walkway to Edison's office, where the inventor greeted them, his hands still grimy from a laboratory project he had abandoned a moment before.

He walked the politicians to the laboratory, where he explained the details of the system, then led them to his library and showed them a big wall map of his proposed lighting district, an area of fifty-one square blocks bounded by the East River and Spruce, Nassau, and Wall Streets. He would build a central station with steam engines and generators, lay conductors under the streets, and run wires into homes and businesses.

The group returned to the laboratory. The tools of the inventing trade had been cleared away, and in their place appeared a long table laid with a feast catered by Delmonico's: roasted and boned turkey, duck, chicken salad, and ham. Champagne, wine, and brandy flowed freely. Edison mixed his wine with liberal doses of cold water before letting it cross his lips. The aldermen drank less cautiously, and by the time they stumbled out to the train at eleven they were promising Edison he would have his permit to dig up the streets of New York and lay his conductors.[14]

AS EDISON REFINED his system in Menlo Park, the Cleveland electrical pioneer Charles Brush was introducing powerful arc lights to American cities from coast to coast. A chemist by training, Brush began following electrical progress in Europe in the early 1870s, and by the end of the decade he had designed his own generator and arc lamp and founded the Brush Electric Company. In 1879 San Francisco became the first city to adopt the Brush system for street lighting, and Boston and Philadelphia followed suit. On December 21, 1880, the same day that the aldermen visited Menlo Park, Brush installed twenty-two lamps on ornamental cast-iron posts along Broadway in New York, from the bottom of Union Square up to Madison Square. The *New York Star* reported that the lamps "made the gaslights look sickly in comparison." The arc lamps were so bright that just twenty-two of them displaced 500 gas streetlights.[15]

Brush installed dozens of lights in textile mills and steel plants, while for other customers the appeal of the lamps lay in their novelty.

Charles Brush's arc lamps lit up Broadway in Manhattan in December 1880.

Brush sold one of his first lamps to a Cincinnati dentist, who installed it on his balcony to attract patients. Many small-town Americans saw their first arc lamps inside the bigtop of P. T. Barnum's traveling circus. John Wanamaker bought twenty lights to dazzle customers at his new department store in Philadelphia, and for a time Coney Island's arc lights were more popular than the rides. By the end of 1880, Brush had installed more than 5,000 lamps.[16]

Before the late 1870s, Americans experienced electricity in only a few ways: as pulses zipping through telegraph wires or as surprising shocks administered by doctors or carnival showmen. When arc lamps appeared—slender wires carrying enough energy to produce blinding light—they heightened the sense of mystery surrounding this unknown force. Whenever arc lamps appeared in a city, people gathered to experience this latest manifestation of electricity's powers.

Brush arc lamps arrived in Buffalo, New York, on the night of July 13, 1881. After noticing a strange glow in the sky that night, many city residents walked across the Michigan Street bridge into an industrial area known as the Island, at the edge of Lake Erie. As they turned left onto Ganson Street, the Island's main thoroughfare, they saw an iron pole, twenty feet high, topped by a blazing white light. Two wires ran from that post to another, a few hundred feet beyond, and beyond that was still another pole—twelve in all, stretching alongside the wooden plank roadway for more than a mile. A small sun rested atop each pole, buzzing and sputtering and throwing a hard, silvery glare on the industrial landscape of planing mills, grain elevators, and dry docks. By the time a group of reporters and leading citizens arrived, their carriages could barely force a passage through the crowd of thousands who milled about in the road, blinking and shielding their eyes.[17]

When the lights were shut off late that night, the people strolled back across the bridge and made their way home, but word spread around town, and each night more people crossed over to the Island to bask in the glare. The Brush company, eager for good publicity, allowed visitors to enter the rough wood-frame building that held the steam engines and dynamos. After inspecting the machinery, some of the visitors—up to a dozen or so at a time—joined hands in a semicircle, with the person at each end grasping the iron railing surrounding the generators. A small amount of electricity leaked from the dynamos to the iron railing, causing a gentle tingle of electricity to pass through their bodies. It was a descendant of the Leyden jar experiment tried on Parisian monks more than 100 years before, a practice kept alive by traveling showmen with batteries and induction coils. The people of Buffalo found that the old trick worked just as well with this new producer of electric current, the dynamo.[18]

About nine o'clock on the night of August 7, a man named Lemuel Smith entered the lighting works with three friends. They joined hands, touched the railing, and laughed as they felt the mild current pass through their bodies. Smith and his friends left the plant and repaired to a saloon, where they drank a lot of beer; Smith then returned, alone, to the lighting works. When he reached over and tried to touch one of the dynamos, G. W. Chaffee, the plant's manager, ordered him away, and a police officer sent Smith tumbling out the door. He returned, and again the officer pitched him out. Thinking he had solved the problem, the officer took a stroll down Ganson Street. Smith, who had been lurking around the corner, once more entered the plant. This time Chaffee had his back turned.

Smith leaned over the railing and grasped one of the poles of the generator with his right hand, hoping to feel the tingle again. He felt nothing. He reached with his left hand and took hold of the other pole—thus completing a circuit and sending an enormous surge of electricity through his body. Smith gave one convulsive gasp and collapsed across the railing. Bystanders dragged him away from the machine and laid him on his back on the rough wooden floorboards. He was dead.[19]

According to the *Buffalo Morning Express*, "It was a lightning death, and a painless one."[20]

CHAPTER 6

Wiring New York

IN FEBRUARY 1881 Edison moved his family to the Lennox Hotel in Manhattan and established a new headquarters at Sixty-five Fifth Avenue, just below Fourteenth Street. The brownstone held Edison's offices, but it also served as a showplace. A steam engine and dynamo were installed in the basement, and the windows of the building blazed with electric lamps. Edison threw open the house to visitors every night until midnight, and New York's elite flocked to see the light and meet the great man. They learned that, among Edison's other skills, he could hit a spittoon with pinpoint accuracy.[1]

Not long after he moved to New York, Edison tried to persuade some of the Edison Electric Light Company's major stockholders to invest in manufacturing companies that would build the components–generators, lamps, conductors–of the new central station. The stockholders, who had yet to see a return on the money they invested in Edison's lighting system two years before, had no interest in sinking more money into the project. They saw Edison Electric as a patent-holding corporation: It would avoid risky manufacturing enterprises and make its money simply by licensing Edison's patents to outside companies.

A dapper Edison with cigar.

In the absence of investors, Edison was forced to sell off much of his stock in Edison Electric and finance the manufacturing concerns himself, and he drafted his top Menlo Park lieutenants to run the new companies. John Kruesi, the master machinist, headed the Edison Electric Tube Company, established on Washington Street in Manhattan to build and lay the underground conductors (which were encased in iron pipes, or *tubes*). Kruesi's Menlo Park assistant, Charles Dean, ran the Edison Machine Works on Goerck Street on the Lower East Side, which built the new generators. Francis Upton was in charge of the Edison Lamp Company, the lone holdout in Menlo Park. Sigmund Bergmann, another old friend of Edison's, operated Bergmann & Company, manufacturer of the smaller elements of the Edison electrical distribution system, such as sockets, switches, fixtures, and meters.[2]

. . .

THE LAYING OF underground conductors—the feeders and mains—under the streets of New York proved to be a monstrously difficult challenge. It was also an avoidable one. Telegraph services and arc light companies ran their wires above ground, on poles or across rooftops. Had Edison followed suit, building his system would have been relatively easy. But back in 1878, when he made his first premature announcement of success in incandescent lighting, he said that the wires would be "laid in the ground in the same manner as gas pipes," and he never wavered from this plan. As an ex-telegraph man, Edison knew that overhead lines could be felled by storms or cut by unscrupulous competitors, causing disruptions in service and loss of faith in the new technology. Edison wanted his light to be as reliable as gas and thus felt that his current had to run underground, as gas did.[3]

The work started in the spring of 1881. The city allowed Edison to lay conductors only at night, to avoid disrupting traffic during the day, and it also required him to hire five city inspectors at five dollars a day. "We watched patiently for those inspectors to appear," Edison said. "The only appearance they made was to draw their pay Saturday afternoon." It was classic Tammany Hall graft, and Edison happily paid it to avoid interference.[4]

To lay the conductors underground, a gang of men first tore up the heavy granite cobblestones and stacked them on the curb. Another gang dug a trench a few feet deep and laid the mains. Each main consisted of two copper conductors wound with rope to separate them from each other and slipped inside an iron tube. Then the insulating compound—kept bubbling hot in cart-mounted cauldrons—was pumped into the tube. At every street corner was a "safety box" containing a lead fuse that melted if the circuit became overloaded. Problems—such as air bubbles in the insulation that allowed electricity to leak—cropped up constantly and had to be fixed on the fly. Once the mains were laid, another crew punched holes into cellar walls and connected street mains to the buildings.[5]

Edison hired fifteen "wire runners"—most of them telegraph line-

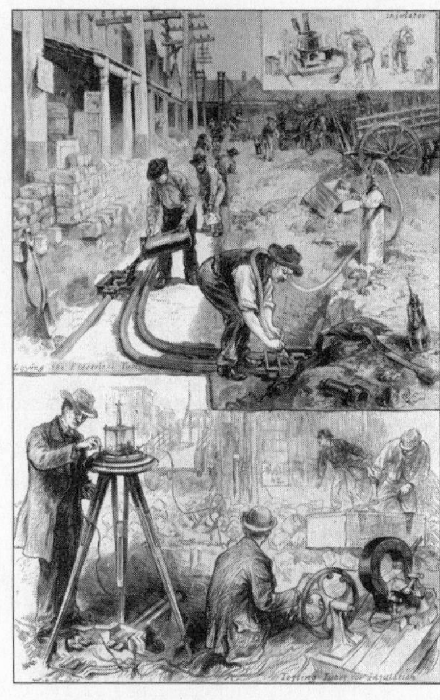

Workers laying underground conductors for Edison's first lighting district in lower Manhattan.

men or burglar alarm installers—to wire, free of charge, the homes and offices of every customer in the district who agreed to try the electric light. He promised them that they would be enjoying the lights within a few months, by the summer of 1881. But in December—a year after he displayed his Menlo Park model system to New York's aldermen—only about a third of the tubes had been laid. Then frost locked up the earth, halting work until spring.[6]

OVER THE WINTER, the Machine Works became the inventor's favorite haunt, where he received visitors and worked on the generators late into the night—this despite the plant being located in a Manhattan slum so dangerous that Edison Electric sent its house detective to protect Edison whenever he ventured there. At two or three in the morning he would repair to a tiny all-night restaurant, where, he said, "for the clam chowder they used the same four clams during the whole season, and the average number of flies per pie was seven. This was by actual count."[7]

The Machine Works was busy building small dynamos to feed the explosive growth of a business that Edison earlier had resisted: "isolated lighting" plants, in which users bought generators and produced their own electricity, such as those that lit the offices at Sixty-five Fifth Avenue and the SS *Columbia*. By early 1881 Edison Electric had received more than 3,000 requests for such plants from around the globe, but Edison initially turned nearly all of them down for fear they would distract from central station work. He believed that isolated plants were too expensive to be anything more than a niche market, and that central stations were the only cost-effective way to bring his light to the world.[8]

As the Manhattan campaign encountered delay after delay, however, the isolated plant business grew too lucrative to ignore. Some of the first customers were lithography plants and cotton mills, which were too fire-prone for illuminating gas but needed good light to distinguish colors. Major cultural institutions–Boston's Bijou Theater, the Academy of Music in Chicago, and Milan's La Scala Opera House–also bought Edison plants, both to improve lighting and to cut the risk of fire from gas lighting. Among New York's upper crust, an electric lighting plant became the latest status symbol. Cornelius Vanderbilt installed lights in his Fifth Avenue mansion, but his wife insisted that it be stripped out after crossed wires scorched the wallpaper. J. P. Morgan's electric wiring sparked a fire that charred his library walls and carpets, but he remained enthusiastic about the lights.[9]

Rather than hampering central station work, the isolated lighting business proved to be a godsend, demonstrating the brilliance of the Edison system in the United States, Europe, and Latin America. It also created a market for the products of the Edison manufacturing companies, allowing them to remain solvent as they refined manufacturing techniques. This was especially true for Bergmann & Company, maker of the new lamp sockets, which were a big improvement over the crude wooden ones used in the *Columbia* installation. According to legend,

Edison had a flash of inspiration while unscrewing the cap of a can of kerosene, and the screw socket was born. Like many of Edison's inventions, the design seemed so simple and obvious that people wondered why no one else had thought of it. The screw socket allowed even the inexperienced user to seat the bulb firmly in its base without jarring the fragile filament. Bergmann transformed Edison's original wooden screw socket into an elegant plaster of Paris design, and he worked similar magic on other electrical components. In addition to practical elements like switches and safety catches, Bergmann & Company designed and sold ground-glass shades, wall brackets, and the elaborate ceiling fixtures known as *electroliers.* Bergmann also made the electrical meter required for Edison's central station business. The meter, perfected by the late spring of 1882, worked on the principle of electroplating: Zinc plates were placed in a solution of zinc sulfate, and a tiny, fixed fraction of the electricity entering a home was shunted into the solution, causing the zinc from the solution to be deposited on the plates. Edison employees "read" the meters by weighing the plates.[10]

Although the Machine Works churned out smaller dynamos for the isolated lighting plants—ones that powered about sixty bulbs—most attention centered on machines big enough to power thousands of lamps from a central station. The first such generator was destined for use not in New York but at the Paris International Electrical Exposition. Edison and his staff completed it in September 1881, just hours before it was set to be shipped to Europe. They knocked the machine apart and packed it into 137 crates, and a convoy of horse-drawn trucks rattled to the wharves, its path cleared by an army of bribed police officers. The crates were loaded aboard a steamship that recently had delivered P. T. Barnum's elephant, Jumbo, to New York. Newspaper writers linked the big beast to the big machine, which henceforth was known as Jumbo.

The Jumbo dynamo and the rest of the Edison system took the Paris exhibition by storm, earning top honors and showing the world that Edison was miles ahead of Hiram Maxim and other competitors. Early in 1882 a second Jumbo was shipped to England, where a model central station lit up hotels and offices in the heart of London, earning still more acclaim for Edison's light.[11]

For a time it seemed that Edison might follow the Jumbo dynamo across the Atlantic. In January 1882 the family doctor warned Edison that his wife, Mary, "seems very nervous and despondent and thinks that she will never recover," and he advised Edison to take her to Europe. Mary suffered from a medley of ill-defined health problems, and marital neglect did not help the situation. Edison's first love was always his laboratory, with home serving primarily as a source of meals. The couple's daughter later said, "Seeing my Father on Sunday was not enough for Mother," but often she was lucky to see him even then. Her health declined further after the 1878 birth of her third child, Will—who weighed in at twelve pounds—but her husband remained as sparing with his sympathy as with his time. In one lab notebook Edison doodled his wife's name several times, transforming her maiden name—Stilwell—into "Stillsick." The doctor's warning early in 1882, however, had frightened Edison. A European vacation seemed excessive to the inventor, but he agreed to take Mary and the children to Florida for a couple of months, and her health improved.[12]

Edison and his family left Florida and moved back to Menlo Park just as the spring thaw was making it possible for his work crews to start laying feeders and mains again. Reports of the Edison system's triumphs in London appeared in the American press, calling further attention to the delays in New York. The London station buzzed to life so quickly because it was a temporary exhibition only, and Edison therefore allowed the conductors to be strung along a viaduct rather than laboriously buried underground. Most New Yorkers, though, knew only that the Londoners were enjoying the new technology while they were not. Residents of the lighting district grumbled that for more than a year the interiors of their homes had been festooned with useless electric wires, "objects of neither ornament nor utility," as the *New York Times* described them.[13]

. . .

EDISON WAS STILL CONFIDENT that the decision to place conductors underground was the right one, but his reasons had shifted. Originally motivated by a desire for reliability, he soon became convinced that safety was an equally important concern.

In Edison's propaganda battle with the gas lighting companies, the safety issue was an important weapon. Even gas companies had to admit that electric light was cleaner and steadier than gaslight, so they played up electricity's ability to start fires, such as those at the Vanderbilt and Morgan mansions. In response, the Edison Electric Light Company's *Bulletin*, distributed to stockholders and the press, highlighted the dangers of gas: house fires, explosions at gas plants, and asphyxiation by leaking gas pipes.[14]

Before long, many of the warnings issued by Edison Electric referred not to illuminating gas but to electric arc lighting. While the Edison system used only about 100 volts, arc lights required as much as ten times that, and this potent current had begun to claim lives. A carpenter died in Lyon, France, in 1879, and the following year two more men died from arc light shocks–one aboard a Russian yacht and another in Birmingham, England. In 1882, a year after Lemuel Smith in Buffalo became the first American to be killed by a dynamo, there were electrical deaths in a Pittsburgh iron mill and a Cleveland steel plant, and a lineman for an arc lighting company died dramatically atop a pole on Canal Street in Manhattan.[15]

When newspapers began to warn of the dangers of electric lighting, the Edison company insisted that its own system was harmless. Because its wires were underground, there was no danger that they would cross with overhead arc light cables and allow high voltages into homes and offices. Edward Johnson, the old friend of Edison's who served as vice president of Edison Electric, took the lead in promoting the safety of the company's current. Showing off for reporters one day, Johnson allowed the Edison current to flow through his body and reported that "there was no appreciable sensation whatever." Johnson was being less

The exterior of the Pearl Street station. Coal is being dumped from a cart through a sidewalk vault into the cellar.

than honest—a shock of 100 volts would have been not only palpable but painful—but he was certainly right to point out that the Edison current was unlikely to kill.[16]

Preparations in New York reached a fever pitch in the spring and summer of 1882, particularly at the four-story building on Pearl Street that would serve as the hub of the lighting system. Butting against the Third Avenue elevated train line and within smelling distance of the Fulton Fish Market, the site was chosen because it was cheap and near the center of the lighting district. One side of the Pearl Street building was set aside for offices and sleeping quarters for Edison and the station attendants, the other for the power plant itself. The two floors that would hold the heavy machinery were reinforced with wrought iron. By April the boilers were installed, and the steam engines had been received from the manufacturers. As the Jumbo dynamos neared completion, the Machine Works was so busy that it was bursting at the seams. Edison paid off the local Tammany boss to let him work on the sidewalk, and soon he ran power belts through the shop windows and set up engine lathes on the streets.[17]

The laying of feeders and mains was proceeding at a rate of about 1,000 feet a night. Edison spent as many as four nights a week down in the ditches, solving technical problems with the conductors and lending another strong back to the cause. By July the entire network had been laid—more than fourteen miles in all.[18]

One morning in July black coal smoke streamed from the two steel stacks that towered eighty feet above the Pearl Street building—evidence that the system was being tested for the first time. The station was arranged vertically, with every bit of space, from basement to top story, in use. The process started on the street. A workman heaved open the metal door of the sidewalk vault, and a teamster tipped his truck and sent a load of coal rumbling down into the cellar. Screw conveyers, powered by a twenty-horsepower engine, carried the coal up one story to the boiler room (and more screw conveyers delivered the ashes to barrels back in the basement). Workmen shoveled the coal into stoke holes of the four Babcock and Wilcox boilers, which fed steam to the six steam engines in the floor above. The shaft of each engine was coupled directly to an adjacent Jumbo dynamo, each of which stood six feet tall and produced enough current for 1,200 lamps. Electricity from the dynamos fed into heavy copper conductors along the walls, which terminated in a switch that linked them to the feeders that took current into the streets. For testing purposes, the current could be diverted one floor up, to a bank of 1,000 lights mounted on the wall.[19]

After conducting test lightings of a few selected buildings, Edison believed he was finally ready for a public unveiling, and he set the date for September 4, 1882—almost exactly four years after he had first started to work on the incandescent lamp. When Edison arrived at the Pearl Street station at about nine that morning, he was wearing a frock coat and white, high-crowned derby, but he immediately threw off his coat and collar and plunged into a final check of his system. In midafternoon he put his coat and collar back on and synchronized watches with the station engineers. Trailed by a gaggle of

Jumbo dynamos at the Pearl Street station.

reporters, he set out for the Wall Street offices of his bankers, Drexel, Morgan & Company.

For the moment, the lighting system served only fifty-nine customers with 1,284 lamps, but those few customers were an influential group that included not only Drexel, Morgan but also the buildings of several newspapers. Edison arrived at J. P. Morgan's offices and greeted the board of directors of the Edison Electric Light Company, who had gathered to witness the culmination of the work they had financed. The men chatted nervously as they waited for three o'clock sharp, the hour appointed for the test.

Back at Pearl Street, the head electrician closed the switch and sent current flowing through the streets. At the Morgan office, Edison, pocket watch in hand, advanced toward the switch controlling a large electrolier and turned it.[20]

The lamps glowed, but they seemed a bit dim—it was midafternoon on a late-summer day, and electric light could not outshine the sun. As darkness fell, the virtues of the lamp grew more apparent. The throng of Brooklynites heading down Fulton Street to the ferry were stopped in their tracks by the new lights shining in the windows of the shops. Instead of tongues of flame, they saw slender threads glowing within glass globes. According to the *Herald*, "From the outer darkness these points of light looked like drops of flame suspended from the jets and ready to fall at every moment."[21]

Edison told the gathered reporters, "I have accomplished all that I promised."[22]

CHAPTER 7

The Hanging Ritual

ONE RAINY AFTERNOON a few days before the official start-up of the Edison system, a policemen ran into the Pearl Street station and said there was trouble at the corner of Ann and Nassau Streets. Edison rushed to the spot, where a wet patch of pavement was giving electric shocks, and a large crowd had gathered to await the next unwary passerby. Just as Edison arrived, an old horse approached, pulling a fruit peddler and his cart. When the mare reached a certain spot on the cobblestones, she reared like a circus stallion and galloped off at a frantic clip, nearly capsizing the cart. The crowd roared with laughter, but a policeman put a stop to the fun by blocking off the street. After Edison and his men cut the current, they dug up the street and found that a company laying steam pipes had damaged the electric main, allowing electricity to escape into the wet ground.

The newspapers played the story for laughs, and so did Edison, who said that a horse seller requested a dynamo for his lot so "he could get old nags in there and make them act like thoroughbreds."[1]

At the low voltages of the Edison current, accidental shocks were a source of humor, but the inventor understood that higher voltages could be deadly. The three arc lighting fatalities in 1882 had raised public concerns about the safety of electricity; they also revived interest in electricity as a method of deliberate killing. In late August 1882, a week or so before the cart horse incident, Edison received a letter from a woman who wanted to know if electricity could kill livestock humanely. He told her that a dynamo could "kill instantly."[2]

Others claimed that electric currents might be a good way to kill human beings. In 1879 the *New York Herald* had run several articles on executing prisoners with electric currents, and a year later the trade journal *Manufacturer and Builder* proposed placing a prisoner in "an arm chair—the seat of death" and using electricity to "shock the victim into the next world." The 1870s and 1880s saw many proposals for new ways of killing condemned prisoners. The suggestions arose from the growing sentiment that hanging—the standard method of capital punishment in America—was barbaric and therefore unsuitable for a civilized nation.[3]

HANGING RITUALS and other elaborate methods of execution arose alongside the modern state. In medieval Europe justice had been carried out on a private basis: When a person was murdered, his family members sought vengeance upon the killer. That practice held sway as long as political power was fragmented, but by the fifteenth century a few rulers had asserted control over large regions of Europe and found that personal feuds were not conducive to orderly government. In order to gain a monopoly on violence, they decreed that justice would be a public rather than a private matter—the state, rather than the victim's family, would kill the murderer. Authorities began to stage executions as public ceremonies in which criminals were branded, whipped, beheaded, burned, drawn and quartered, or hanged. These were not gratuitous displays of violence; they were civics lessons, intended to teach lessons on the perils of lawbreaking.[4]

The colonists who settled in America brought with them the English justice system, including hanging practices. As in England, the American theater of execution was strictly scripted: prayers in the jail cell, a formal procession through the streets to the gallows, a confession by the condemned, more prayers, and, finally, the drop. By taking a man's life, the magistrates demonstrated the power of the state, while the ministers gave divine sanction to the proceedings. The prisoner, by confessing, assured the authorities and the audience that the execution was just.[5]

The members of the crowd played the most important role, because the ceremony was performed for their benefit. Watching a hanging was considered a wholesome activity, even for women and children. In their execution sermons, ministers described the condemned not as a moral monster, utterly unlike the spectators, but as a common sinner, exactly like them. All people bore the stain of original sin; all were guilty of lust and anger—the man on the scaffold had just traveled a bit farther down the well-trodden path of iniquity. As they watched the execution, citizens faced the burden of sinfulness that was the fate of all men. The crime of murder destroyed families, sowed distrust among neighbors,

William Hogarth's *The Idle 'Prentice Executed at Tyburn,* 1747. Riotous crowds at public hangings were common in England and America in the eighteenth and early nineteenth centuries.

and ripped communities apart; the execution ceremony brought them together again. A public hanging was intended to serve as a civic ritual of retribution and reconciliation.[6]

ON AUGUST 24, 1827, a man named Jesse Strang was hanged in Albany, New York, for the murder of his lover's husband. On the morning of the execution, the Reverend Lacey, an Episcopal minister, prayed with Strang in his cell for several hours. At precisely one o'clock, the condemned was ushered out of the prison. He wore black gloves, a long white shroud trimmed in black, and a white cap, also trimmed in black—traditional garb for the gallows. The minister walked on one side of Strang, the sheriff on the other. The jailer and other civil officials joined the procession, which was led by two black horses drawing a wagon that bore an empty coffin. The Albany Republican Military provided an escort.[7]

The militia escort was not a formality. Steamboats and ferries arriving in Albany were swarmed with passengers, and the roads into the city were choked with carriages and farm wagons. Visitors thronged the streets and sidewalks, bringing normal business to a halt. With great difficulty the militiamen managed to part the crowds to allow the procession to make its way to the execution site. The gallows had been constructed next to a bend in the Ruttenkill, a small river about a quarter mile from the state capitol building. From the valley floor steep hills rose abruptly on three sides, forming a natural amphitheater, and more than 30,000 people crowded the hills. Earlier in the day, when the crowds first started to swell, thirteen companies of volunteer militia had marched to the execution site and formed themselves three deep in a large circle, keeping the spectators from surging toward the scaffold.

A little before half past one the circle of soldiers parted to allow entry to the hearse, the civil and religious officials, and the condemned man. Strang mounted the steps to the gallows, and the sheriff invited him to address the crowd. In a voice loud enough to reach the hilltops, Strang expressed sorrow for his crimes and hope that his death might atone for his sins. He urged those present to "reflect upon the effects of sin and lust" and avoid his terrible fate. The sheriff granted the condemned man another hour of prayer to make his peace with God, but Strang said he was ready to die. Reverend Lacey read the burial service of the Episcopal Church: "In the midst of life we are in death: Of whom may we seek for succour, but of thee, O Lord, who for our sins art justly displeased." Strang joined fervently in the prayers.[8]

When the service was ended, Reverend Lacey commended Strang to his maker and resigned him to the civil authorities, and the prisoner drew the cap down over his own face. The sheriff bound Strang's legs, pinioned his arms behind his back, and adjusted the noose around his neck. The sheriff gave the signal, the drop fell, and Jesse Strang "was launched into eternity."[9]

In its formal aspect–the procession, the gallows confession, the prayers–Jesse Strang's 1827 hanging resembled the execution ceremonies in Massachusetts Bay a century before, but the similarities ended there. In his gallows speech, Strang held up a pamphlet titled *Confession of Jesse Strang* and declared that "every word that it contains . . . is true." The condemned man used the last minute of his life on Earth to plug a lurid account of the romance and murder that led him to the noose. Strang's endorsement delighted the hucksters who were working the huge crowd to sell the pamphlet, along with other goods. Thirty thousand spectators required food and drink, and Albany's tavern keepers and grocers happily supplied rum and ale, bread and meat. Rather than the Puritans' solemn ceremony, hanging day had become a public holiday. In the nineteenth century there was probably no single event that attracted as many spectators as a hanging, and the men and women were not always on their best behavior. According to a newspaper account of Strang's hanging, "scenes of the most disgraceful drunkenness, gambling, profanity, and almost all kinds of debauchery" had occurred around the gallows, "even at the very time the culprit was suffering."[10]

The type of disorder witnessed at Strang's hanging was not unusual, and it was beginning to make America's ruling classes nervous. Sheriffs were staging hangings as their predecessors had a century before, but the United States had become a very different place. The country's population had grown enormously, from 1.5 million in 1750 to 13 million in 1830. Many of the newcomers crowded into cities, where the nature of work and social life was being altered. Before the Revolution America had been a paternalistic society. Most goods were produced in small shops, and owners worked alongside their employees and kept a close eye on them, both when they were at work and in their leisure hours. The emergence of large factories and the growth of a market economy changed that. By the 1820s the United States was home to tens of thousands of masterless men, who put in their hours at work and could do what they pleased in their free time: get drunk, go to theaters, gamble, visit brothels. They also had a tendency to express their views through group violence. In the years before the Civil War, there were riots over elections, abolition, Catholicism, Mormonism, brothels, race, immigration, working conditions, theatrical performances, and a dozen other issues. The first American police forces were created in these years specifically to address the problems of mob violence.[11]

The working class revealed a reluctance to behave as their betters wished, and this independent streak revealed itself at public executions. The sheriffs and clergy tried to teach lessons about state power and the price of sin, but members of the audience, bent on drinking and gambling, proved to be reluctant pupils. Sometimes the condemned man chose to "die game"—according to the phrase at the time—refusing to play the contrite role and defying the authorities to the end. In such cases the crowd often treated the condemned man as a hero and the authorities as villains. "An hundred persons are made worse, where one is made better by a public execution," one man wrote in 1826. The mob had wrenched control of the ceremony out of the hands of the authorities and given it a subversive new meaning.[12]

AUTHORITIES TRIED various methods of reasserting control over hanging day. To limit the size of crowds, New York City officials in the late 1820s tried moving hangings to islands in the harbor, or starting the procession early in morning, or simply quickening its pace. In 1829 a newspaper reported that an execution procession moved "with such rapidity as to prevent the rabble from keeping pace." The public execution had become, in the words of one legal scholar, "a spectacle in flight from its audience."[13]

Many states took this flight a step farther: They privatized executions, moving them from public spaces to the isolation of the prison yard, where only a select few could watch. Pennsylvania abolished public executions in 1834, and a year later New York, Massachusetts, and New Jersey followed suit. By 1845 every state in New England and the Mid-Atlantic region had privatized executions.[14]

Although the fear of riotous hanging day mobs was the most immediate cause of the new laws, private executions also reflected a more general move away from criminal punishments that involved the public infliction of pain. Over the course of the eighteenth century, American officials gradually made fewer crimes subject to the death penalty and began to be more sparing in their use of corporal punishments such as whipping. The practice of gibbeting–hanging the criminal's body in public for weeks or months after the execution–also stopped. In the 1780s and 1790s American states started building penitentiaries, where, in theory at least, criminals enjoyed the solitude to consider their crimes, grow penitent, and transform themselves into virtuous citizens. Rather than punishing the criminal's body, the state embarked on reshaping his soul.

The United States eased physical punishments earlier than its European counterparts. In the early nineteenth century, England operated under the "Bloody Code," a set of laws that rendered more than 200 crimes capital offenses. People–almost exclusively poor people–were hanged for petty offenses such as forging bank notes or stealing a few spoons. Similar codes prevailed in American colonies for much of the eighteenth century, although executions for these lesser offenses were much more rare in America than in England. After the Revolution most American states abolished the death penalty for crimes such as burglary and sodomy, retaining it only for murder and treason. Having fought a war to be free of the tyranny of King George III, Americans had no desire to institute his brutal system of punishments in their pure young nation. In 1794 Pennsylvania lawmakers invented the concept of *degrees* of murder, in which only first-degree–premeditated–killing earned the noose. The new ideas about criminal justice also became enshrined in the federal Constitution, in which the Bill of Rights protected accused criminals and banned "cruel and unusual punishments." That phrase, borrowed from the English Bill of Rights of 1689, guarded against a return to such barbarous European practices as burning alive or drawing and quartering. Capital punishment itself was not seen as cruel and unusual.[15]

Some people, however, believed that the logical conclusion of the move away from physical punishment was the abolition of the death penalty. From the 1820s through the 1850s a vigorous anti-capital punishment movement emerged in the northern states,* supported by, among many others, the poet William Cullen Bryant and the newspaper editor Horace Greeley. The death penalty foes were met by an equally vigorous group of death penalty defenders.

The battle turned on competing views of human nature. Religious conservatives believed that humans were deeply depraved, and that only harsh laws rigidly enforced could keep anarchy at bay. Civil government was the earthly representative of divine government, and biblical justice demanded death for murderers: According to Genesis 9:6, "Whoso sheddeth man's blood, by man shall his blood be shed." Opponents of the gallows emerged from the same school of thought that created the penitentiary. In their view, human nature at heart was moral, reasonable, and capable of reform. People's characters were formed not by the corruption of original sin but by their environments, so convicted criminals needed not punishment but proper

*The movement made little headway in the South, whose citizens believed capital punishment to be necessary for controlling the slave population.

training. It was the government's duty to reform criminals, not to kill them. Abolitionists tended to belong to the more liberal Christian churches–Unitarian, Universalist, Quaker–while most vocal death penalty supporters were Congregationalist or Presbyterian. Walt Whitman, founder of a Brooklyn anti-death penalty group, denounced this ministerial support for the gallows in an 1845 essay: "When I go by a church, I cannot help thinking whether its walls do not sometimes echo, 'Strangle and kill in the name of God!'"[16]

The debate over capital punishment also took place on utilitarian grounds. Supporters believed executions deterred others from committing murder, and that outlawing the gallows would "unkennel the bloodhounds of disorder . . . and perhaps overturn the very foundations of political existence." Death penalty abolitionists considered this argument ludicrous. "Imagine a man who would like to kill another, sitting down and balancing the relative gravity of hanging or imprisonment," one lawyer wrote, "and ending it all by giving up the formation of the purpose because of capital punishment, or nursing and maturing it because of imprisonment for life!" On the other hand, many of those sentenced to life in prison managed to avoid serving the full sentence. Although successful appeals of capital sentences were rare in the nineteenth century, executive pardons were not uncommon, especially for well-heeled convicts or those who had gained the public's sympathy. "In point of fact imprisonment for life would mean confinement for a few years and then liberty to commit another crime," the *New York Herald* wrote. "Political influence and bribery would be brought to bear and the prisoner would simply laugh at his sentence as meaning next to nothing."[17]

Nearly every northern state saw legislative action to repeal the death penalty, and some efforts succeeded. In the wake of a hanging day riot in 1837, Maine passed a law that effectively (although not explicitly) ended the death penalty. In 1846 Michigan's government became the first English-speaking legislative body anywhere in the world to abolish capital punishment officially, and Rhode Island and Wisconsin followed within a decade. These successes, though, were unusual. More typical was the case of New Hampshire. When the state held a referendum on the issue in 1844, the vote ran two to one in favor of the death penalty.[18]

DESPITE THE LAWS that privatized hangings in the 1830s and 1840s, executions still drew a large audience, through the medium of the newspaper. The press expanded coverage of executions after the privatization laws went into effect, allowing people to satisfy in print an appetite forbidden in person. For the authorities, this seemed to be a perfect situation. The hanging rituals could assert their deterrent effect, but now the audience was dispersed—at home, on the streetcar, in the park. People could absorb the lessons of the spectacle without participating in it or disrupting it in any way.

In some counties, however, sheriffs flouted the privatization law by appointing hundreds of friends and constituents as "special deputies" with viewing privileges. Admission tickets distributed by sheriffs quickly found their way onto the black market, and those without tickets often climbed hills overlooking the prison yard. At the Tombs prison New Yorkers clung to chimneys and railings of nearby buildings to get a view of the sufferings of the condemned.[19]

Technically, there should not have been any suffering, because hanging methods had been changed over the years in an attempt to make executions more humane. The earliest gallows were two posts with a crossbar—the victim either walked up a ladder, which was then removed, or stood upon a cart, which was driven away. In either case, the prisoner fell just a few inches and died of strangulation. By 1800 the trapdoor gallows was common, and a few decades later some states adopted a new design: Rather than dropping the prisoner through a trap, these gallows used a suspended weight that, when released, jerked him up into the air. Sheriffs hoped that the force of the fall—or of the hoisting up—would break the neck and cause rapid death. Professional hangmen created a formula in which rope length was a function of the prisoner's weight: the heavier the victim, the shorter the drop. But many times these delicate calculations of anatomy and gravity failed to add up. Sometimes the drop was too short, and the prisoner strangled. Sometimes it was too long, and those assembled found themselves witness to a decapitation rather than a hanging.[20]

There is no way of knowing how often these problems occurred earlier in the nineteenth century, because most descriptions of hangings did not include such information. The lack of detail reflected the feeling that what was important about hanging day—the ritual procession, prayers, and confession—took place before the drop. Around 1850, however, newspapers began printing details of the victim's sufferings at the end of the rope. They described the prisoner's body convulsing and twitching, legs twisting and kicking, throat gurgling, eyes bulging, face turning purple. (The press was too delicate to note that most hanged men also urinated, defecated, and ejaculated.)[21]

Although the American people were largely deaf to pleas that capital punishment was unjust on principle, death penalty opponents had one strong card to play: the suffering of the condemned. Edmund Clarence Stedman, a foe of capital punishment, published long articles with explicit depictions of botched hangings and criticized "the dreadful, the inconceivable *physical* agonies of men who are hanged." He hoped to play on the public's sympathies, and the strategy worked. The *New York Times* claimed that bungled hangings were "accountable for most of what opposition exists among us to capital punishment." Maine in 1876 officially abolished capital punishment after a botched hanging outraged the public. Death penalty opponents, it seemed, had found the issue that might carry the day.[22]

But this strategy contained a flaw. In arguing that the problem with the death penalty was the suffering of the prisoner, Stedman and others suggested that the death penalty without pain would be unobjectionable, and they implicitly challenged death penalty advocates to find a better way to kill.

CHAPTER 8

The Death Penalty Commission

THE CONCERN WITH SUFFERING that characterized the death penalty debates was a relatively new phenomenon. For most of human history violence and pain had been unremarkable aspects of everyday life. In premodern societies minor insults led easily to duels or fistfights, which regularly escalated into bloody feuds and vendettas. In medieval England homicide was as common among noblemen as among peasants, producing a murder rate twice as high as in the modern United States. Many of those who did not succumb to violence fell victim to accident and disease: Women died during childbirth; illness routinely felled infants and children; men were maimed and killed in warfare and farm accidents; plague and famine killed indiscriminately. Pain provided the texture of everyday life, and people accepted it as inevitable. Christians saw it as punishment for sin, or even as on opportunity to draw closer to the divine by sharing the suffering of Christ. Physical suffering was routine, and compassion was a precious resource, easily exhausted and grudgingly dispensed.[1]

By the end of the nineteenth century the situation had changed. William James, the great psychologist and brother to Henry, noted in 1901 that in the past century a "moral transformation" had "swept over our Western world. We no longer think that we are called on to face physical pain with equanimity." Compassion was now extended to all of humanity, and cruelty became the worst of sins. An 1891 advertisement for a laxative expressed the new mood in rhyme: "What higher aim can man attain than conquest over human pain?"[2]

The intellectual origins of this revolution in human feeling lay in the late seventeenth century, when Anglican clerics, rejecting both the angry God of the Puritans and Thomas Hobbes's dark view of human nature, redefined God as benevolent and human nature as intrinsically compassionate. Inspired by these attitudes, a generation of writers—the philosopher Francis Hutcheson, novelists Sarah Fielding and Samuel Richardson—gave birth to a "humanitarian sensibility" and spread the gospel that cruelty was repugnant and sympathy the highest human emotion. The word *civilization* arose alongside the humanitarian sensibility and was tightly bound with it. Those who inflicted pain were "barbarous"; those who did not were "civilized." People widened their circle of sympathy to embrace not only family members and fellow villagers but a broad swath of humanity. Practices that had been unquestioned for centuries—child labor, slavery, torture, bull-baiting—came to be criticized as heartlessly cruel. The U.S. Constitution's prohibition of "cruel and unusual punishments" was an expression of this compassionate sensibility, as was the decision by many states to limit the number of crimes punishable by death.[3]

This way of thinking was restricted to the upper classes of England and America in the years before the American Revolution. In the nineteenth century the middle classes embraced humanitarianism as a sign of respectability and applied it to a wide range of social reforms. Some of the first targets were the miserable conditions in the slums of industrializing cities. Massachusetts, the first American state to industrialize, enacted its first laws regulating child labor in 1842. Other humanitarian reforms included the education of the blind and deaf, treatment of the mentally ill, public health measures, and better housing for the poor. The most famous and most significant movement was abolition, undertaken in part out of sympathy with slaves. These were complex movements driven by many motivations, but the desire to decrease the measure of human suffering was prominent among them.[4]

Nothing better illustrates the revulsion at pain than the discovery and lightning-quick adoption of medical anesthesia. Surgery before anesthesia was an exercise in torture. Doctors, like orthodox Christians, traditionally viewed pain as inevitable, or even as a necessary part of the healing process. But as fear of suffering grew, fewer doctors found this attitude viable. In 1846, a Boston dentist demonstrated that ether gas could prevent the pain of surgery. Within a few years doctors discovered that chloroform and nitrous oxide produced similar effects. Some doctors resisted because of the health risks of anesthesia, or because they continued to believe that suffering was good for the patient. Most, though, were thrilled to have a way to ease their patients' suffering. No medical discovery had ever gained general acceptance so quickly. Anesthesia both reflected and reinforced the culture's aversion to pain—now that it could be relieved, it became even more repugnant.[5]

Not only humans benefited from the humanitarian sensibility. In 1866 Henry Bergh, the wayward and wealthy heir to a shipbuilding fortune, founded the American Society for the Prevention of Cruelty to Animals. The names of the men who assisted Bergh with the project—John Jacob Astor Jr., Horace Greeley, Peter Cooper, Ezra Cornell—indicated the social constituency of the cause. Bergh's organization was modeled after the Royal SPCA in London, but the American humane movement soon outstripped its British counterpart. SPCAs were established in Philadelphia, Boston, Buffalo, and other cities. The societies had the power to bring legal action against animal abusers, but their primary mission was not prosecutory but pedagogic: teaching people, especially children, that it was wrong to make animals suffer.[6]

The SPCA drew its membership almost entirely from the upper and middle classes, and there was good reason for this. The poor—trapped in the lethal squalor of industrial America, working jobs with high fatality rates—lived lives of premodern brutality and therefore assigned a low priority to the suffering of animals. Faced with a daughter who had lost her arm in a textile factory, who cared about a cart driver beating his horse? Members of the SPCA considered this indifference to animal cruelty as unfortunate for animals and dangerous to society.

Humanitarians railed against a common game—known as "spinning the cockchafer"—in which children pinned a beetle to the end of a string and spun it in the air so they could enjoy the loud whirring noises it made. Reformers objected not out of sympathy with the animal's suffering but because they saw it as the first step down the slippery slope of cruelty. What started with beetles would then lead to dogs, and to people: "Cruelty to animals predisposes us to acts of cruelty towards our own species," ASPCA founder Henry Bergh explained. In 1870 the Pennsylvania SPCA noted two cases in which men arrested for cruelty to animals later committed murder. In an SPCA journal, an illustration titled "The Labor Problem" showed a factory in flames and the arsonist, a union man, shot dead by a militia. The caption read, "Shall it be this—or humane education of rich and poor?" The humane movement was as much about taming the lower orders as about protecting animals. In the peculiar vision of some anticruelty reformers, social problems sprang not from vicious exploitation of the poor but from torturing beetles or kicking dogs. SPCAs lobbied hardest against the cruelties that were most public—especially the beating of carriage horses—because such spectacles "tend to brutalise a thickly crowded population."[7]

This same fear—of the contagiousness of cruelty—motivated much of the opposition to hanging. Executions whipped spectators into such a frenzy that they committed violent crimes themselves. It was reported, for example, that a man who attended Jesse Strang's 1827 hanging in Albany committed murder eleven days later. Hanging produced a "demoralizing effect upon society," one writer said. "We would put an end to capital punishment, for the sake of the law-abiding classes;

just as the abolition of Slavery was wisely urged for the benefit of the white man." The statement reveals the self-interest that often lay at the heart of humanitarian sentiment. Reformers were concerned less with the suffering of victims than with the social consequences of that suffering. Spectacles of cruelty—the whipped slave, the beaten horse, the man dangling at the end of a rope—were thought to produce yet more acts of cruelty.[8]

Although a few citizens cited the suffering of hanged men as an argument in favor of abolishing the death penalty, not many were willing to take this step. Supporters of capital punishment, however, were not untouched by the anticruelty movement; they wanted to achieve the goals of the death penalty without fraying the moral fabric of society. According to one capital punishment advocate, a painless execution method would "deprive those who have the bad manners to argue against the death penalty, of one suggestion by which they operate on the nerves of others."[9]

THE PIONEERS IN THE FIELD of scientific killing were individuals connected to anticruelty societies. In the 1850s Benjamin Ward Richardson, a distinguished British physician and expert in anesthesia, constructed a "lethal chamber" for killing unwanted animals with carbon monoxide, and for the next thirty years he advocated this method of "humane destruction." In 1874 the Pennsylvania SPCA built a special brick room for killing dogs with carbon monoxide, becoming the first American organization to move beyond shooting and drowning as methods of killing unwanted animals. Instructions for building a lethal chamber were published in *Popular Science Monthly*.[10]

Soon after the Pennsylvania SPCA built its lethal chamber, a Philadelphia physician suggested that the same method be used for condemned criminals, because it produced "the easiest and quickest death known to science." This was one of many proposals for making the death penalty more humane. In 1847 a Brooklyn murderer was knocked cold with ether before he was hanged, on the theory that he would then not suffer during the drop. In the 1870s the New York Medico-Legal Society formed the Committee on Substitutes for Hanging and considered carbon monoxide, poisons, the garrote (a brass collar fitted with a screw that was turned until it crushed the spine), and the guillotine, before finally deciding that hanging remained the best option, because it shed no blood and produced death with reasonable certainty. Another physician came to the same conclusion after conducting an experiment on himself. He had a friend strangle him near the point of death with a towel and at the same time prick him with a knife so that he could judge his level of sensation. After reviving, the physician reported that he lost consciousness in eighty seconds and felt no discomfort at all—not even from the "knife thrusts he was inflicting upon my hand." He concluded that, even if the neck did not break, hanging was not cruel.[11]

Despite these reassurances, many argued that science could offer a method more palatable than hanging, and electricity soon became a favored candidate. Benjamin Franklin's killing experiments were well known, as were the 1869 tests on pigeons, rabbits, and frogs by lethal chamber inventor Benjamin Ward Richardson. Also in 1869, a writer in *Putnam's Magazine* predicted that "with new scientific knowledge, a painless mode of killing may be discovered,—as by an electric shock." *Scientific American* claimed in 1873 that killing with electricity would be "certain and painless." After interviewing prominent physicians and electricians in 1879, the *New York Herald* concluded that electricity offered a "humane, effective and impressive" method of executing condemned criminals.[12]

The discussion of electrical execution in the *Herald*, like the earlier one in *Scientific American*, mentioned batteries, Leyden jars, and induction coils as sources of electricity. By the early 1880s a more potent generator of electricity had become common: the dynamo.

Although arc lighting accidents had claimed four lives in America by the end of 1882, the only one of those deaths to receive

Alfred Porter Southwick

serious scientific attention was that of Lemuel Smith, the man who died after grasping the poles of a dynamo in the Buffalo Brush lighting plant in August 1881. The morning after the accident, Joseph Fowler, Buffalo's coroner, conducted an autopsy. Smith's lungs were congested with blood, and the blood appeared to be somewhat thinner than usual, but there was no obvious cause of death. If he had not known the circumstances of the case, Dr. Fowler said, he would not have known what killed the man. He was so intrigued by the case that he returned to the morgue for a second autopsy later in the day. After peeling back the skin from Smith's arms and chest, Dr. Fowler discovered the path of the current: a line stretching from shoulder to shoulder, about two or three inches in width, where the flesh was a little darker than the tissue on either side. A force potent enough to fell a man in seconds left only a delicate tracing upon the flesh.[13]

Reports of the autopsy caught the attention of Alfred Porter Southwick, a Buffalo dentist who had been following the debates on capital punishment and execution methods. After hearing about Smith's death, Southwick began to collect stray dogs and kill them with electric shocks. As a supporter of the death penalty, Southwick wanted to preserve capital punishment by removing the strongest objection to it:

the pain it wrought on its victims, and the damage inflicted upon society by the spectacle of suffering. "Civilization, science, and humanity demand a change," he explained.[14]

Born in Ohio in 1826, Southwick had moved to Buffalo after finishing high school and found work on the steamboats, eventually becoming chief engineer of the Western Transit Company. Restless in that profession, he studied dentistry and opened a practice in 1862. Southwick helped found the state dental society, later becoming its president, and also helped organize the Dental Department at the University of Buffalo, where he taught dental surgery. His one published scientific article was titled "Anatomy and Physiology of Cleft Palate." Southwick's two professions–steamboat engineering, which familiarized him with power production, and dental surgery, which involved the use of anesthesia–provided a background of sorts for the new art of humane killing with electricity.[15]

Southwick proved to his own satisfaction that electricity provided the most humane method of capital punishment, but he kept the results private for a few years. Then, in a speech early in 1885, New York governor David B. Hill stated that the "present mode of executing criminals by hanging has come down to us from the dark ages" and proposed that science could furnish a "less barbarous" method of execution. In those words, Southwick sensed an opportunity, because he believed that electricity was precisely the method Governor Hill sought. Southwick also had good political connections: His friend Daniel McMillan was a state senator. At Southwick's urging, McMillan introduced a bill in the state legislature to create a commission to investigate "the most humane and approved method" of execution. The bill, which passed the state legislature and became law in 1886, named three men to the commission, and one of them was Alfred Southwick.[16]

Southwick's fellow commissioners were Matthew Hale and Elbridge T. Gerry. Hale was an obscure Albany lawyer, but Gerry–chairman of the commission–was a man of note. The grandson and namesake of a signer of the Declaration of Independence, Gerry

Elbridge T. Gerry

moved in the highest circles of New York society. At the time of his appointment to the death penalty commission, he was commodore of the New York Yacht Club, and he was also a famous philanthropist. Gerry served as legal counsel to the American Society for the Prevention of Cruelty to Animals, and in 1874 he had founded the Society for the Prevention of Cruelty to Children, the first organization of its kind in the world. He made frequent appearances in the pages of New York newspapers, presiding over regattas in Newport and staging raids on theaters employing child actors. Gerry was widely recognized as a humanitarian, and his presence on the panel lent considerable weight to the inquiry.[17]

The commissioners began their work by sending a circular to attorneys, physicians, and public officials, requesting their opinions on capital punishment. Elbridge Gerry hired a staff of nine assistants and set them to work in his private law library, poring over historical and legal books for information about execution methods.[18]

Southwick, meanwhile, had an opportunity to conduct further experiments. In the summer of 1887 packs of dogs roamed the streets of Buffalo, and the city council fixed a bounty of twenty-five cents for each stray brought to the pound. Local boys, quick to spot a business opportunity, rounded up dogs by the dozen and deposited them at the pound, which became overwhelmed. Concerned about the animals' welfare, the local chapter of the Society for the Prevention of Cruelty to Animals assumed operation of the pound, including the destruction of unwanted animals. The usual method, shooting, was rejected as cruel. The society instead tried asphyxiating the dogs in a carbon monoxide lethal chamber, but that proved unreliable.[19]

The SPCA then discovered that a citizen of Buffalo was already adept at dog killing and would be happy to share his expertise. On July 16 Alfred Southwick and a friend, the Buffalo physician George Fell, constructed a pine box, lined it with a zinc plate, and filled it with an inch of water. They ran an electrical line from the nearest arc light cable, connecting one pole to the zinc plate, the other to a muzzle with a metal bit. A small terrier was fitted with the muzzle and led into the box. The *Morning Express* reported what happened next: "A simple touch of a lever—a corpse." Twenty-seven more dogs followed. At its next meeting the SPCA reported that the dogs died "instantly and seemingly without pain" and urged that all unwanted dogs be destroyed with electricity.[20]

A *Scientific American* illustration of Southwick and Fell's dog-killing cage, used in Buffalo in the summer of 1887.

. . .

THE DEATH PENALTY commissioners completed their work in January 1888 and submitted a report to the New York legislature. Although the report did, at the end, propose a new method of capital punishment, the bulk of it was devoted to a catalog of death, in which the commissioners described—in alphabetical order and exquisite detail—every method of execution they had discovered.

Southwick and his colleagues began with *auto da fe* (literally "act of faith," the Spanish Inquisition's ceremonial execution of heretics, usually by burning), then proceeded to beating with clubs and beheading, duly noting national differences in decapitation practice among the English, Chinese, and Japanese. Next came blowing from a cannon, in which the victim was either lashed to the cannon's mouth or stuffed into the barrel: "Here is no interval for suffering," the report stated, because "no sooner has the peripheral sensation reached the central perceptive organ than that organ is dissipated on the four winds of heaven." Executions by boiling employed not only water but also melted sulfur, lead, and oil. Breaking on the wheel, in which the victim was lashed to a large wheel and beaten viciously with clubs, was common for a time in western Europe (a blow to the head that brought death and ended suffering was known as a *coup de grace*—"stroke of mercy"). Death by burning was familiar to students of European history, but the commissioners did not content themselves with such pedestrian examples. They discovered a Persian practice known as "illuminated body," in which the victim was bound to a slab and "innumerable little holes were bored all over his body. These were filled with oil, a little taper was set in each hole and they were all lighted together."

The death commission's report marched on, through burying alive, crucifixion, decimation, dichotomy (splitting the body in two), dismemberment, drowning, exposure to wild beasts (in some cases a victim was sewn "in a sack alive, venomous serpents with him, and sometimes a dog, a monkey or the like were added"), flaying alive, flogging, garrote, guillotine, hanging, *hara-kiri*, impalement, iron maiden, poisoning, pounding in mortar, precipitation (throwing the victim off a cliff), pressing to death, the rack, running the gauntlet, shooting, stabbing, stoning, strangling, and suffocation.[21]

The commissioners were aware that there was something strange about their exhaustive inventory of death. They explained apologetically that "brief mention of these monstrosities" was included to "indicate the thoroughness of the research." But the descriptions were not brief, and they indicated more than the researchers' desire to display their industriousness. In part, the commissioners hoped that compared to such barbarism, their own mercy would shine all the brighter.[22]

The catalog of death also revealed the dark underbelly of the humanitarian sensibility. During the same years that pain became unacceptable, the public grew more fascinated with violence and death. Edgar Allan Poe was only the most famous of hundreds of nineteenth-century writers who dwelled with delight on blood, murder, dissection, and the putrefying corpse. At dime museums, catch-all repositories of nineteenth-century popular culture, the public could see waxwork reproductions of famous murder scenes and jars containing body parts of executed murderers. The anticruelty movement may have curtailed public executions and blood sports, but it only fed the public's appetite for violent death. Horror writing had not existed in the premodern world, when physical torment was an accepted part of everyday life. But when suffering became obscene, the stage was set for a pornography of pain.[23]

The death commissioners made great efforts to present themselves as thoroughly rational, but they had ventured into territory not easily reducible to the cold logic of science. In their efforts to make executions more civilized, the commissioners found themselves pushed and tugged by the dark allure of violent death.

IN THE SURVEY they sent out as part of their research, the commissioners had asked judges, district attorneys, sheriffs, and a number of physicians to comment on three methods of execution currently in use in "civilized" nations—hanging, guillotine, and garrote—and two novel ones, electricity and poison. Of those who replied, eight supported poison, five the guillotine, four the garrote. Eighty wanted to retain hanging. According to the commissioners, "eighty-seven were either decidedly in favor of electricity, or in favor of it if any change was made." This vague phrase left it unclear how many of those eighty-seven actually preferred hanging to electricity.[24]

But Southwick and his colleagues were not holding a referendum; they were marshaling evidence to buttress their own recommendation. Although the survey revealed that hanging had many defenders, the commissioners did not seriously consider retaining the gallows. They noted that the public disapproved of hanging female prisoners, which led to obviously guilty women being acquitted by juries or pardoned by governors. The commissioners asserted (without evidence) that people objected "not so much to the *execution* of women as to the *hanging* of women," and therefore would embrace killing women by some other method. The commissioners also recounted many instances of bungled hangings, involving broken ropes, faulty trapdoors, slow strangulations, and decapitations. Considering these problems, the commissioners believed that support for the gallows was a product of blind conservatism and could therefore be discounted.[25]

The report revealed that there were issues at stake besides pain. Carbon monoxide, the method of choice for some SPCAs, was not considered because death from the gas, while painless, could take several minutes. Anxious to avoid a prolonged killing process, which they associated with torture, the commissioners insisted that the death be "instantaneous." They rejected the spine-crushing garrote because it was too slow and because it disfigured the body. The mutilation complaint disqualified the guillotine as well. The French—who as late as the 1780s occasionally burned criminals at the stake and broke them on the wheel—had adopted the guillotine as a humanitarian gesture. New York's death penalty commissioners conceded that the method was quick and painless, but they objected to the blood.* Considering "the fatal chop, the raw neck, the spouting blood," such executions "cannot fail to generate a love of bloodshed among those who witness them." Given that the awful details would be "presented to the public in the journals of the week," the harmful effects would spread throughout society.[26]

The problem was not so much the suffering of the condemned, the commissioners explained, but the bloody spectacle. Like the SPCA leaders who focused first on public cruelties, the death commissioners wanted to protect the public from the brutalizing spectacle of suffering. Their task, then, was double: to find a painless death, and an unspectacular one.

Rejecting hanging, guillotine, and garrote, the commissioners were left with two options: poison and electricity. Poison advocates recommended a hypodermic injection of prussic acid (a form of cyanide) or morphine. Critics claimed that poisons were an unreliable form of killing, because people differed in their reactions to them. Also, the hypodermic syringe was a relatively new tool in medical practice, and physicians feared that using it for executions would create a prejudice against it "among the ignorant."[27]

That left electricity. The commission report contained a lengthy description by Dr. George Fell of the electrical dog killings he and Southwick had conducted for the Buffalo SPCA in the summer of 1887. The commissioners quoted from electrical experts who were familiar with accidental deaths from electricity and printed excerpts from survey responses that were in favor of electrical executions. The commissioners concluded by recommending electrical executions, explaining that they would be "instantaneous and painless" and "devoid of all barbarism."[28]

---

*Southwick had changed his mind about the painlessness of the guillotine. A year before the report was issued, he told a newspaper reporter that "the guillotine is a barbarous mode of execution. Death does not ensue instantly, as it should in such cases. In fact, a condemned man and I can agree upon certain eye and mouth signals before his head is laid on the block, and I can communicate intelligently with the severed head for some time after the execution by means of such signals."

The unanimous conclusion of the report masked a furious debate that took place behind the scenes. Alfred Southwick had favored electricity from the start, and fairly early on he convinced Matthew Hale to support it as well. But Elbridge Gerry—the most famous and well-respected man on the panel—believed that an injection of morphine would be the most humane method. Without Gerry's support, electricity did not stand a chance. At the last moment, however, he abandoned his preference for poison and came out in favor of electrical execution.[29]

It was not Alfred Southwick who changed Gerry's mind. On December 19, 1887, just a few weeks before the commission issued its report, Thomas Edison wrote a letter to the commission advocating electrical execution. Edison's opinion shifted Gerry's vote from morphine to electricity. The great inventor was an "oracle," Gerry later explained. "I certainly had no doubt after hearing his statement."[30]

CHAPTER 9

George Westinghouse and the Rise of Alternating Current

IN THE LATE summer of 1882, about the time the Pearl Street station began delivering light to lower Manhattan, Edison took a lease—at an exorbitant $400 a month—on a home on Gramercy Park. The move to New York had been proposed by his wife, Mary, who was anxious to escape the isolation of Menlo Park. At first she thrived on city life, hosting receptions and teas in her home. A year later, though, her health deteriorated so much that she had to give up housekeeping, and the family moved into a hotel.

Mary's health improved somewhat after a Florida vacation in the spring of 1884, but something was still not right. "I am so awfully sick," she wrote in April. "My head is nearly splitting and my throat is very sore." In the summer the couple once again left the city for the comfort and familiarity of Menlo Park, but Mary's condition worsened, and she unexpectedly died early in the morning of August 9. Dot awoke that day and found her father "shaking with grief, weeping and sobbing so he could hardly tell me that mother had died in the night." Mary Stilwell Edison was twenty-nine years old. The doctor attributed her death to "congestion of the brain."[1]

Mary's death so upset Edison that it brought about (at least temporarily) a realignment of his concerns. A thirty-seven-year-old widower, for the first time he began paying attention to his children: eleven-year-old Marion, eight-year-old Tom, and five-year-old Will. After her mother's death, Marion became her father's chief companion. He took her to the theater and afterward to Delmonico's, where he smuggled her into the men's dining room and kept her there with him until past midnight. He also bought his daughter a horse and a parrot. The bird never learned to speak, and the inventor complained that it had "the taciturnity of a statue, and the dirt producing capacity of a drove of buffalo." The horse was more useful. At least once a week they hitched it to a carriage and took a drive in the country, with Marion at the reins (her father could never be trusted to control a horse). "It seems wonderful . . . that Father had so much time for me," Marion later recalled. "He was interested in my clothes, diary, the novel I was going to write." As the ultimate mark of approval, he allowed her to assist at the laboratory. Her entry into this all-male preserve demanded a new nickname, so he took to calling her "George." In a lab notebook he left a doodled love note: "Dot Edison angel Miss Marion Edison Sweetest of all."[2]

WHILE DISCOVERING his sentimental side at home, Edison grew more aggressive in his business dealings, because he was dissatisfied with the progress of the industry. The Edison lighting empire was organized as a three-tiered system. The Edison Electric Light Company held the patents and licensed them to the independent Edison manufacturing companies (the Lamp, Tube, and Machine Works). The manufacturing companies sold the equipment they produced either to individuals who installed isolated lighting systems for private use in one office building or factory, or to local electrical utilities that funded the construction of central stations and then sold electricity to customers in their areas. The directors of the parent company, Edison Electric, hoped that the Pearl Street station in New York would prove the reliability and economy of electric light and lead investors in other cities to organize and capitalize local utilities.[3]

Because the New York station was long delayed and over budget, however, cities that had been contemplating central stations shied away. The Edison Electric directors were content to bide their time and make money from the sale of isolated lighting plants. Edison, however, did not want his lights to dot the night landscape in a factory here and a hotel there; he wanted to light whole cities, to make the entire nation glow with the radiance of his lamp. He believed that the parent company should invest in more central stations to demonstrate the system's viability. The problem, as he saw it, lay not with his system but with Edison Electric's failure to promote it—yet another example, he thought, of "the characteristic timidity of capital."[4]

"If the business is to be made a success it must be by our personal efforts and not by depending upon the officials of our companies," Edison declared. "I am going to take a long vacation in the matter of inventions. I won't go near a laboratory."[5]

He became, as he put it, "a regular contractor for electric light plants." He did not have the resources to take on something as big as the New York station, so he concentrated his efforts on smaller towns. He plowed his own money into a new company called the Thomas A. Edison Construction Department. The company's agents recruited local investors, who then formed utility companies and paid Edison—in cash and stock—to install central stations. In the next year Edison installed central stations in nearly twenty towns. But local investors were slow with their payments, and the Construction Department's up-front costs were high. Most of the money was coming directly out of Edison's pocket, and he soon ran low on funds. In the fall of 1884 he disbanded the Construction Department and allowed the parent company to take over its assets and operations.[6]

Edison decided that if the officers of Edison Electric would not promote the business, he would change the officers. When Edison Electric was founded, he owned five-sixths of its stock, but he had sold most of his stake in order to finance the manufacturing firms. Edison, though, had more influence than the average minority stockholder. In a proxy battle at the stockholders' meeting in October 1884, he gained enough support from other investors to oust the board of directors and install a new one that included several of his closest friends, men who saw the business as he did. Francis Upton and Charles Batchelor won positions on the board, and the new president was Edward Johnson, Edison's longtime friend and formerly the vice president of the company.[7]

"I have given a perfect system, and I want to see it sold," he told a newspaper reporter. "I have worked eighteen and twenty hours a day for five years, and I don't want to see my work killed for want of proper pushing."[8]

Now that his friends ran the company, Edison pursued strategies that the previous board had avoided. Edison Electric filed patent suits against infringing rivals and gave Edison's lighting system the promotion it deserved. The results were striking. Edward Johnson pushed the expansion of the central stations in New York and Boston, but most of the growth came in smaller towns. This "village business," as Edison called it, relied on one of his latest patents. Because the high cost of copper remained the greatest stumbling block to his light's success, Edison deployed the "three-wire system," a new circuit arrangement that required 60 percent less copper than the feeder-and-main arrangement used in New York. He installed the first three-wire central station in Sunbury, Pennsylvania, in the summer of 1883, and in the next few years the system spread across the country.[9]

Before Edison installed the new board of directors, there had been only eighteen Edison central stations in the United States. A year later there were thirty-one, and by August of 1886 there were fifty-eight.

During the same period the number of isolated lighting plants nearly doubled. As the companies prospered, Edison recovered his investments and became a wealthy man. Just as important, his dream of seeing his light spread across the country was beginning to come true.[10]

EDISON'S LIGHTING SYSTEM inspired an electrical craze. A New York bartender mixed an "electric cocktail" by caramelizing sugar with an electric light wire, then adding alcohol. The *Electrical Journal* reported in 1882 that a Richmond man "cooked the first four eggs boiled in water by electricity, the first piece of beefsteak, and probably the first bacon" (the rival claimant for the bacon prize was not identified). A physician used the Edison current for the "eradication of a mustache from a woman's lip." J. P. Morgan's daughter appeared at a ball with tiny glowing bulbs in her hair, and the Electric Girl Illuminating Company, incorporated in 1884, adorned servants with incandescent lamps and hired them out, promising "girls of fifty-candle power each in quantities to suit householders."[11]

Although Edison's light had captured the nation's imagination, in 1886 it remained a luxury service, available only to the relatively few people who had purchased isolated lighting plants or who lived in the central districts of some cities. Arc lamps, because they were placed on public streets, had a more immediate impact.

Brush Electric remained the leading arc lighting firm, but a strong competitor had entered the field. In the late 1870s Elihu Thomson and Edwin J. Houston, two science teachers at Philadelphia's Central High School, designed an arc lamp system, located investors in Connecticut, and founded a manufacturing company there. Originally known as the American Electric Company, in 1883 the firm was reorganized as the Thomson-Houston Electric Company. Brush's supremacy in arc lighting was challenged not only by Thomson-Houston but also by the United States Electric Lighting Company, Fort Wayne Electric, and about twenty smaller firms.[12]

Because competition was so intense, arc lighting manufacturers marketed their systems aggressively, with salesmen for the companies essentially acting as promoters. In each city, they located investors, helped them organize a local lighting utility, secured a franchise from the local government, arranged for the purchase of equipment, and assisted with setup. Installing an arc lighting system, like installing an Edison incandescent system, involved creating a central station with one or more dynamos, then running conducting wires to carry electricity to individual street lamps, but the similarities ended there. An incandescent lighting system reached into the interior rooms of houses and offices and required the wiring of thousands of individual lamps. An arc lamp central station, by contrast, served just a few dozen street lamps. Even more important, most arc lamp companies avoided the most laborious part of Edison's system installation: digging up the streets to lay conductors underground. Arc lighting firms, like operators of telegraph services, strung their wires on poles over city streets.[13]

Because they were relatively easy—and therefore inexpensive—to install, arc lighting systems spread more quickly than incandescent. The number of arc lamps in service in the United States jumped from 6,000 in 1881 to 140,000 in 1886. New York City followed the national trend during that period, with the number of arc street lights rising from 55 to 700, illuminating more than thirty miles of the city's avenues and major cross streets.[14]

Whereas the arc lighting industry was crowded with competitors, Edison enjoyed a virtual monopoly in incandescent lighting. Through the end of 1886, he controlled four times the market share of his closest competitor, the United States Electric Lighting Company. The mistake of U.S. Electric and Edison's other early rivals was in trying to compete with him directly. They could try to copy his system, but they did not have the Edison brand name and could not match the years of experience of Edison and his men. The competitors offered pale imitations of the Edison system, and most customers were wise enough to choose the original article. Late in 1886, however, a formidable new challenger arrived.[15]

. . .

George Westinghouse was born in 1846 (a year before Edison) in the upstate New York town of Central Bridge. When he was nine, his father opened a shop in Schenectady for making farm machinery and steam engines. The boy found his father's shop much more interesting than school, so he cut class and spent his days designing model powerboats, waterwheels, and other bits of "trumpery," as his father called them. Aside from his deft mechanical touch, young George was known for a volcanic temper and a sometimes misguided devotion to efficiency. A story was told about the day his father whipped George for misbehavior. After the wooden switch broke twice, George choked back his sobs long enough to point out a leather whip that would do the job more competently.[16]

In 1863, at the age of seventeen, Westinghouse enlisted as a cavalryman in the Union army. Early the next year he transferred to the navy and put his mechanical skills to use on steam-powered gunboats. After the war he enrolled at Union College in Schenectady but did not last long. "He was my despair," one of his teachers recalled. "While the other boys were struggling with German syntax or French pronunciation, he would amuse himself making pencil drawings on his wristbands. His sketches were always of locomotives, stationary engines, or something of that sort." The school dismissed Westinghouse after three months, which was just as well with him, for he could then work on machines rather than draw them on his cuffs. Soon after leaving school, he invented a device for putting derailed trains back on track, as well as a new type of "frog," the mechanism that allowed trains to move from one track to another. Westinghouse persuaded two local investors to finance the manufacture of the devices, and before long he was a partner in a thriving industrial operation.[17]

While returning home from a business trip one day, Westinghouse was delayed by an earlier head-on collision between two freight trains. He wandered up to inspect the damage and fell into conversation with the chief of the wrecking crew. It was a straight stretch of track, the

George Westinghouse

man explained, and the engineers saw each other, but they could not halt in time. "You can't stop a train in a minute," the man said.[18]

When an engineer needed to halt his train, he used a whistle to signal to the brakeman, who climbed to the roof of a car and twisted a wheel that tightened a chain that pressed the car's brake shoe against the rail. Then he jumped to the next car and repeated the maneuver. It was a slow process even in routine circumstances, and trains often overshot stations and had to back up to the platform. In an emergency, such as when two trains were hurtling toward each other on the same track, they were usually stopped not by brakes but by impact.[19]

Westinghouse decided to build a better brake. He first tried using steam lines running the length of the train: The steam pressure drove pistons that instantaneously would press all the brake shoes against the rails. The invention failed miserably, however, because the steam condensed before it could do its work. After reading a magazine article about mining drills driven by compressed air, Westinghouse adapted his steam brake to use compressed air and patented the design.

Like Edison with the electric lamp, Westinghouse was attacking a problem that many had attempted before. There were already more than two dozen patents covering automatic braking systems, including one using compressed air. But Westinghouse was the first inventor to build a practical system and the first to organize for its manufacture. After trials proved the worth of his system, a few wealthy railroad men formed the Westinghouse Air Brake Corporation in 1869. They built a factory in Pittsburgh, capitalized the company at $500,000, and named George Westinghouse president. He was twenty-two years old. In the decade after its invention, the Westinghouse air brake became standard equipment on passenger trains around the world.[20]

In the early 1880s both Westinghouse and Edison were wealthy industrialists in their early thirties, but the two men played the role very differently. Unlike the smooth-cheeked Edison, Westinghouse cultivated the extravagant facial hair common at the time: bushy side whiskers and a handlebar mustache. Whereas Edison was boyish and playful, Westinghouse, over six feet tall and barrel-chested, was an imposing, stern figure. Most strikingly, Westinghouse avoided granting interviews or being photographed. "When I want newspaper advertising, I will order it and pay cash," he once said. "If my face becomes too familiar to the public, every bore or crazy schemer I meet in the street will insist on buttonholing me." Although the company bore his name, Westinghouse did not attempt to sell himself along with his products. An engineer who worked in the Pittsburgh factory said Westinghouse never struck him "as being a wizard, but he seemed to be a plain human being with lots of initiative, with nerve to attempt difficult things."[21]

Having conquered the railroad brake industry, Westinghouse went looking for new challenges. According to Thomas Edison's secretary, a fit of pique drove Westinghouse into the lighting business. When Westinghouse tried to interest Edison in a steam engine he had invented, Edison supposedly replied, "Tell Westinghouse to stick to air brakes. He knows all about them. He doesn't know anything about engines." Westinghouse, so the story goes, decided to avenge the insult by competing with Edison.[22]

The true story is more mundane. Westinghouse first met Edison when he visited Menlo Park while looking for an isolated lighting plant for his Pittsburgh house. Edison showed him around the lab, and Westinghouse later recalled the visit warmly. He entered the light business not out of spite but for the same reason so many others did: He saw profit and glory in it. Early in 1884 he met a young electrical expert named William Stanley, who had worked for U.S. Electric and other electric lighting companies, and brought him to Pittsburgh to develop an incandescent lighting system. In 1885 and 1886 Westinghouse and Stanley installed isolated lighting plants in hotels in New York and Pittsburgh and built a central station in Trenton, New Jersey. But Westinghouse ran into the same problem that stymied U.S. Electric: His system was too similar to Edison's to be competitive with it.[23]

DESPITE ITS SUCCESS, Edison's light suffered from a serious weakness. The feeder-and-main and three-wire distribution systems lowered copper costs somewhat, but Edison still could not serve areas more than a mile or so away from the generating plant. There was, though, an inexpensive way to cut copper usage and broaden the service area of a central station: by boosting the voltage.

The behavior of electricity can be thought of in terms of water flowing through pipes. A pump (voltage) drives the flow, the diameter of the pipe determines resistance, and the rate of flow (amperage) is determined by voltage as limited by resistance. If there are two pipes, one with double the capacity of the other, both will carry the same amount of water if the pressure in the smaller is twice that in the larger.

Similarly, to deliver a given amount of energy, engineers could either couple a thick (lower-resistance) conductor with a low-voltage current or couple a thin (higher-resistance) conductor with a high-voltage current capable of overcoming the resistance. Given the high cost of copper, the choice was clear: A high-voltage system with thin wires was much more economical.

Because domestic lighting required low pressures of about 100 volts, an ideal system would allow a company to transmit through slender wires at high voltages (over 1,000 volts), then lower the pressure to 100 volts for use in the home. Charles Brush, the arc lighting pioneer, was the first American to exploit the economies of high-voltage transmission. In 1882 he built an incandescent lighting system in which electricity was transmitted over longer distances at over 1000 volts, then fed into batteries near the site of consumption. Charged at high voltage, the batteries then discharged electricity at much lower voltages to operate lamps. Brush's battery system, though, was clumsy and inefficient, and it never caught on.[24]

In 1885 Stanley and Westinghouse began to work on a lighting plan that, like Brush's storage battery system, involved transmitting at high voltages, then lowering the voltage and distributing into homes and offices at low voltages. Their system, though, had one significant difference: Rejecting direct current, which flowed continuously in one direction, they chose to use alternating current, which changed direction back and forth many times a second.

It was a radical step. Rotating a coil of conducting wire between the poles of a magnet naturally produced alternating current, because the relationship between the coil and the north and south poles of the magnet reversed with each half turn (see figure 1). The earliest generators in the 1830s produced alternating current, but, because this current had no apparent uses at the time, instrument makers added switching devices that converted alternating-current generators into direct. From the 1830s through the 1870s, electroplating and electrochemistry—both of which required direct current—were the primary commercial use of electrical generators, and as a result most available dynamos were direct-current machines. When inventors took up electric lighting in the 1870s, they patterned their generators on the direct-current models common at the time. Charles Brush's arc lights used direct current of 1,000 or more volts, and early incandescent light systems, like Edison's, used direct current of 100 volts or so.[25]

In the early 1880s a few inventors realized that alternating current had one distinct advantage over direct, an advantage related to the principle of induction, discovered independently by Michael Faraday and the American scientist Joseph Henry in the 1830s. The principle stated that a variation—starting, stopping, or reversing—of the current

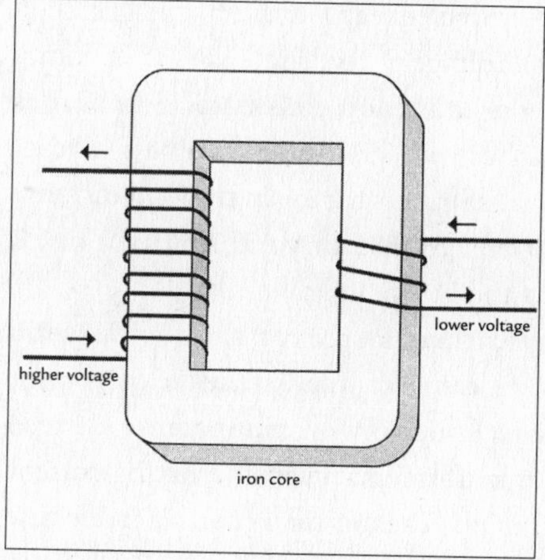

Figure 3: A transformer consists of two distinct coils of wire wound around an iron core. The input current enters the primary coil, producing a magnetic field in the iron core that repeatedly switches on and off. The core transfers this field to the secondary coil, where it induces an output current. The degree of change in the voltage depends upon the ratio of the number of windings of the coils: This transformer steps the current up or down three times.

in one coil of conducting wire will induce a current in a second coil placed in close proximity to (but not touching) the first. Direct current, which flowed continuously, without variation, produced no induction effect. But alternating current did, because it reversed direction many times a second. An alternating current in one coil of wire (the primary circuit) caused a current to flow in another coil (the secondary circuit) placed close by.[26]

Induction allowed the conversion of low voltage to high or high voltage to low. The degree of change depended on the ratio of the number of coils of conductor in the primary circuit to the number in the secondary circuit. Given nine coils of wire in the primary circuit and three coils in the secondary circuit, a current of 300 volts in the primary would induce a current of 100 volts in the secondary. Operated in reverse, the arrangement raised voltage instead of lowering it. One of Edison's favorite toys, the induction coil, operated on this principle: Low voltage from a battery in the primary circuit induced a higher voltage in the secondary, which was then used to administer shocks.

In the early 1880s the French inventor Lucien Gaulard applied the principle of induction to the problem of transmitting electrical energy.

He would transmit alternating current cheaply at high voltages, then use induction coils to reduce the voltage to levels safe for use in homes and offices. Used in this way, the induction coils became known as converters or *transformers*. Together with his British business partner John Gibbs, Gaulard designed transformers for use in the long-distance transmission of alternating current. In 1884 they transmitted along a fifty-mile circuit between Lanzo and Turin, Italy.

Gaulard and Gibbs displayed their system at the 1885 London Inventions Exhibition, where it came to the attention of George Westinghouse, who purchased on option on the patent. He also ordered several Gaulard-Gibbs transformers and an alternating-current generator made by Werner Siemens, the German pioneer in electric lighting.[27]

The engineers on Westinghouse's staff opposed these moves. Like most experts, they believed alternating current could never compete with direct because transformers wasted too much energy. An editorial in a leading electrical journal explained that the best minds in electrical science had tested and abandoned alternating current, and that these men would never "have allowed this subject to dwindle away to mere nothingness had there been a chance of bringing the matter to a successful issue."[28]

Westinghouse disagreed, and his enthusiasm for alternating current was shared by William Stanley. Starting early in 1886, Stanley redesigned the Gaulard-Gibbs transformer and tested it in a new system. He installed an alternating generator and transformer in an old mill in Great Barrington, Massachusetts, strung a mile of thin copper wire into the center of town, installed another transformer in the basement of a store, and ran wire from that basement to lamps in a hotel and several stores. Stanley used the first transformer to step up a generator's voltage from 500 to 3,000 volts, transmitted the current to the center of town, then used the other transformer to step it back down to 500 volts.[29]

"All the converters are under lock and key, so that no one knows anything about them," Stanley wrote to Westinghouse after a successful test in March. "I might say a great deal about the system, but briefly, it is all right."[30]

Despite the continued objections of his other engineers, Westinghouse moved ahead with alternating current. He developed a system that transmitted current at 1,000 or 2,000 volts, then stepped it down to 50 volts for distribution within homes and offices. After further refinement of the system in Pittsburgh, Westinghouse Electric installed its first commercial alternating-current plant in November 1886 in Buffalo, New York, where, at the same time in the same city, Alfred Southwick was in the midst of his research on killing with electricity. Orders for twenty-five more plants arrived within a few months, and Westinghouse Electric moved into bigger quarters in Pittsburgh to handle the business.[31]

THE TRANSFORMER ALLOWED high-voltage transmission of alternating current over long distances and low-voltage distribution near the point of consumption. Since Edison's low-voltage system required such thick copper conductors, he had to build several different direct-current generating plants to serve an area of a few square miles. But the same area could be served by a single large alternating-current central station, which allowed lower initial capital costs. Changes in the international copper market also helped Westinghouse. In 1886 copper sold for about ten cents a pound, but the following year a French syndicate cornered the world's copper supply and doubled the price. This made the alternating system, with its thinner conductors, even more attractive.[32]

Direct current had a few points in its favor. Most important, its dynamos converted mechanical energy into electrical energy with a loss of only about 10 percent, whereas alternating generators had losses closer to 25 percent. Direct current also offered the versatility of supplying power as well as light. A lightbulb had no preferences as to current type–its filament heated to incandescence whether the current flowed in one direction or alternated back and forth. But electrical companies needed a way to earn income from their plants during the day, when there was little demand for light. Direct current could run electric motors, allowing Edison to sell electricity to users of sewing machines, elevators, and other machines. There was no motor for alternating current, so Westinghouse's plants would sit idle during the day. There was also no way of measuring usage. The only electric meter–Edison's–worked on the principle of electroplating, strictly a direct-current phenomenon. Without a meter, Westinghouse would have to charge a flat rate, and this would lead to inefficient use of his power.[33]

EDISON FOLLOWED Westinghouse's work with great interest. Late in 1886–around the time Westinghouse's Buffalo plant went on-line–he wrote a thirteen-page memo to Edison Electric president Edward Johnson about alternating current, professing to be not a bit worried. The inventor pointed out that the generators and transformers wasted energy, and that the lack of meter and motor was a problem. Edison also worried about the dangers of sending high voltages through wires over city streets and into people's homes. Whereas direct-current lighting used just 100 volts, the Westinghouse system, Edison explained, "uses 2000 volts alternate–This gives a difference of 4000 volts (!) (HOLY MOSES). . . . Suppose W[estinghouse] uses 2000 and one leg gets crossed, the first man that touches a wire in a wet place is a dead man." The inventor then made a prediction: "Just as certain as death Westinghouse will kill a customer within 6 months after he puts in a system of any size."[34]

Edison's evaluation of alternating current was accurate, given the state of the industry in 1886. But Westinghouse was placing a bet that the efficiency would improve, that meters and motors would be invented, that the dangers of high voltage could be controlled. Just as Edison had done in his work on the incandescent lamp, Westinghouse chose to gamble that he and his engineers could do what most experts said was impossible.

. . .

THROUGH HIS FLAIR for promotion, Edison had created a demand for electric light that he could not meet. Most Americans still lived in sparsely populated areas, and the limited range of the Edison system left them in the dark. Westinghouse filled the gap. His high-voltage system delivered affordable electricity to homes and businesses outside the reach of the Edison system.[35]

When Edison wrote his memo to Edward Johnson, he admitted, "One thing that disturbs me is the fact that Westinghouse is a great man for flooding the country with agents and travellers." It was an astute assessment. Unlike Edison, Westinghouse shunned publicity and built his business not through the power of personality but through aggressive marketing. Although at first Westinghouse was content to promote his system in rural areas Edison could not reach, before long he was competing head-to-head with Edison for customers.[36]

In April 1887 Westinghouse moved into New Orleans, selling light at a loss in order to undercut Edison's operation. "They are robbing our business pretty badly," Edison's local representative complained. The numbers bore out the complaint, in New Orleans and across the country. Westinghouse sales, less than $200,000 in 1886, jumped to $800,000 the following year. By the fall of 1887, just a year after the first alternating central station went on-line in Buffalo, there were 68 Westinghouse central stations in operation or under contract. Edison, who had been in the business several years longer, had only 121.[37]

At the end of 1887, Edison agents from across the country sent panicked letters to the head office in New York, begging for a way to compete with Westinghouse in smaller towns and the outlying districts of cities. "There is an enormous pressure everywhere for a system to cover distances," one agent explained. An agent in Tacoma was more blunt: "Are we going to sit still and be called 'old fashioned,' 'fossils,' &c., and let the other fellows get a lot of the very best paying business?"[38]

Edward Johnson told Edison in December that he was going to Chicago to talk to the agents there and "stiffen them up a bit—they have been completely demoralized by Westinghouse." Johnson also warned Edison that until they came up with a new system to compete with Westinghouse's, they would have to accept the fact that they would "do no small town business, or even much headway in cities of minor size."[39]

In November 1887, at the height of the panic over the Westinghouse competition, Edison received a letter from Alfred Southwick concerning the work of the death penalty commission. Southwick asked Edison to name "the necessary strength of current to produce death with certainty in all cases," and also to specify the best equipment for the task.

In reply, Edison told Southwick that he was opposed to capital punishment and refused to offer any advice.

On December 5, Southwick wrote again. "The question does not arise do we as individuals believe in capital punishment," Southwick told Edison, because the practice "has existed by law in all ages and in all nations and perhaps will for all time to come." This being the case, Southwick continued, "it appears to me that science and civilization demands [*sic*] some more humane method than the rope. The rope is a relic of barbarism and should be relegated to the past." The dentist also admitted that he was seeking more than technical advice from Edison. "The reputation you have as an electrician," Southwick wrote, would be invaluable in persuading the legislature to abolish the gallows and substitute electricity. Southwick closed by begging Edison, "Change your mind on the subject and give us the benefit of your knowledge."

The plea worked. Edison wrote a letter to Southwick on December 19, 1887.

> Your points are well taken, and though I would join heartily in an effort to totally abolish capital punishment, I at the same time

realize that while the system is recognized by the State, it is the duty of the latter to adopt the most humane method available for the purpose of disposing of criminals under sentence of death.

The best appliance in this connection is to my mind the one which will perform its work in the shortest space of time, and inflict the least amount of suffering upon its victim. This I believe can be accomplished by the use of electricity and the most suitable apparatus for the purpose is that class of dynamo-electric machine which employs intermittent currents.

The most effective of these are known as "alternating machines," manufactured principally in this country by Mr. Geo. Westinghouse, Pittsburgh.[40]

CHAPTER 10

The Electrical Execution Law

EDISON'S LETTER convinced Elbridge Gerry that he should abandon his preference for morphine executions and throw his support behind electricity. The three commissioners, Southwick, Gerry, and Hale, added some final touches to the commission report–including an extract from Edison's letter–and delivered it to the New York State Legislature in January 1888. Gerry, who was experienced in legislative matters, drafted a new capital punishment bill and attached it to the end of the report.[1]

The commission's bill changed the method of execution but did not stop there. Executions would take place not at the county jail but within the walls of three state prisons: Auburn, not far from Syracuse; Clinton, in northern New York near the Canadian border; and Sing Sing, along the Hudson River about thirty miles north of Manhattan. Fewer than twenty witnesses would be allowed at each execution: a Supreme Court justice, the sheriff and district attorney from the county of conviction, two physicians, and "twelve reputable citizens" selected by the prison warden. The bill included an unusual press restriction: "No account of the details of any such execution, beyond the statement of the fact that such convict was, on the day in question, duly executed according to law at the prison, shall be published in any newspaper."[2]

The most surprising aspect of the bill had to do with the disposition of the prisoner's corpse. Doctors would conduct a postmortem examination, after which the prisoner's body was to be either turned over to a medical school for dissection or buried "with a sufficient quantity of quick-lime to consume such body without delay." Under no circumstances could a relative or friend claim the corpse, and "no religious or other services shall be held over the remains." Although the commissioners had chosen a killing method that would not mutilate the victim's body, as guillotine and garrote would, they recommended dissecting the corpse and obliterating it with quicklime. The "criminal classes . . . are certainly very indifferent as to the infliction of death," the commissioners explained, so capital punishment by itself would serve no deterrent purpose. But those same classes were also "superstitious" and would avoid committing a crime if "they were certain that after execution their bodies were to be cut up in the interest of medical science."[3]

Dissection was, in fact, widely feared. A corpse was considered sacred, and strong taboos forbade desecrating it or denying it proper burial. When gibbeting and dismemberment fell from favor a century earlier, officials settled on dissection as an alternative way to add horror to a capital sentence. (Dissection laws also provided medical schools with corpses, always a scarce commodity.) American judges, sensitive to the public horror it provoked, rarely sentenced the condemned to dissection. The death penalty commissioners hoped to remove the judges' discretion by making dissection mandatory.[4]

The quicklime poured into the coffin would prevent the corpse from becoming a commercial attraction. After an 1878 hanging in Pittsfield, Massachusetts, the victim's father-in-law collected the body and put it on exhibit, charging admission of ten cents a head. A New York dime museum claimed to own the pickled head of Charles Guiteau, executed assassin of President James Garfield, while a competing museum advertised "The Head and Right Arm of Anton Probst, The Murderer of the Deering Family, amputated after Execution." After one New York execution, armed guards stood watch at the man's grave. "I'm not going to have any dime museum get ahead of me," a prison official vowed. "The graves will be watched until there is nothing down below."[5]

Electrical execution devices would be so expensive and complex that they could not be provided at every county jail, as gallows were. The commissioners, however, provided two reasons for moving executions to the three state prisons, and neither involved technology. First, they explained that the new location would make escape more difficult, county jails being notoriously insecure. And, because the site of execution would be far removed from the scene of the crime, the prisoner's friends would be less likely to gather outside the jail and express sympathy for the condemned, a common practice that was, the commissioners believed, "discreditable to public decency and dangerous to public peace." Like so much of American society, executions would become centralized and bureaucratized, the dangers of local variation removed.[6]

The death commissioners saw their main duty as controlling what they called the *criminal classes.* The term, usually synonymous with *the poor,* had been thrown around for decades, but it had a particular resonance in 1880s America. With the emergence of big factories and sprawling cities, the myth of the United States as a classless society became harder to sustain. The new economy had given birth to a permanent working class, and many of the workers were new immigrants from Germany, Russia, and Italy, who brought unfamiliar languages, strange customs, and dangerous political ideas.

An event in Chicago became the touchstone for the era. On May 3, 1886, police fired into a crowd of striking workers in Chicago's Haymarket Square, killing four. The following day a group of anarchists held a peaceful protest meeting. When the authorities tried to break it up, someone threw a bomb among the police, killing eight officers. The bomb thrower was never identified, but eight anarchists were convicted on conspiracy charges and sentenced to death. By all accounts the trial was a farce: The prosecution proved that the men held radical ideas but not that they had anything to do with the bombing.

In nineteenth-century America, labor violence was not unusual, nor was the gross miscarriage of justice. What made Haymarket unusual was the hysteria it provoked. The *New York Times* prescribed the Gatling gun as the only remedy for "an acute outbreak of anarchy," while the *St. Louis Globe-Democrat* announced that "there are no good anarchists but dead anarchists." A Cincinnati paper recommended lynching the prisoners, and–although legal procedures technically were followed–that is more or less what happened. Despite abundant evidence of innocence, the U.S. Supreme Court turned down a final appeal, and four of the men were hanged on November 11, 1887. All of the prisoners remained defiant to the end, and one shouted "Hurrah for anarchy!" from the scaffold. After the executions, the families held one of the largest public funerals Chicago ever saw, with 20,000 people marching in the funeral procession and 200,000 more lining the streets. One reporter noted that to the crowd the executed men were "martyrs in the cause of the poor against the rich."[7]

The Haymarket tragedy exposed the deep fears that pervaded the United States in the 1880s. As cities grew uncontrollably large and the working class demanded its rights, many of America's leaders felt as if the country was slipping out of their grasp and into the hands of the rabble. Mob violence, it appeared, had to be met by the measured violence of capital punishment. But the death penalty commissioners recognized that state violence had to be applied carefully, lest it promote rather than quell disorder. They reported that at the funerals of executed men "evil deeds are glorified into acts of heroism." A member of New York's legislature defended the clause forbidding funerals by noting "the insurrectionary funeral of the Chicago Anarchists." The *World* noted approvingly that with electrical execution there would be "no glory left in execution–nothing, in short, but a cold and quiet death, such as dogs meet in the public pounds."[8]

Similar motives underlay the press gag provision of the proposed law. "The newspapers are extensively read by the criminal classes," the commissioners wrote, "who glory in the description of the courage shown by their colleagues undergoing the sentence of death." The result was that "the execution, instead of operating as a deterrent, . . . has been known even to stimulate others to the commission of crime." If newspapers did not report on executions, this danger was removed.[9]

The execution bill Gerry drafted tried to complete the movement that began with the shift to private executions in the 1830s. Because sheriffs had found ways to skirt the law–and because newspapers always reported on executions–private hangings were not private at all. The hanging ritual still seemed to subvert rather than buttress the social order. By moving executions to central locations and providing for dissection, corpse destruction, and press restrictions, the bill was intended to destroy the old execution rituals and the dangers they posed to society. The invention of a more humane form of killing became subordinate to the larger goal of controlling the passions of the poor.[10]

IN MARCH 1888 the New York State Assembly's Judiciary Committee reported favorably on Gerry's execution bill, and the full Assembly passed it. The Senate Judiciary Committee, however, subjected the measure to a curious revision. It eliminated the clause substituting electricity for hanging while leaving in place the other provisions. A member of the committee considered electrical execution "so radical" that he thought it deserved another year of study. The full Senate, however, restored the excised provision and passed the bill by a vote of 87–8. The version of the bill passed by the two houses was essentially the same one Gerry drafted, with a few notable exceptions: The dissection clause was omitted; religious services were not forbidden; and the victim's family could claim the body. But the commissioners got most of what they recommended. Electrical executions would take place at three state prisons, with fewer than twenty witnesses, and unclaimed corpses would be buried in quicklime. The press gag clause survived. On June 4 Governor David B. Hill signed the bill into law.[11]

The bill encountered surprisingly little opposition from lawmakers in its path through the legislature, and most newspapers backed it as well. Some observers did worry that a humane method would destroy the deterrent effect of capital punishment. Gerry later admitted that killing with morphine was rejected in part because it might "make death somewhat agreeable" and "rid it of its terrors." Electricity seemed to strike the proper balance between humanity and terror. In Gerry's view, "Criminals would infinitely more dread a silent going away–to be deliberately killed by a terrible but silent force to them unknown." Park Benjamin, a prominent electrician, agreed, explaining that "the instant extinction of life in a strong man by an agency which it is impossible to see, which is unknown, may create in the ignorant mind feelings of the deepest awe and horror, and prove the most formidable of all means for preventing crime." The new execution law tapped into the sense of mystery surrounding this invisible force that no one could properly explain. Lightning from the heavens splintered trees, and the artificial lightning of battery and dynamo carried telegraph messages, produced blazing light, and cured illnesses. The power to kill was the latest manifestation of electricity's magical powers.[12]

Although terrifying, electrical execution was also said to cause no physical suffering, therefore offering more evidence of the glorious progress of civilization. "The state of New York may pride herself in the fact that the gallows is to be banished and a more humane and scientific method of executing criminals is to be instituted," *Scientific American* wrote. According to the *New York Times*, "It will be creditable to the State of New-York to be the first community to substitute a civilized for a barbarous method of inflicting capital punishment." With

Olympian certitude the *Tribune* insisted that "no right-minded person can fail to approve the enactment of the law."[13]

A FEW VOICES nonetheless declined to approve. "Talk . . . about 'the dark ages' and 'barbarism' is nonsense," the *Buffalo Express* wrote. "We know what hemp will do, but we don't know what electricity may do. This movement is a pure scientific experiment, in which criminals are to be killed to test an open question."[14]

The *Express* realized that—beneath a veneer of scientific authority—the death penalty commission's report was full of holes. In support of its recommendation, the report provided three sorts of evidence: testimonials from electrical experts such as Thomas Edison, none of whom had personally investigated electricity's ability to kill; reports of accidental deaths from electricity, which had not occurred under controlled conditions; and the dog-killing experiments of Southwick and Fell, who were largely ignorant of electrical technology, and who had failed to record the technical details of their tests.[15]

The law decreed that death must be "inflicted by causing to pass through the body of the convict a current of electricity of sufficient intensity to cause death." No one had yet specified the type of generator, voltage, amperage, or how to attach electrodes to the body.[16]

"The technical difficulties of killing a man with electricity . . . are considerable," a medical journal stated, "and apparently have not been by any means completely studied out by the learned Commission." According to the *Herald*, Dr. William A. Hammond, former U.S. surgeon general, described the commission's arguments as "the weakest he had ever seen in an official paper." The *Herald* went on to observe that the new law seemed to be "somewhat senseless if not, indeed, very idiotic."[17]

All the bluster about progress and civilization and science could not hide the fact that no one knew how to kill a man with electricity. The *New York World* recognized the deficiencies in the commission's report. Rather than simply point out these facts, the newspaper decided to investigate, and a reporter persuaded Thomas Edison to conduct some experiments.[18]

BY THE SUMMER of 1888, calling on Thomas Edison required a journey to Orange, New Jersey, where the inventor had a new home, a new laboratory, and a new wife.

Like many nineteenth-century widowers, Edison did not dawdle in his quest for another bride. Early in 1885—about six months after Mary died—Edison traveled to an industrial exposition in New Orleans in the company of his daughter, Dot. As much as he loved machines, Edison found himself distracted by Mina Miller, a nineteen-year-old, dark-haired beauty from Ohio who was attending the exposition with her father, Lewis Miller, a wealthy inventor of farm machinery. Edison could not pursue his attraction in New Orleans, but opportunity arose again soon, when he traveled to Boston and stayed at the home of Ezra Gilliland, a friend and business associate. Conveniently enough, Mina Miller attended school in Boston and was a friend and frequent visitor of the Gillilands.[19]

Edison became so infatuated with Mina that he lingered in Boston, eager for more encounters. Neglecting the rough labors of the laboratory, he sank into the refined world of the Gilliland household and joined in the sorts of activities—polite literature, boating parties, parlor games—that he normally disdained. He recorded his experiences in a diary, the only time in his life that he kept one. He read Hawthorne, Goethe, and Rousseau, and was not much impressed with any of them, and he found some of the outdoor activities even less congenial. One day after supper he and Dot threw a ball back and forth. Amazingly, Edison claimed it was the first time he ever played catch: "It was as hard as Nero's heart—nearly broke my baby-finger—gave it up." Dot read him the outline of her proposed novel, which was about "a marriage under duress." Her father told her "that in case of a marriage to put in bucketfulls [*sic*] of misery. This would make it realistic."[20]

For a man with such a pessimistic view of the wedded state, he seemed terribly anxious to enter into it again. Mina Miller became "a sort of yardstick for measuring perfection," the inventor wrote, and the mere thought of her drove him to distraction. While walking the streets of Boston, he "got thinking about Mina and came near being run over by a street car." As the ultimate sign of his love, the famously slovenly dresser tried to please his beloved by improving his wardrobe: "For the first time in my life I have bought a pair of premeditatedly tight shoes–These shoes are small and look nice."[21]

Dot, to her dismay, found that she had been supplanted in her father's affections by a girl just six years older than she. Even Edison recalled that his constant talk about Mina "makes Dot jealous, she threatens to become an incipient Lucretia Borgia." Dot later recalled that this time was "the most unhappy of my life."[22]

Not much troubled by his daughter's sadness, Edison devoted himself to winning over Mina's devoutly Methodist family. Although her father, Lewis Miller, earned his money in farm equipment, he gained fame as the cofounder of the Chautauqua Institute, a new type of religious enterprise in western New York. Nineteenth-century Protestants harbored deep suspicions about most leisure activities. Virtue lay in hard work, they believed, while reading novels, attending plays, or going to the beach simply provided occasions for sin. At the Chautauqua Institute, Lewis Miller and a minister friend created an educational summer camp where middle-class believers could come together to listen to lectures on science and literature–as well as swim, boat, and fish. Chautauqua made vacationing safe for Protestants.[23]

Although the Millers were religious liberals by the standards of the time, they were far more church-oriented than Thomas Edison. To the extent that he thought about religious matters at all, he was a deist: He thought the design of the universe indicated a creator, but he did not believe in a personal God and never went to church. "My conscience seems to be oblivious of Sunday–it must be incrusted [*sic*] with a sort of irreligious tartar," he wrote in his diary. Later in the summer he traveled to Chautauqua, where he had an opportunity to charm the Miller family. Lewis Miller needed little convincing–what

Mina Miller Edison at about the time of her marriage to Edison.

better match for his daughter than a wealthy inventor and industrialist like himself? Lewis's wife, Mary, was worried by Edison's ungodliness, but she accepted his explanation that he avoided church because he was hard of hearing. She was also won over by Edison's storytelling skills, which became legendary in the Miller family. "Our folkes try to tell storys," Mary Miller wrote in a letter, "when they gett along a little ways they will stop and say O if Mr Edison was hear he could tell it."[24]

As for Mina's two young brothers–ages ten and twelve–Edison knew the way to a boy's heart. He sent them a telegraph set and a telescope, as well his own favorite: the induction coil. Edison gave elaborate instructions for making the battery ("be sure in pouring the sulphuric acid that you do not let any of it spatter into your eyes") and explained that turning the crank built up an electrical charge that could be administered to unwitting victims. "The wheel should be turned about 200 times per minute for a black cat and 199 1/2 for a cat with a sanguine temperment [*sic*]," Edison told the boys. "This coil is *very powerful*. I tried it on a Dutch carpenter today and it knocked him down instantly."[25]

After his stay at Chautauqua, Edison persuaded Mina to join him and the Gillilands on a trip into the White Mountains in New Hampshire. Edison taught Mina Morse code, and on long drives through the mountains the pair conversed secretly by tapping messages onto each other's hands. On one of the last drives, Edison asked her to marry him. She tapped out "yes" in reply.[26]

The pair was married in Akron, Ohio, in February 1886, a year after they first met. On their honeymoon the couple traveled to Chicago, Atlanta, and Jacksonville, Florida, before arriving in Fort Myers, where Edison was building a winter home and laboratory. While in Florida he sketched plans for telegraph and telephone improvements, a new cotton-picking machine, a cream separator, and a hearing aid. He and Mina tried some more fanciful experiments as well. "Shock an oyster see if it won't paralize [*sic*] his shell muscle & make the shell fly open," he wrote in a lab notebook. "Dead failure," he scrawled below.[27]

For their northern residence, the couple chose a house in an area called Llewellyn Park in Orange, New Jersey, not far from Newark. Llewellyn Park was the first planned residential suburb in the United States, an exclusive neighborhood of private streets and grand houses on landscaped, wooded lots. The Edison house, known as Glenmont, was enormous—a twenty-three-room mansion in the Queen Anne style with stained-glass windows and carved wood paneling. A clerk in a dry goods firm had built and furnished the house at a cost of about $400,000, but his employers soon discovered that he had embezzled most of that money. The clerk fled to the West Indies, his firm assumed ownership of the property, and Edison bought it, furnished, at the fire-sale price of $235,000. Glenmont became home to Edison's three children from his first marriage, and over the next decade Edison and Mina would add three children of their own—a daughter, Madeleine, born in May 1888, and two sons, Charles and Theodore.[28]

GLENMONT SAT at the crest of a hill, commanding sweeping views of the Orange River Valley, which was mostly undeveloped at the time. "See that valley?" Edison asked his secretary one day as the pair enjoyed the scenery.

"Yes, it's a beautiful valley," the man replied.

"Well, I'm going to make it more beautiful. I'm going to dot it with factories."[29]

Edison's first project in the valley was not a factory but a new laboratory, which he began building in the summer of 1887. At that time, his lighting business was thriving, with more than 100 central stations across the country. In Manhattan Edison had expanded from the financial district uptown all the way to Forty-second Street, and there were also major stations in Brooklyn, Chicago, Boston, Detroit, New Orleans, Philadelphia, and St. Paul. All of the central stations purchased bulbs, dynamos, fixtures, and other equipment from the Edison manufacturing companies: the Lamp Company, Machine Works, Bergmann & Company, and Tube Company. Since Edison owned majority stakes in all of these companies, a large proportion of the companies' growing profits went directly into his pocket.[30]

As a result, the inventor could afford to build himself an extravagant workshop at a cost of more than $150,000 (the equivalent of nearly $3 million today). Constructed on a fourteen-acre plot half a mile from Glenmont, the laboratory looked like a tidy factory complex. It consisted of a main building, three stories high and 250 feet long, as well as four one-story buildings set perpendicular to it. The first smaller building held sensitive electrical equipment, the second a chemistry lab, the third a chemical storehouse and carpentry shop, and the fourth a metallurgy lab. Copper wires ran underground from a powerhouse—or dynamo room, as it was known—to all of the other buildings, supplying 100-volt direct current for lighting and power. The grandest space was the library, at one end of the main building, which had parquet floors, Oriental carpets, tall tropical plants, and polished wood bookcases holding one of the world's best private collections of scientific books and journals.

Visitors entering the heavy machine shop on the first floor encountered a chaotic scene. Leather belts whirred across the ceiling to spin the drive shafts that provided power; enormous machines roared as they planed, bored, rounded, and cut iron and steel; grimy workmen hammered and filed the new metal castings of the devices under construc-

Edison's Orange, New Jersey, laboratory complex.

tion. Up a steep staircase lay another room as large as the first, with a similar range of activities, only here the noise was a few octaves lower, the machines more dainty, the workmen a bit cleaner. This was the precision machine shop, where men worked on the more delicate equipment, such as the intricate brass workings and delicate needles of the phonograph. Edison's private experimental room was on this floor. Nearby were a drafting room, where spectacled men turned Edison's rough sketches into precise plans for new machines, and various experimental rooms. The third floor held more experimental rooms and a large space used as a lecture hall and phonograph recording studio.

One of the most remarkable parts of the laboratory was its stockroom, which, as Edison had planned it, held "almost every conceivable material of every size," including shark teeth, walrus hide, narwhal horn, tortoise shells, ostrich feathers, peacock tails, printer's ink, pumice stone, mica, oats, buckwheat, dried grasses, gums, spices, drugs, chemicals, sheet metals, ice-cream freezers, and wheelbarrows. "The most important part of an experimental laboratory," Edison explained, "is a big scrap heap." When two workmen were stranded at the lab during the great blizzard of 1888, the experimental storeroom doubled as a pantry, providing buckwheat for pancakes, maple syrup, pemmican, and macaroni with olive oil. In fair weather Edison's young sons lingered near the storeroom and begged pieces of gum chicle to chew. The men at the lab, including the boss, were given to chewing not only chicle but also samples of asphalt and tar.[31]

The range of equipment–from heavy machine tools to the most sensitive devices for measuring electricity–was unrivaled, as was the staff. Edison employed more than 100 men, including some of the country's most skilled craftsmen, chemists, and engineers. One of Edison's top researchers described the staff as "about as interesting an aggregation of learned men, cranks, enthusiasts, plain 'muckers' and absolutely insane men, as ever forgathered under one roof." Even more so than Menlo Park, the Orange laboratory was an invention factory, designed to let ideas flow smoothly from the library to the experimental areas to the drafting room to the machine shops. Between his men, material, and machines, Edison boasted, his new lab could "build anything from a lady's watch to a Locomotive."[32]

Although Edison personally had financed construction of the laboratory, he paid its operating expenses by conducting experiments and developing new products for Edison Electric, the Edison Machine Works, the Edison Lamp Company, and Bergmann & Company. The laboratory began operations at the end of 1887, just as the panic over the Westinghouse competition was peaking. As a result, most of the early experimental work was devoted to testing alternating-current systems, improving the efficiency of dynamos, motors, and lamp filaments, and developing better wire insulation. Assisting Edison with these experiments was a new hire, Arthur E. Kennelly, a quiet, serious, twenty-six-year-old Englishman with a thick mustache and thinning hair. Although self-educated, Kennelly was one of the few electricians of the day who had mastered the complex mathematical knowledge required for modern electrical engineering. He felt lucky to be hired by Edison. "The laboratory is just heaven," he wrote to a friend. "It is certainly one of the finest in the world, and the finest in the States."[33]

In Edison's laboratory Kennelly began the pioneering research in

Arthur E. Kennelly

electrical theory that would eventually earn him international renown and a professorship at Harvard. Some of his first experiments were also his most unusual.

IN JUNE 1888, when the reporter from the *New York World* asked Edison for help answering questions about electrical executions, the inventor was happy to oblige. He instructed Kennelly and Charles Batchelor to prepare the machinery and scheduled an experiment for the afternoon of Thursday, June 21. The *World* procured a dog for the occasion. About twenty people, Edison among them, gathered outside the dynamo room at three in the afternoon. Kennelly laid a wooden board on the ground and topped it with a sheet of tin, one corner of which was attached by wire to an alternating-current dynamo running at 1,500 volts. Beside this tin sheet, but not touching it, Kennelly placed a metal pan filled with water and insulated from the ground with two strips of rubber. Another wire from the dynamo was attached to the pan. When the dog drank from the pan of water, its body would close the circuit between the tin sheet and the pan.[34]

Kennelly looped a rope around the animal's neck and tugged it toward the tin plate. The dog yanked violently and broke the rope, but it was quickly recaptured and tied with a stouter cord. Kennelly pulled the dog onto the tin plate, but it refused to drink. When it made another frightened leap, though, a front paw touched the pan of water while a back paw was still on the tin sheet. "There was a quick contortion," the *World* reporter noted, "a smothered yelp, and the little cur dog fell dead."

"How quickly will electricity kill a man?" the reporter asked.

"In the ten-thousandth part of a second," Edison replied.

When the reporter asked how to attach the electrodes to the condemned man, Edison picked up pencil and paper and sketched two hands manacled, with a chain attached to each manacle. The officers of the law could handcuff the condemned man in his cell and lead him to the place of execution, Edison explained, where a wire from a dynamo would be attached to each handcuff. "When the time comes, touch a button, close the circuit, and"—Edison snapped his fingers—"it is over."

"The current," Edison made sure to tell the reporter, "should come from an alternating machine."

CHAPTER 11

"A Desperate Fight"

ALTERNATING CURRENT was much on Edison's mind during the summer of 1888. The French syndicate's grip on the world copper supply had not loosened, which meant that the price of conductors for Edison's system remained high. The Thomson-Houston Electrical Company, which previously sold arc lighting and direct-current incandescent equipment, began selling an alternating system as well, giving Edison a second major competitor in incandescent lighting. George Westinghouse continued to win most of the business in small towns, and he was doing well in cities, too. An Edison agent in one town wrote to describe "a desperate fight between the Westinghouse Co. and Edison" to win a lighting contract. Francis Upton, Edison's longtime associate, urged him to build a new system that could transmit greater distances and therefore compete "in places where now we have no chance, or where the Westinghouse Alternating System will be used."[1]

Among the drawbacks of the alternating system were the lack of a motor and meter. By March 1888, however, word leaked out that Westinghouse engineers had produced a meter, which would allow their current to be sold more efficiently. Even more alarmingly, in May the

U.S. Patent Office issued five patents for an alternating-current motor to Nikola Tesla, a brilliant young Serbian inventor who earlier had worked for Edison. George Westinghouse snapped up the patents and put Tesla on his payroll. The motor was not yet ready for market, but the Edison interests worried that soon enough they would lose another major advantage over the competition.[2]

Early in 1888 Edward Johnson, Edison Electric's president, printed hundreds of copies of a long, scarlet-covered pamphlet titled *A Warning from the Edison Electric Light Company* and mailed them to reporters and officials of local lighting utilities that had bought or were considering buying equipment from Edison's rivals. Johnson cited numerous violations of Edison patents—including the incandescent lamp—by Westinghouse, Thomson-Houston, and other companies, and he cautioned buyers of these rival systems that they might get stuck with worthless equipment if Edison's patents were upheld. The Edison system, he added, was far more efficient than its alternating-current rivals. Johnson also issued a grave warning concerning the dangers of alternating current. "Death-dealing" high-tension currents had killed dozens, Johnson reported, and he reprinted excerpts from newspaper articles about electrical accidents. The Edison system, on the other hand, had a "glorious record" without "a single instance of loss of life."[3]

Manufacturers of alternating current rose to its defense, and in the late spring of 1888 electrical societies staged heated debates between partisans of each system. "It is no longer a question of discussing the pros and cons in amicable conclave," the journal *Electrician* reported, "but of fighting tooth and nail." "The battle of the currents," as it became known, had begun.[4]

Just as the battle was heating up, George Westinghouse made a gesture of peace by writing to Edison. "I believe there has been a systematic attempt on the part of some people to do a great deal of mischeaf [*sic*] and creat [*sic*] as great a difference as possible between the Edison Company and The Westinghouse Electric Co., when there ought to be an entirely different condition of affairs," Westinghouse wrote. Perhaps, he proposed, Westinghouse Electric could make "some sort of arrangement with the Edison Company whereby harmonious relations would be established."[5]

Edison wrote a terse letter of reply. "My laboratory work consumes the whole of my time and precludes my participation in directing the business policy" of the company, he explained. This was not true, for Edison was actively involved in Edison Electric policy—he simply saw no reason to accept Westinghouse's offer. Because Edison believed his own lighting system was safer and more efficient, he expected victory and saw no reason to call a truce.[6]

In the spring and summer of 1888 the Edison and Westinghouse companies argued bitterly over the relative efficiency, reliability, and versatility of the two systems: which converted mechanical to electrical energy with the least loss; which was less liable to break down; which could provide power to sewing machines and elevators as well as lamps; which could serve houses in far-flung districts as well as urban centers. These were arcane and technical matters that remained confined to the dry pages of technical journals. But one issue—danger—had distinct popular appeal. In 1888 the daily newspapers reported an increasing number of accidental deaths from high-voltage alternating current. A Cleveland man died while adjusting a lamp at a theater, and in upstate New York a light company manager died while fixing the carbon on an arc lamp. In New York City, one lighting utility engineer received a fatal shock while oiling a dynamo, a second while cutting a wire that interfered with fire-fighting equipment, a third while demonstrating to a friend that shocks did not affect him.[7]

Most of those who died worked in the industry, but electricity posed a threat to the public as well. To see the source of the danger, pedestrians in New York City needed only to look up in the air. By the late 1880s Manhattan hosted nearly a dozen electric light utility companies, which included Harlem Lighting, Manhattan Electric, Mt. Morris Electric, North New York Lighting, Brush Electric Illuminating, United States Illuminating Company, Safety Electric, and Ball Electrical Illuminating. Edison Illuminating remained the city's primary supplier of incandescent lighting, its system now stretching as far north as Fifty-ninth Street, while the other companies were primarily in the business of supplying high voltage to arc lamps on city streets. In addition to the lighting firms, there were countless providers of telegraph, telephone, stock quotation, and fire alarm services. Whereas all of Edison's wires were buried underground, the wires of the other companies were strung overhead on tottering poles—some of which carried more than a dozen cross-arms and several hundred wires—or looped across the facades and over the rooftops of buildings.

Bad insulation made many of these wires unsafe. One electrician explained that the most common wire-coating material provided "as much value for the purposes of insulation as a molasses-covered rag." The better insulations were compounded of rubber, gutta-percha, tar, pitch, asphalt, linseed oil, and paraffin, but not even these could long survive the rigors of wind, rain, and high-pressure current. Some arc light cables carried 6,000 volts, far more pressure than the insulation could bear. Poorly insulated high-voltage lines were strung across tin roofs—creating hundreds of square feet of lethal metal surface—or placed just under windows, within easy reach of building occupants. About a third of the overhead wires were "dead," abandoned by their owners and left for years, stripped of insulation and draped across live wires. A drooping bare wire could saw through the insulation of even a brand-new line, and when metal touched metal, whatever the wires carried—human voices on the telephone, telegraphic dots and dashes, or sizzling electric current—veered off on a new path.[8]

Major cities in Europe required that all electric light wires be placed in underground conduits, and Chicago followed the same course. New York, however, allowed what an electrical journal called "aerial freebooting," in which wires were strung with "absolutely no official supervision." In 1884 the New York State Legislature passed a law much like Chicago's, requiring that all wires be placed in underground conduits. The lighting and telegraph companies, however, ignored the law, so the following summer the legislature created the

In the late 1880s thousands of telegraph, telephone, and electrical wires ran above New York's streets, posing a danger to the public.

Board of Electrical Subway Commissioners to enforce the burial of wires. As with much governmental action at the time, incompetence and corruption delayed the work. The law granted an exclusive contract to build the underground conduits to the Consolidated Telegraph and Electrical Subway Company, which was controlled by friends of the subway commissioners. The company's officers dithered in the digging of trenches, and their cronies on the subway commission were not inclined to hurry them along.[9]

The delays pleased the lighting utilities. They resisted placing their wires underground because they would have to rent space in the conduits, and because underground wires required more expensive forms of insulation. This relentless pursuit of cost savings also discouraged the companies from conducting routine maintenance or installing safety devices that would have protected the public from overhead wires. In 1887 the legislature tried to force the issue with a stronger law that replaced the subway commissioners with the Board of Electrical Control and named New York's mayor as a member. After underground conduits were ready, companies would have ninety days to put their wires in them. When that time expired, the electrical board was to cut down any overhead wires that remained. Finally, by the spring of 1888, a few telegraph and light wires began to go underground.

But there was one notable holdout. The United States Illuminating Company, an arc lighting firm that used alternating current, refused to bury its wires or even maintain them properly. The firm challenged the law in court, insisting that its lines were perfectly safe and required neither burial nor regulation.[10]

That position became increasingly difficult to defend. In April 1888 a boy peddler died on East Broadway after touching a downed telegraph wire that fell across a high-tension light wire. Later that month a young clerk died in front of his uncle's store after touching a low-hanging arc lamp wire. In May a lineman for Brush Electric died from a shock while working outside a building on Broadway. All of those fatal wires carried alternating current.[11]

When asked his opinion of the best method of executing criminals

Harold Pitney Brown

with electricity, Thomas Edison replied, "Hire them out as linemen to some of the New York electric lighting companies."[12]

WITH THIS CLUSTER of three deaths in New York, the debates over high-voltage current jumped from the pages of electrical journals to the mass-circulation daily newspapers. On June 5—one day after Governor Hill signed the electrical execution bill into law—the *New York Evening Post* printed a letter from an obscure electrician named Harold Pitney Brown. Just thirty years old, Brown had a decade of experience in the electrical industry. In 1877 he had started working for Western Electric in Chicago, which manufactured a wide variety of electrical devices, including medical apparatus, telegraph and telephone equipment, and an "electric pen"—a sort of early mimeograph—invented by Edison. During the early 1880s Brown installed arc lighting plants for Brush Electric, and by 1888 he was describing himself as an independent electrical consultant. Brown's boss at Western Electric had been a man named George Bliss, a close associate of Edison's who became a fierce opponent of Westinghouse in the battle of the currents. Brown, it turned out, agreed with his former boss.[13]

In his letter to the *Post* Brown defended those who wanted to ban high-voltage overhead lines in New York, but he also opened a more general attack on alternating current. He echoed the arguments of Johnson's *Warning*, but he expressed them in incendiary language. Alternating current was so dangerous to human life, Brown claimed, that it could "be described by no adjective less forcible than *damnable*." The only reason to use it was to save on copper costs: "That is, the public must submit to *constant danger from sudden death* in order that a corporation may pay a *little larger dividend*." An alternating-current wire above a New York street, Brown warned, "is as dangerous as a burning candle in a powder factory." Brown advised New York City's Board of Electrical Control to limit the transmission of alternating current to no more than 300 volts—a restriction that would effectively ban it altogether, for without the advantages afforded by high-voltage transmission, it could not compete effectively with direct current. Brown was not troubled by this possibility, because he considered direct current far preferable: "a continuous current of 'low tension,' such as is used by the Edison Company for incandescent lights, is perfectly safe."[14]

Manhattan's Board of Electrical Control suddenly found itself at the center of a nasty fight over electrical safety. At a July meeting of the board, defenders of alternating current—most of them employees or allies of Westinghouse—vilified Brown. One described Brown's letter as "a villainous budget of perversions of fact," while another claimed Brown was "entirely ignorant" of electrical technology. A Westinghouse vice president cited Edward Johnson's circulars and Brown's letter as evidence of the "desperation of opposition companies." Brown's attack, one critic claimed, was "made rather with a view to injuring rivals than with a purpose of protecting the 'dear public.'"[15]

But the critics did not stop at impugning Brown's knowledge and motives. Whereas Brown had claimed that direct current was much safer than alternating, they argued that just the opposite was true. "The alternating current," one claimed, "while disagreeable, is far less dangerous to life than a direct current of the same tension," and another insisted, "I have myself received 1,000 volts without even temporary inconvenience." As long as the current is less than 1,100 volts, one man said, "no fatality can occur, or serious inconvenience result." One electrician went so far as to claim that a shock could be good for people. When a person was felled by a lethal direct-current shock, he said, "the passage of an alternating current through the body has been very efficacious in restoring life."[16]

These electricians did not explain why they believed alternating current was less dangerous. Brown and his allies, on the other hand, put forth a theory to justify their stance. The dangers of electricity, they said, lay not in the current itself but in its interruptions. Direct current, flowing smoothly in one direction, did little damage, while alternating current, which changed directions many times a second, ripped and tore at the body's tissues. "It is the rapid succession of shocks that kills," Brown wrote in his letter to the *Post*, "while a single steady impulse of the same intensity would do little damage."[17]

Brown's letter and the responses to it revealed a dispute that had not yet been resolved by experimental evidence: At a given voltage, which was more dangerous, alternating current or direct?*

To prove the alternating forces wrong, Brown needed allies and equipment. "I therefore called upon Mr. Thos. A. Edison, whom I had never before met," Brown later explained, "and asked the loan of instruments for the purpose, which could not be obtained by me elsewhere. To my surprise, Mr. Edison at once invited me to make the experiments at his private laboratory."[18]

EDISON'S EXPERIMENT for the *World* reporter in June was an unsystematic demonstration, with no records kept. In July experiments of a different sort took place. The key participants were Brown, Arthur Kennelly, and Dr. Frederick Peterson, a specialist in nervous

*The danger of an electric current actually depends on the combination of volts and amperes; a high-voltage current is not dangerous if the amperage is low. Published reports on these debates, however, usually noted only the voltage, omitting the amperage measurements.

diseases who would later become the president of the American Neurological Association. The experiments took place under carefully controlled conditions, and Kennelly kept meticulous records in the official laboratory notebooks. Over the next year several dozen animals would die at the Edison laboratory to test the dangers of electricity. These experiments constituted the first scientific researches into killing with electrical generators.[19]

On July 12 Kennelly and Brown wired a fox terrier to a direct-current dynamo, and the dog survived momentary shocks of 400, 600, and 800 volts. Finally, at 1,000 volts, it died. The second subject, a half-breed bulldog, died almost immediately after receiving a two-second shock of 800 volts alternating current. A few days later a half-breed shepherd was given 1,000 volts direct current, then, in succession, shocks of 1,100, 1,200, 1,300, and 1,400 volts. At each shock, Kennelly jotted in his notebook, the "dog yelped once but not much hurt." The shepherd was released. Next a "black mongrel" was subjected to increasing jolts with alternating current, and Kennelly recorded the grim results: at 300 volts, "dog howled for about 1 minute & struggled violently"; at 400 and 500, "dog yelped and struggled"; at 600, "dog yelped and groaned. Died in 90 seconds."[20]

On July 17 Kennelly killed a "yellow mongrel" with 500 volts direct current, with the electrodes attached to the dog's legs, as they had been in all the earlier experiments. Kennelly then applied 300 volts alternating current to the ears of a bull terrier. The dog showed little effect while the current was on, but when it was cut off, the animal bled from the eyes and appeared to be in great pain. Kennelly then "hastened to apply the whole power of the machine in order to terminate his sufferings."[21]

The dogs that died in these experiments had been bought from neighborhood boys at twenty-five cents a head. The Orange area was soon depopulated of strays, and Edison went looking for a new supply. Back in April he had received a letter from Henry Bergh, the president of the ASPCA, who had heard about the Buffalo SPCA's use of electricity to kill strays and wanted Edison's advice on the matter. The

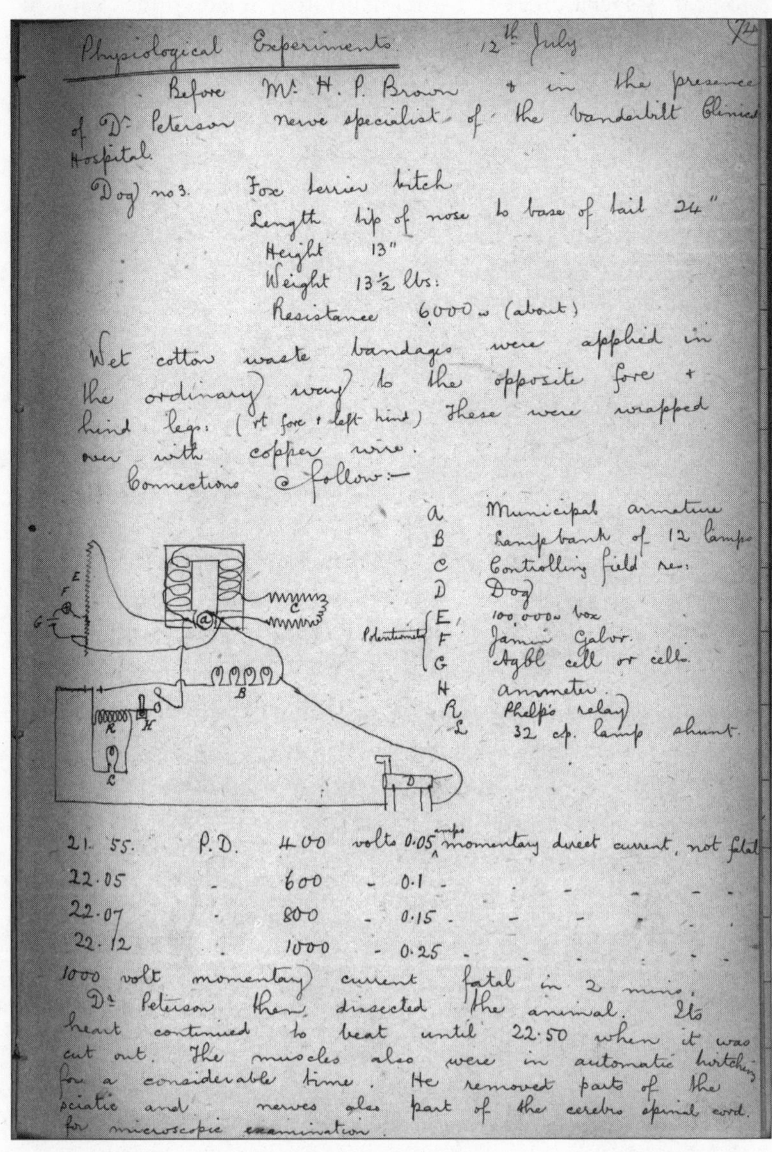

Physiological Experiments. 12th July

Before Mr. H. P. Brown + in the presence of Dr. Peterson nerve specialist of the Vanderbilt Clinical Hospital.

Dog no 3. Fox terrier bitch
Length tip of nose to base of tail 24"
Height 13"
Weight 13½ lbs:
Resistance 6,000ω (about)

Wet cotton waste bandages were applied in the ordinary way to the opposite fore + hind legs: (rt fore + left hind) These were wrapped over with copper wire.

Connections @ follow:—

A Municipal armature
B Lampbank of 12 lamps
C Controlling field res:
D Dog
Potentiometer { E 100,000ω box
F Jamin Galvr.
G Agbl cell or cells
H ammeter
R Phelps relay
L 32 cp. lamp shunt

21. 55. P.D. 400 volts 0.05 amps momentary direct current, not fatal
22.05 " 600 " 0.1 " " " " "
22.07 " 800 " 0.15 " " " " "
22.12 " 1000 " 0.25 " " " " "

1000 volt momentary current fatal in 2 mins.

Dr. Peterson then dissected the animal. Its heart continued to beat until 22.50 when it was cut out. The muscles also were in automatic twitching for a considerable time. He removed parts of the sciatic and nerves also part of the cerebro spinal cord for microscopic examination

A page from the laboratory notebook of Arthur Kennelly, Edison's chief electrician, showing a circuit arrangement for one of the experiments on dogs.

The Edison laboratory's dynamo room, where the dog-killing experiments took place.

inventor had recommended "a small alternating machine." In July it was Edison's turn to ask a favor.[22]

"I have lately been trying various experiments on dogs with a view to finding how great a pressure and quantity of electricity it takes to kill them," Edison wrote to Bergh. Although he knew the pressure "within certain rough limits," Edison explained, he wanted to fix it more precisely. "Can you therefore aid me to obtain some goodsized animals for these experiments?"[23]

Bergh indignantly rejected Edison's request as "antagonistic to the principles which govern this Institution." The SPCA's goal was to use enough current "to produce instantaneous and merciful death," whereas Edison's efforts to determine the minimum lethal voltage were "calculated to inflict great suffering upon the animals." Edison, aware from his own experiments of the truth of Bergh's statement about suffering, quickly apologized.[24]

The experiments continued with dogs obtained in other ways. Most of the tests took place late at night, between ten and midnight. Edison normally welcomed visitors, but he kept these experiments as secret as possible. The lab workers, though, knew what was happening, alerted by the agonized howls of dogs. Edison's men stood outside the dynamo room and peered in through the windows to catch glimpses of the experiments.[25]

BROWN AND KENNELLY believed they were ready to prove that alternating current was more dangerous than direct, so they scheduled a public demonstration. On the evening of July 31, 1888, a crowd of seventy-five gathered in an auditorium at Columbia College in Manhattan. The audience included the best and brightest of New York's electrical community: members of the city's Board of Electrical Control, officers of the Edison and Westinghouse companies, the editors of *Electrical World* and *Electrical Review*, and the secretary of the American Institute of Electrical Engineers. Brown, assisted by Kennelly and Dr. Peterson, produced a seventy-six-pound black mongrel. Although it suffered dreadfully, the dog survived successive jolts of 300, 400, 500, and 700 volts direct current. When given 500 volts alternating current, it "gave a series of pitiful moans, underwent a number of convulsions and died." Brown and Kennelly then prepared to experiment upon another dog, but one of Henry Bergh's ASPCA officers stopped the proceedings. The officer said that "if it was necessary to torture animals in the interest of science it should be done by the colleges or institutes and not by rival inventors."[26]

Undeterred, Brown scheduled a second demonstration for the evening of August 3. This time Kennelly was not present, nor were the representatives of the Edison or Westinghouse companies. To avoid further interference from the ASPCA, Brown enlisted the assistance of two physicians licensed to practice vivisection. Brown used alternating current to kill three dogs: a black mongrel (killed by 272 volts), a Newfoundland (340 volts), and a Newfoundland-setter mix (234 volts). The doctors then dissected each dog to note the physiological effects of the current.[27]

"If this sort of thing goes on, with the accidental killing of men and the experimental killing of dogs," the *Electrician* wrote, "the public will soon become as familiar with the idea that electricity is death as with the old superstition that it is life."[28]

"We made a fine exhibit yesterday," Harold Brown wrote to Kennelly. "It is certain that yesterday's work will get a law passed by the legislature in the fall, limiting the Voltage of alternating currents to 300 Volts."[29]

CHAPTER 12

"Criminal Economy"

THE *Electrical Review* thought so little of Harold Brown that it published a poem, "Mr. Brown and the Dog," which included the sarcastic couplet "No company employed him, and his motives he felt sure,/ Were thoroughly unbiased, philanthropic, and most pure." No one could prove that Edison Electric employed Brown, but the logic of such an arrangement was obvious: If Brown succeeded in outlawing high-voltage alternating current, Edison Electric would gain an enormous advantage.[1]

That goal appeared to be increasingly unattainable, since even more modest efforts to regulate electricity had been stymied. United States Illuminating continued to fight the state law requiring it to move its Manhattan wires from overhead poles to underground conduits, and in the fall of 1888 the company scored another victory. The state supreme court granted a preliminary injunction preventing the city from removing its wires. As Harold Brown and Arthur Kennelly were killing animals to prove the dangers of alternating current, lighting companies were winning the right to keep that current sizzling overhead.[2]

While Thomas Edison stayed in the background, Kennelly became an outspoken advocate of execution by alternating current. The editors of electrical journals regarded Brown with deep suspicion, but they took note when Kennelly spoke in defense of the dog-killing tests at the Edison lab. The *Electrician*, which had harshly criticized Brown's claims, changed its tune after hearing from Edison's chief electrician. "[A] letter from Mr. A. E. Kennelly," the *Electrician* noted, "completely rehabilitates Mr. Brown's position, and in our estimation invests the experiments with an importance which they did not before possess." Although the journal remained noncommittal on the relative dangers of the two currents, Kennelly had caused some advocates of alternating current to rethink their positions.[3]

Kennelly also took the fight to the popular press. When the *New York World* reported that, since the dog experiments started, "the faith of experts in electricity as an executioner has steadily decreased," Kennelly took it as an affront to himself and to his boss and immediately sent a long letter of rebuttal. "Having carried out those experiments under Mr. Edison's instruction," Kennelly explained, he had proven that he could "kill a dog instantly by electricity . . . and without sigh, moan, or struggle." Alternating current was "beyond all doubt more fatal than the continuous current."[4]

According to Brown, these dog experiments assessed the relative

Drawing cables into an underground conduit. Many companies resisted state laws requiring that all wires be placed under the street rather than on utility poles.

danger of alternating and direct currents for lighting purposes and had nothing to do with electrical execution. Most people failed to grasp the distinction. Brown's avowed goal of preventing accidental deaths and New York State's interest in intentional killing were two sides of the same coin: Both required knowing what type of electrical current could kill a human being.

In November 1888 Thomas Edison granted his first public interview on electricity's dangers since his dog experiment for the *World* in June. He told the *Brooklyn Citizen* that electrical execution was "a good idea. The man will be killed with a current of the proper number of volts in the tenth of a second." He explained how he came to this conclusion: "I did quite a lot of experimenting with currents on dogs. It was funny. At first, we used continuous currents. After the electricity charged the dog he stood still without the slightest change in his appearance. He did not move, and his eyes retained the same expression they usually wore. Then after a minute, or a minute and a half, he would collapse and tumble over, dead. Finally, we tried alternate currents. . . . Then it was found that this shock of one-tenth of a second killed the dog." Edison said that one dog was granted a reprieve after surviving a shock of 1,400 volts (current type unspecified): "He is walking round to-day as good as ever after his shock. We call him Ajax, because he played with the thunder-bolts."[5]

Edison used the occasion of this interview to tackle one of the puzzles of the debate: Although its advocates described electrical execution as painless, electric shocks were known to hurt quite a lot. Arthur Kennelly himself accidentally took a 1,000-volt alternating-current shock and vividly described the sensation as "violent rending pain." Defenders of the new execution method would have explained that Kennelly suffered only because he had not been killed. Nerve sensation, they said, traveled at a rate of about 100 *feet* per second, whereas electricity sped along at 160,000 *miles* per second. In the victim of lethal electric shock, a mismatched race was staged between the sensation of pain and the destructive force of electricity. By the time a nerve impulse from the victim's arm or leg meandered its way to the brain to announce the presence of pain, electricity had gotten there first and destroyed consciousness. As Edison put it, "There won't be time for these sense-bearing nerves to telegraph the news that he is hurt to his brain before he will be dead from the shock."[6]

The explanation sounded reassuringly mathematical, but it skirted some important issues. If only a fraction of a second separated the sensation of pain from the end of life, it was crucial to understand how and when death occurred. But doctors could not agree whether electricity killed by stopping the heart, cutting off the breath, destroying the blood, or some other mechanism. In an experiment described in the death penalty commission report, Dr. Fell anesthetized a dog and removed its chest wall so he could observe its heart. As soon as the current was applied, the heart "ceased its action and became a mere mass of quivering flesh," Fell wrote. "Not the least resemblance to a rhythmical movement was observed after the current was made." As Fell saw it, this irregular heart movement—what doctors today would call fibrillation—was proof that the dog had died instantly. Fell either did not know or did not admit that such heart failure did not cause immediate unconsciousness and often was accompanied by intense pain.[7]*

New York's Medico-Legal Society decided to find answers to the many questions surrounding electrical execution. The private society, which was dedicated to improving relations between the medical and legal professions, took up the matter of its own accord and had no official status with New York State. Its membership simply recognized that New York State needed expert guidance, and that a physician's knowledge could apply just as well to ending life as to saving it. The death penalty commission had rejected the injection of poison as a method of execution in part because of "the difficulty of finding any medical man who would act as a public executioner." As it turned out,

*A critic of electrical execution noted that George Fell shared a surname with a figure from a nursery rhyme—"I do not like thee, Doctor Fell / The reason why I cannot tell"—and commented, "The clue appears now perhaps to be within our grasp."

electrical execution relied heavily on medical knowledge, and state officials had no trouble finding physicians willing to kill. Many doctors saw their primary responsibility as serving not individuals but the state. The death penalty was "a question of therapeutics," in the *Medical Record*'s view. "If society can better preserve and protect itself against criminals by their strangulation or electrothanatosis, why, then, these measures should be employed. No question of first principles or inherent rights or sentiments is involved."[8]

In September 1888 the Medico-Legal Society appointed a committee to investigate the technical requirements of killing with electricity. One of the committee's members was Dr. Frederick Peterson, who had taken part in the Edison lab's dog experiments. In a report presented at the society's November meeting, Peterson and his fellow committee members recommended a current of 3,000 volts. On the question of type of current, the committee wavered: "Either a continuous [direct] or an alternating current may be used, but preferably the latter."[9]

Rather than adopt the committee's report, the Medico-Legal Society tabled it until its December meeting. A few members still needed to be convinced that alternating current was indeed more dangerous than direct. Others objected because the report was based on tests involving dogs, which were considerably smaller than humans. To reassure the doubters, the committee turned for assistance to Thomas Edison, who again agreed to serve as host for experiments in electrical killing. Arthur Kennelly and Harold Brown made the necessary arrangements.

Kennelly hoped to hold the tests at night because "an experiment of this kind by daylight naturally attracts very much curiosity and renders the necessary privacy and care difficult to obtain." He did not get his wish. On the afternoon of Wednesday, December 5, 1888, Thomas Edison welcomed a group of observers, including members of the Medico-Legal Society, newspaper reporters, and—most important—Elbridge Gerry. Because Gerry was the chairman of the death penalty commission and the author of the new law, his views carried great weight.[10]

The guests arrived around noon, but equipment problems delayed the experiments until after three. Brown and Kennelly connected an alternating generator to a transformer, which they hoped would boost the current up well over 1,000 volts. The first victim was a calf weighing 124 pounds. They clipped the hair from two areas, between the calf's eyes and beside its spine just below its shoulders, and attached two electrodes, both covered with sponges soaked in a conducting solution of zinc sulfate. They ran the dynamo up to 1,100 volts and closed the circuit, but "the animal fell uninjured," according to Kennelly's notes. He determined that the problem lay with the transformer, so he disconnected it. He wired the calf directly to the dynamo, ran the current up to 770 volts, and closed the circuit for eight seconds. "Death was instantaneous," Brown claimed in his published report. Next came another calf, weighing 145 pounds, which died after receiving 750 volts for five seconds.[11]

To eliminate all doubt that electricity could kill animals larger than dogs and calves, Brown led in a horse weighing 1,230 pounds. This time he and Kennelly tried a different electrode arrangement that had been proposed by Edison, who believed that prisoners should be executed by placing an electrode on each hand, so that the current passed through the chest. To test this theory, Brown and Kennelly attached an electrode to each of the horse's forelegs. They applied 600 volts for five seconds, but the animal did not die, and a second shock of fifteen seconds also failed to kill it. Finally, 700 volts for twenty-five seconds proved fatal.

Kennelly's notes of the experiments indicated that the calves and horse suffered, just as many of the dogs had. But in a report of the experiments published in the *Electrical World*, Brown insisted that the deaths of all three animals were "instantaneous and painless." He concluded that "the alternating current is the best adapted for electrical executions."

The well-publicized experiments at the Edison lab caught the public imagination. A Brooklynite wrote to Edison suggesting that electrical execution be combined with hanging, by weaving metal electrodes into the hemp noose such that one pole would contact the spine, the other the throat. Combining the noose with an electric shock, the

The December 1888 horse-killing experiment at the Edison laboratory, as illustrated in *Scientific American*.

man wrote, would be "more *wholesome* than screwing the poor victim into a vise of electrical horrors." In Connecticut P. T. Barnum allowed electricians to give nonfatal shocks to animals in his menagerie, including a monkey, baboon, hippopotamus, and elephant. The monkeys screamed in agony, but the elephant, named Tom Thumb, reportedly "squealed with delight." When Chief, a rogue circus elephant in Cincinnati, had to be euthanized, his owners considered killing him with electricity, but they ended up strangling him instead, thus robbing alternating current's foes of an ultimate demonstration of electricity's killing power.[12]

The Medico-Legal Society, meanwhile, continued its efforts to ensure that condemned criminals would not die by the rope. At the society's annual meeting on December 12, the committee on electrical execution recommended that the state should give condemned prisoners a shock for fifteen to thirty seconds with an alternating current of 1,000 to 1,500 volts. The full society unanimously adopted the report and transmitted it to state authorities.[13]

The lessons of the tests were not lost on the press. The *Daily Tribune* noted that the Edison lab experiments "showed that less than half the pressure used for electric lights in our streets is sufficient to produce instant death. Evidently the danger from electric light wires has not been over-estimated."[14]

GEORGE WESTINGHOUSE disagreed with the *Tribune*'s conclusion, and a few days after the experiments he published a response in several New York papers. Westinghouse argued that the Edison lab experiments were so carefully calibrated to cause death—with large electrodes placed at the most sensitive portions of the body—that they had no relevance to the possibility of accidental electric shock. "It has been found that pressures exceeding 1,000 volts can be withstood by persons of ordinary health without experiencing any permanent inconvenience," Westinghouse wrote. He described an episode in which a man "held his hand in contact with the wires [bearing 1,000 volts] for a period of three minutes without fatal results—in fact, was able to go on with his work after a short period." Not only could the shocks be survived, but "the alternating current is less dangerous to life" than the direct.[15]

As Westinghouse saw it, the real reason for the tests at the Edison laboratory had to do with the fact that Westinghouse sold a better, cheaper product and was destroying Edison's business.

> It is generally understood that Harold P. Brown is conducting these experiments in the interest and pay of the Edison Electric Light Company; that the Edison Company's business can be vitally injured if the alternating current apparatus continues to be as successfully introduced and operated as it has heretofore been, and that the Edison representatives, from a business point of view, consider themselves justified in resorting to any expedi-

ent to prevent the extension of this system. . . . We have no hesitation in charging that the object of these experiments is not in the interest of science or safety, but to endeavor to create in the minds of the public a prejudice against the use of the alternating currents.[16]

Westinghouse could not prove his charges against Brown, but one claim in his letter was undoubtedly true: Westinghouse Electric was badly hurting Edison Electric's business. The Edison market position had deteriorated further since summer, as Westinghouse continued to win most of the lighting contracts in smaller towns and to expand much more quickly than Edison.[17]

Edison Electric remained silent on Westinghouse's charges, but Harold Brown proffered a dramatic challenge to Westinghouse in a letter to the *New York Times*. "First, allow me to deny emphatically that I am now or ever have been in the employ of Mr. Edison or any of the Edison companies," Brown wrote. He claimed that it was Westinghouse who was motivated by selfish business interests: To protect his alternating-current system, he was denying solid scientific evidence of its dangers. Because Westinghouse had insulted his honor, Brown proposed a duel: "I therefore challenge Mr. Westinghouse to meet me in the presence of competent electrical experts and take through his body the alternating current, while I take through mine a continuous current. . . . We will commence with 100 volts, and will gradually increase the pressure 50 volts at a time, I leading with each increase, until either one or the other has cried enough, and publicly admits his error." An electrical journal suggested a way to make the duel even more dramatic: It published an illustration of two men fencing with electrified foils. To the regret of many in the industry, the duel never took place.[18]

Brown found other angles of attack. In late December he sent a circular to city governments, insurance executives, and prominent businessmen in every town in the United States with a population of more than 5,000. He enclosed a copy of the Medico-Legal Society report, explaining that it supported alternating current's "adoption for EXE-

CUTION PURPOSES at exactly the pressure used for 'safe and harmless' electric lighting." Calling for laws limiting the current to 300 volts, Brown urged public-spirited citizens, "Lend me your aid in securing legislation which will keep this EXECUTIONER'S CURRENT out of our homes and streets and prevent reckless corporations from saving their money AT THE EXPENSE OF THE LIVES OF THOSE DEAR TO YOU." The appeal had its desired effect. Anxious citizens from across the country wrote to Brown, and in response he sent more information about the dangers of alternating current. Some letters brought requests for Brown's personal services, so he traveled to many towns to conduct tests on the local alternating systems, sometimes persuading mayors and city councils to support the call for a 300-volt limit.[19]

The December circular, a three-and-a-half-page typewritten document, offered a mere sketch of Brown's case against Westinghouse. In the early months of 1889 he published a more sustained attack, a sixty-one-page booklet titled *The Comparative Danger to Life of the Alternating and Continuous Electrical Currents*. It was a narration of Brown's crusade against alternating current, starting with his letter to the *Evening Post* in June 1888 and continuing through his experiments with Kennelly and Edison and his work with the Medico-Legal Society. Brown once again charged that Westinghouse was endangering the public for the sake of higher profits. "So many lives have been sacrificed by criminal economy in electric lighting that legislative control of the subject is imperatively demanded," he wrote. The pamphlet reprinted many of the relevant documents, including newspaper articles and letters, reports of accidents in various cities, detailed descriptions of the Edison lab experiments, and the Medico-Legal Society report. The pamphlet was professionally printed and mailed to the press and government officials across the country.[20]

WHILE BROWN and Westinghouse wrangled publicly, the electrical execution law quietly took effect on January 1, 1889. Austin Lathrop, the superintendent of the New York State prison system, appointed a commission to finalize the technical plans. Dr. Carlos F. MacDonald, the chairman of the state's Lunacy Commission and a professor at Bellevue Hospital Medical College, headed the panel, and he was joined by a fellow physician, A. D. Rockwell, the coauthor with George Beard of the standard text on electrical medicine and a cofounder of the American Neurological Association. Late in February Lathrop and MacDonald met with Harold Brown in New York and told him they wanted to conduct further experiments. Once again Thomas Edison offered the use of his laboratory.[21]

On the afternoon of March 12, Brown, MacDonald, two surgeons, and a professor from the University of Pennsylvania convened in Orange. As in the earlier experiments, Kennelly handled the electrical apparatus and kept careful records in a lab notebook. By the end of the day, four dogs, four calves, and a horse had died. Since the new state commission had accepted the Medico-Legal Society's recommendation of alternating current, no direct current was used. The main purpose of the experiments was to determine the proper placement of electrodes on the bodies of the victims. Kennelly and MacDonald tried different arrangements: foreleg to hind leg, head to hind leg, back of the neck to hind leg, back of the neck to heart. The easiest deaths were suffered by those animals with one electrode on the head and another on a hind leg.[22]

These experiments "solved any doubts . . . regarding the certainty of quick death by the alternating current," *Scientific American* wrote. "As for the bodies of the slain, they so completely escaped disfigurement that the veal was perfectly suitable for human food, and it was returned to the butcher who had brought the calves to the laboratory."[23]

NOT LONG AFTER these tests, the state prison superintendent awarded Harold Brown the contract to acquire dynamos and other equipment for Auburn, Sing Sing, and Clinton, the three prisons where executions would take place.[24]

The state appropriated $10,000 for the necessary equipment, including three alternating dynamos. Brown had a choice of manufacturers, including Westinghouse, Thomson-Houston, and several European firms. He also could have contracted to have a generator built specifically for the purpose. But Brown insisted on Westinghouse machines, and he was unapologetic about his choice. He explained his reasons in the preface to the third edition of his *Comparative Danger* booklet, published sometime between March and July. "The danger of the alternating current was admitted by all the prominent companies excepting the Westinghouse, which still protested that it was 'safe.' On this account," Brown explained, "its apparatus was purchased for the infliction of death upon condemned criminals." He explained further in a letter to the *Evening Post*: "I determined to educate the public by buying for electrical execution the same apparatus that killed . . . many [people] who made accidental contacts with 'safe' Westinghouse circuits."[25]

For a state expert, it was a startling admission: Brown acknowledged that he would use his official position to punish Westinghouse for his alleged intransigence on the safety issue and to bring alternating current into disrepute.

Although Edison Electric still denied that Brown was making these efforts to harm Westinghouse on its behalf, the company did publicize the new developments. The Edison managers sent out a circular titled *The Deadly Parallel* pointing out that New York State had purchased "three Westinghouse alternating-current dynamos," and that the execution current "will be of the same pressure as that used for electric lighting." The Edison current was described as "absolutely harmless, no matter how one may come in contact with it."[26]

Not surprisingly, George Westinghouse refused to sell his generators for the purpose of execution. Brown said that he managed to obtain them anyway, but he would not reveal how. According to the *Times*, "The news that three of their dynamos have been sold by Mr. Brown to the State will be a surprise to the Westinghouse people."[27]

. . .

As Brown maneuvered to acquire execution dynamos, linguists took up the task of naming the new method of killing. Soon after the law took effect, the *World* noted that "the gallows and the hangman will disappear, to be replaced by what? A machine! That is as far as anybody has yet got in describing the curious contrivance which will play such a ghastly role in future executions." *Death by electricity* and *electrical execution* were clumsy terms, certainly too long for space-conscious headline writers. Some looked to the past for models. The guillotine had been named after Joseph Ignace Guillotin, the strongest advocate (though not the inventor) of the device. Since Elbridge Gerry was the man most identified with the electrical method, some suggested *gerricide*. When the *World* sought Gerry's opinion in February 1889, he instead proposed *electrolethe*, "which is a combination of two pure Greek words. . . . The man who touched the button would be the electrolether, with the accent on the third syllable." Others interviewed by the *World* coined many words using the prefix *electro-*, including *electronirvano*, *electrorevoir*, and *electrosiesta*. An assistant district attorney in Manhattan proposed *virmort* from the Latin for "man" and "death." A warden at the Tombs prison, more conversant in popular slang than classical roots, said, "What's de matter wid callin' her de whizzer?" A fellow warden proposed *razzle-dazzle*.[28]

When the scholarly journal *American Notes and Queries* put out a call for new words, proposals arrived from professors at Cornell, Harvard, Brown, Johns Hopkins, and Yale. Most offered etymologically impeccable but thoroughly impractical words: *electrophony*, *electricize*, *electroctony*, *thanelectrize*, *electrothanasia*, *joltacuss*, *voltacuss*, *electrostrike*, *galvanation*, *galvanification*, *electronation*, and *superelectrification*. A frequent suggestion was *electricide*, modeled after suicide or patricide. When some objected that this word implied the death *of* rather than death *by* electricity (*patricide*, for example, meant the killing of the father), the term was revised to *electrocide* (on the model of *electroplate*, or plating by means of electricity). Forsaking Greek and Latin, one correspondent proposed the Teutonic *blitzentod*–lightning death.[29]

A Johns Hopkins professor threw up his hands in defeat: "The word will be a hideous one, however compounded." Julian Hawthorne (son of Nathaniel and an author in his own right) had the firmest grasp of how new words entered the language: "Very likely some chance inspiration of slang may settle the matter."[30]

The editor of *American Notes and Queries* wrote to Thomas Edison seeking his opinion. The inventor proposed *ampermort*, *dynamort*, and *electromort*, but he preferred to hand over the question to one of his attorneys. "The trouble is that none of us here remember enough latin [*sic*] to inspire confidence in the etymology of these coined words," Edison's secretary told the lawyers. "Mr. Edison would like you to revise them." The lawyer rejected Edison's ideas and proposed a different sort of coinage: "As Westinghouse's dynamo is going to be used for the purpose of executing criminals, why not give him the benefit of this fact in the minds of the public, and speak hereafter of a criminal as being 'westinghoused', [punctuation *sic*] or as being 'condemned to be westinghoused'; or, to use the noun, we could say that such and such a man was condemned *to the westinghouse*. It will be a subtle compliment to the public services of this distinguished man." That suggestion never made it into public discourse. The Edison interests decided instead to throw their support behind *electricide*.[31]

Largely missing from these discussions was the word that would eventually gain currency. The *World*'s February 1889 article on the matter did not even mention *electrocution*. It is unclear precisely when the word was first used, but by the summer it appeared frequently in the *World* and in a few other popular newspapers.[32]

Electrocution was formed by combining *electro* and *execution*. *Execution* derived from the Latin *sequi* (follow), which with the prefix *ex-* means "follow out" or "carry out"—as in executing a sentence of death. Etymologically, *electrocution* meant something along the lines of "the following of electricity." Although state officials, most medical journals, and highbrow newspapers still preferred *electrical execution*, the more concise *electrocution* gradually crept into common usage, and the self-appointed gatekeepers of the English language were appalled. The editors of the *New York Times* wrote, "We pray to be saved from such a monstrosity as 'electrocution,' which pretentious ignoramuses seem to be trying to push into use. It is enough to make a philologist writhe with anguish." The *Buffalo Courier* lodged a "protest against the adoption of the proposed new words, electrocute, electrocution, and electrocuted. There's quite enough bastard philology already." A law journal described *electrocution* as "the very worst barbarism* yet perpetrated by an illiterate neologist."[33]

*To linguists, a *barbarism* is a word formed without respect for etymological meaning.

CHAPTER 13

Condemned

ALTHOUGH the electrical execution law took effect on January 1, 1889, newspapers had to wait until May for the year's first capital murder trial. It took place not in Manhattan, the state's murder capital, but at the opposite corner of the state. Buffalo—the city where Lemuel Smith became the first American to die from the shock of a dynamo, where Alfred Southwick conducted the first tests in electrical killing, and where George Westinghouse installed his first alternating system—was also home to the first man sentenced under the electrocution law.

Mary Reid owned a big square cottage on South Division Street in Buffalo. She lived with her two young daughters at the front of the house and rented four small rooms in the back to William Hort, his wife, Tillie, and their four-year-old daughter, Ella. At about eight o'clock on the morning of March 29, 1889, as Mrs. Reid washed dishes, she heard a shrill scream from the back of the house, then the hollow thwack of someone chopping wood, then a few low moans, then silence. She walked toward the back of the house and called out, but she received no reply. She then walked outside, where she saw Mr. Hort walking toward her, his hands smeared with blood.

"I have killed Mrs. Hort," he said.[1]

Mrs. Reid gathered her children and fled to tell a neighbor, a bookbinder named Asa King, who walked to the Hort residence and opened the kitchen door upon a horrific scene. The tables and chairs were overturned, broken dishes and an uneaten breakfast scattered across the floor, blood spattered across the walls. Tillie Hort was on her hands and knees, rocking back and forth and moaning quietly. Her long hair, matted with blood, swept down to the floor. Not far away was a bloody hatchet. William Hort had returned inside and was standing at the back of the kitchen, wiping his bloody hands on a cloth. King urged Hort to go for a doctor, but he refused. Instead, Hort stepped over his bleeding wife, went out the door, walked down the street to Martin's saloon, and ordered a beer. King, who had followed, told the saloon keeper that Hort had killed his wife. Refused his drink at Martin's, Hort walked down the street to the next saloon, where he had just taken the first sip of a beer when a Buffalo patrolman arrested him.

Mrs. Hort was rushed to Fitch Hospital. She had ugly gashes on her left shoulder, right arm, and right hand, but her head suffered the worst damage. Doctors counted twenty-six cuts in her skull, varying from one to four inches in length. They removed the loose shards of skull and tried to stop the bleeding, but there was little they could do. Tillie Hort died seventeen hours later.

Even before she died, reporters and the police had been piecing together the story of her life. One of the first things they learned was that Mr. and Mrs. Hort were not married. In fact, neither of them was legally named Hort: He was William Kemmler, and she was Tillie Ziegler. About a year and a half before, Kemmler was living in Philadelphia and working as a huckster, selling food from a wagon. One night, while drunk, he married a woman he knew only slightly. Desperate to escape the situation, he sold his horse and wagon and disappeared. He did not leave town alone. Matilda Ziegler, who was related to Kemmler by marriage, was unhappily wed to a man whose "fondness for fast women" had resulted in his contracting "a loathsome disease." She bundled up her young daughter, Ella, and ran away

William Kemmler
and Tillie Ziegler

to Buffalo with Kemmler. They called themselves the Horts and settled down as man and wife. William—known to his friends as "Philadelphia Billie"—took up work as a huckster again and prospered. Before long he owned four horses and three wagons and employed a handful of men and boys who helped him peddle fruit, vegetables, butter, and eggs. He made more than enough money to support his small family, but he drank too much, and his fights with Tillie sometimes kept the neighbors awake.[2]

After his arrest, the police supplied Kemmler with a glass or two of brandy, and he admitted to the crime. Illiterate, he signed his confession with an *X*. When the trial started a month later, on May 7, 1889, the courtroom galleries filled with what a reporter described as "the usual crowd which hastens to a murder trial as to a picnic." Kemmler entered the courtroom looking uncomfortable in a new brown suit. A slender man of twenty-eight, he had been deeply tanned from long days peddling vegetables, but a month in jail had bleached him out. From the moment of the arrest, newspapers made a point of saying that Kemmler took the situation "cooly" and expressed neither defiance nor remorse. He was similarly inscrutable during the trial, spending most of the time gazing at the floor, his elbows on his knees, snapping his thumb as if shooting marbles. Often his head sank so low that it rested on the table in front of him. The *Express* believed that Kemmler's calm indicated not murderous coldness but "mild-eyed imbecility."[3]

If Kemmler's sorry appearance inspired sympathy, the testimony quickly dispelled it. The prosecution took only a day to present its case. Kemmler's motive for murder was jealousy. He believed that Ziegler had become "too familiar" with one of his employees, a Spaniard named John "Yellow" DeBella. Mary Reid and Asa King described the gruesome crime scene and testified that Kemmler had admitted his guilt. A doctor from Fitch Hospital detailed Tillie Ziegler's wounds and handed the jury a paper box containing seventeen skull fragments removed from her head before her death. Dr. Roswell Park, later a noted cancer researcher, had consulted on Ziegler's case. When a defense attorney bizarrely asked whether Tillie Ziegler was "good-looking," Park grimly replied, "Not as I saw her."[4]

Kemmler's attorney did not bother to dispute much of the evidence against him. "We will not ask you to acquit this man," the lawyer told the jury, "for I believe it would be monstrous to turn loose upon society a man with such propensities as his." He argued, though, that Kemmler was a hopeless drunk and therefore incapable of premeditation. If this was so, the defendant deserved life in prison rather than death. A friend testified that he had never seen Kemmler *not* drunk, and an employee reported that while peddling "it was their custom to take in with strict impartiality all saloons on the route." One of Kemmler's employees reported that he got drunk with Kemmler on the afternoon before the murder. "He started in on cider and wound up on whisky," the man said. "We had some eggs to sell, but we were all too drunk to sell eggs." Asked to speculate on Kemmler's condition the following day, the man reported that Kemmler usually was "shaky" when he woke up, and that the two of them "always went for a drink in the morning." The beer Kemmler bought at about half past eight after killing his wife was part of his morning routine.[5]

On May 9, the third day of the trial, the prosecution and the defense attorneys made their closing statements, and the judge sent the jury to its deliberations. At ten that evening, when the court reconvened, the members of the jury had not reached a verdict, so the judge sent them out for a full night of deliberations. The next morning the weary jury members filed back into the courtroom and announced that they remained deadlocked. Rumor had it that the panel was evenly divided between a verdict of first- and second-degree murder. The judge told the jury that to be responsible for his crime a man need not be intelligent; he simply must understand the nature of his act. If a guilty verdict required the defendant to be intelligent, the judge impatiently explained, then "half the human family would be exempt from the consequences of crime." The judge's instructions had the effect he so obviously desired. Within two hours, the jury found Kemmler guilty of murder in the first degree.[6]

Several witnesses testified that immediately after the crime, Kemmler said, "I have done it, and I am ready to take the rope." But New York State, of course, no longer hanged criminals. The court ordered that Kemmler be executed with electricity during the week of June 24, 1889. The prisoner appeared sanguine in the face of death. He told his keeper that he "dreaded the long drop and thought he'd rather be shocked to death."[7]

KEMMLER WAS TRANSFERRED to the state prison in Auburn to await his death. In early June, however, newspapers reported that he would appeal his sentence on the grounds that the new execution method was unconstitutional. His case was to be handled by a new attorney, W. Bourke Cockran, one of the most respected and feared litigators in New York. Born in Ireland and educated in France, Cockran had been practicing law in New York since the late 1870s. His oratorical skills brought him to the attention of New York's Tammany Hall Democrats, who secured him a seat in the U.S. House of Representatives in 1887. Cockran resigned after just one term to return to his lucrative private practice.

Cockran was far too expensive for William Kemmler, and many

Bourke Cockran

newspapers charged that George Westinghouse was paying his fees. It was a logical assumption, because overturning the new law would foil Harold Brown's plan to use Westinghouse generators for executions. Cockran had never worked for Westinghouse before, but it was known that Westinghouse's chief attorney, Paul Cravath, often hired Cockran to handle litigation for him.[8]

Cockran denied any connection. "I wish you would say that I am not retained by the Westinghouse Electric Company on behalf of Kemmler," he told a reporter. "I believe the law to be unconstitutional and inhuman, and deem it due to the honor of the State and for the welfare of justice that it should be tested. No electrical company has retained me, and I am doing this without hope of financial remuneration." Few believed the statement. The *New York Times* bluntly stated that Cockran was motivated not by "a desire to save Kemmler" but by "the objection of the Westinghouse Company to having its alternating current employed for the purpose" of execution.[9]

WESTINGHOUSE HAD ACCUSED Edison of promoting electrical execution to damage the market for alternating current; after Kemmler's trial, Westinghouse found himself accused of attempting to defeat the execution law in order to defend his system. It was clear to all observers that competition between the two firms had grown ferocious since the battle of the electric currents had started a year before.

Westinghouse was not timid about the methods he used to ensure success, and, even by the freewheeling standards of the time, he earned a reputation as an unscrupulous businessman. "We do not like the method of doing business of the Westinghouse Co.," a small-town electrical entrepreneur reported, and many agreed with him. In 1888 the Thomson-Houston company decided to begin producing street-railway equipment. To do so it needed to revise its corporate charter, which required a special act of the Connecticut legislature. Westinghouse interests, fearful of a new competitor in the railway business, lobbied strongly against the bill. Edward Johnson of Edison Electric considered the Westinghouse effort grossly unjust and helped Thomson-Houston win the charter revision. "The methods of the Westinghouse people are, as we know, of the most unfair and undignified character," wrote Elihu Thomson, the cofounder of Thomson-Houston.[10]

A year later Thomson was outraged to learn that George Westinghouse had been awarded a broad patent on an alternating-current meter that he had not even invented. Whereas Thomson's own application for a meter patent—filed much earlier than Westinghouse's—gathered dust in the government office for ten months, Westinghouse's was issued only two months after filing. As Thomson saw it, "the Patent Office could only have allowed this patent through corruption or bribery of some kind." According to a banker who knew all the major players in the industry, Westinghouse "irritates his rivals beyond endurance." Charles Coffin, the president of Thomson-Houston, complained of Westinghouse's "attitude of bitter and hostile competition."[11]

Thomas Edison had a particular reason to resent Westinghouse, who made a habit of appropriating Edison's inventions for his own use. Edison told a reporter, "It . . . always made me hopping mad to think of the pirates in the electric business, not merely stealing the radical inventions which made the lamp possible, but taking advantage gratis of the long line of thousands of experiments which I had made night and day for a couple of years."[12]

Edison Electric was embroiled in two crucial suits with Westinghouse. In one, it defended its basic patent for the incandescent lamp Edison invented in 1879. Although the suit was originally filed against the U.S. Electric Lighting Company in 1885, by the time it came to a hearing in 1889, that company had been purchased by Westinghouse. In the other suit, Edison Electric was the defendant. The Consolidated Electric Light Company had sued Edison Electric, claiming that Edison's lamps violated a patent, owned by Consolidated Electric, on incandescent lamp filaments made from paper. In 1888 George Westinghouse gained control of Consolidated, adopted the suit as his own, and pressed it with vigor. Of the hundreds of lawsuits filed in the first decades of the electrical industry, only these two carried real significance. In both cases, Edison defended his most prized invention—the lightbulb—against infringement by George Westinghouse.[13]

Edison took the matter personally. Early in 1889, as the patent battle intensified, a mutual friend of Edison and Westinghouse tried to broker a peace, urging Edison to visit Westinghouse in Pittsburgh. Edison would have nothing of it, explaining that Westinghouse's "methods of doing business lately are such that the man has gone crazy over sudden accession of wealth or something unknown to me and is flying a kite that will land him in the mud sooner or later."[14]

At the time Westinghouse's kite was flying high, with his business growing far more quickly than Edison's. Some Edison Electric officials believed that they were losing ground to Westinghouse because their company was poorly organized. At the start of 1889 the Edison electrical business was still fragmented into several distinct firms, including Edison Electric Light Company (the patent-holding company) and the manufacturing enterprises (the Tube Company, Lamp Company, Machine Works, and Bergmann & Company). In April these firms were merged to form a new company called Edison General Electric, with Henry Villard as president and Samuel Insull—who earlier had served as Edison's secretary and as manager of the Machine Works—as vice president. (Edward Johnson resigned the presidency of Edison Electric and became the president of the electric railway division of the new company.) Edison General, it was thought, would operate more efficiently and command enough capital to push rapid expansion of the Edison system. A few dissenters within the company suspected that the consolidation failed to correct the company's basic problem: the lack of an alternating system to compete with Westinghouse's.[15]

JUST AFTER the Buffalo jury convicted William Kemmler in May 1889, Harold Brown provided the first public description of the execution apparatus to the *New York Star*: "There will be a strong oaken chair, of the reclining make, in which the condemned will sit, and electrodes for the head and feet. The former of these electrodes consists of a metal cap, with an inner plate covered with a sponge that has been saturated with salt water, which is to be fastened on the condemned man's head by means of stout straps held by another strap around the body under the armpits. The other electrode is simply a pair of electrical shoes tightly laced on the convict's feet." A few weeks later illustrations of the chair appeared in the *New York Daily Graphic* and *Frank Leslie's Illustrated Newspaper*.[16]

Brown apparently believed that revealing these details would help his cause, but they had the opposite effect. The *New York Evening Post* explained that it originally had backed the law in the belief that "the victim would die with the touch of a wire or a knob." As the *Post* saw it, the myth of the simple and tidy execution was exploded by these new descriptions of the apparatus, in which the prisoner would "be seated in a formidable-looking chair, have his feet encased in shoes which contain damp sponges, have another sponge placed on his head, and his head clamped down with metallic bands." The first execution "will doubtless be very interesting to the scientists," the *Post* claimed, but its humanity was in doubt.[17]

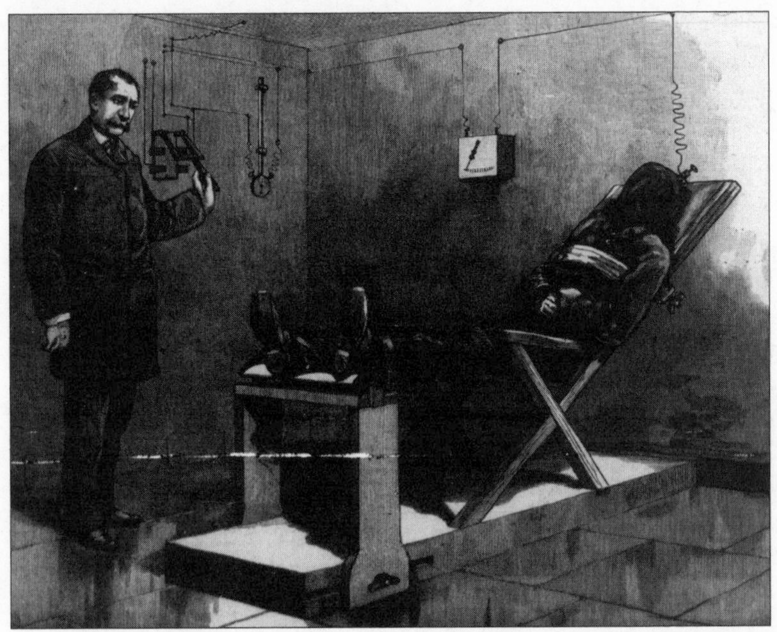

Harold Brown released this illustration of his proposed electric chair in May 1889, not long after winning the state contract to supply execution apparatus.

Electricians and doctors—those who knew most about the balkiness of machines and the resilience of the human body—had first expressed reservations about the law the previous summer, and their doubts had only grown. "The question is really and solely one of aesthetics," the *Medical Record* wrote. Although electricity might kill quickly, it was not "humane, or civilized, or even scientific, to strap a condemned prisoner to a chair and throw him into a convulsion, or possibly burn the top of his head off." Although the popular press at first had praised the new law as a great leap forward for civilization, many editorial writers got cold feet as the date of the first execution approached. The *Buffalo Express*, *Philadelphia Bulletin*, *St. Paul Pioneer-Press*, *Albany Argus*, *New York World*, and *New York Herald* formed a chorus of opposition to the new law. A *World* story noted Harold Brown's links to the Edison laboratory under the headline "Thomas A. Edison Said to Be Behind the Plan to Use the Alternating Current in the Death Penalty in Order to Bring it into Disrepute." The *Tribune*, which originally had supported the law, now predicted that the first execution would be "distressingly bungled." The *Sun* went a step farther: "The only thing to do is to repeal the law." Because many suspected that the law would be declared unconstitutional, murder trials across the state were put on hold pending a judicial decision.[18]

The growing opposition to electrical execution pleased Bourke Cockran. In June he argued Kemmler's case before S. Edwin Day, the Cayuga County judge whose courtroom was just a few blocks away from the cell where Kemmler awaited his death. Far from being a humanitarian reform, Cockran said, electrical execution constituted cruel and unusual punishment and therefore violated both the state and federal constitutions. No one had ever killed with electricity deliberately, Cockran explained, and so no one really knew what would happen to the first victim: "We hold that the state cannot experiment upon Kemmler."[19]

By the summer of 1889 there were so many charges of backhanded dealings and shoddy science that most people did not know what to think of electrical execution. The courts charted a deliberate course. In response to Kemmler's appeal, Judge Day ordered hearings to gather evidence on the question of whether electrical death violated the constitutional prohibition of cruel and unusual punishments. Supporters and opponents of the law, who had sniped at each other in the press for over a year, now would meet in open debate, to be examined and cross-examined by attorneys for the state and for William Kemmler. From the crucible of the adversarial legal system, it was hoped, the pure light of truth would emerge.

CHAPTER 14

Showdown

THE HEARINGS ON the constitutionality of electrocution opened on Tuesday, July 9, 1889, in Bourke Cockran's legal offices in lower Manhattan. Presiding was Tracy C. Becker, a Buffalo attorney whom Judge Day had appointed as the referee. The proceedings stretched on for more than three weeks, and among the four dozen witnesses were Thomas Edison, Elbridge Gerry, two future presidents of the American Neurological Association, and three future presidents of the American Institute of Electrical Engineers.

New York's daily papers gave the hearings a great deal of space in their pages, with special emphasis on the performance of Cockran. Just under six feet tall, the lawyer had a wide chest, bushy mustache, massive round head, and drooping jowls. He spoke with an Irish accent and was famous for his sharp wit and withering cross examinations. During the hearings Cockran pitched his message not only to the appellate judges but also—through the medium of the press—to the public and the state lawmakers. Although Kemmler's appeal might fail in the courts, the legislature could repeal the law if public opposition to it continued to grow. As a veteran Tammany politician, Cockran knew how to appeal to the masses, and a proper show involved something more than a parade of experts.[1]

Cockran called to the stand a portly man named Alfred West, who offered a tale of his own encounter with electricity. While strolling one recent summer day along the Palisades, high above the Hudson River in New Jersey, West sought shelter from a thunderstorm under a large tree. "I heard a shock which seemed to be an explosion," West testified, "and that ended my experience." The lightning bolt threw him twenty feet and blew off his pants, underclothes, socks, and shoes. When he woke up, his friends gave him some brandy and sent for medical help. The physician who came to the scene subscribed to the popular theory that a dangerous amount of electrical fluid lingered in the bodies of shock victims, rather like an overdose of a drug. He prescribed a folk remedy to "draw off" the electricity: "Put his feet in warm water, and just pull on his toes the same as you milk a cow," the doctor said.

"They done so," West explained to Cockran, "and I could feel the electricity go out of me."

"How did you feel it go out?" asked Cockran.

"The same as though you would strike your elbow and you could feel it down the fingers."

West had spent three weeks in bed after the accident, and he still bore its scars. The bolt scorched the hair off his chest and groin, and a four-inch-wide swath of burned skin ran from his chest down his right leg, then curled up around his left leg like a ribbon. His chest bore a delicate tracery of pale lines, like the veins on a leaf.[2]

West was important to Cockran because he was still alive. When Cockran examined electrical experts, he asked them whether a dynamo could produce enough current to throw a man twenty feet, or strip his clothes off, or shatter even a broomstick, much less a tree. The answer was always no. As Cockran saw it, if an electric charge could be powerful enough to toss a man like a doll and strip him naked—and still not kill him—then the apparently weaker electric force of an execution dynamo could not be counted on to kill.[3]

Cockran delivered the same message by trotting out men who allegedly survived severe electric shocks. "I have had five or six, where I know it was 1,000 volts, right from the machine," said Carpenter Smith, a Westinghouse manager. On various occasions he had accidentally grasped the bars of a live switchboard, touched the poles of a dynamo, and stepped on uninsulated wires. According to Smith, taking the shocks felt like being "kicked by a mule" or "struck with a bundle of loose rods."

Despite the pain, Smith said, such shocks were not a major concern. He told of the time when one of his employees took a big shock that burned his left palm to a crisp black. "I gave him fifty cents," Smith testified, "and told him to go and get a big drink of whisky and call at the doctor's on the way home."

"He took the first part of your advice, I suppose?" Cockran asked dryly.

"He did," Smith replied. "Then he came back after getting the drink of whisky. He wanted to come back and go to work.

"We don't count a 1,000 volt shock anything in the business," Smith told Cockran. "I don't mean by that a man would go and take hold of it, but I don't know a lineman in my employ who hasn't had it time and time again." His fellows felt the same way. "If a man gets a shock like that, he doesn't lay up. He shakes his fingers and thinks he was foolish to get in the road and goes on with his work."[4]

ALSO TAKING THE STAND was Charles Tupper, who ran a restaurant on Eighth Avenue in Manhattan and lived above it with his wife and their dog, Dash, a St. Bernard–collie mix. One day early in July he had been in front of the restaurant watching two Western Union men cut dead telegraph wires down. Mrs. Tupper, who was upstairs, sent Dash down with a newspaper to deliver to his master. His task completed, Dash ran along the sidewalk and stepped on a wire that was not dead. Stripped of insulation, it had draped across an electric light wire, with its ends dangling to the street. "When he put his paw on the wire he jumped about two or three feet in the air," Tupper said. Dash gave a piercing yelp and fell onto the wire. His owner started forward to pull him away, but the lineman held him back, explaining that if Tupper touched the dog, he would be shocked himself. In any case, the lineman said, Dash was already dead.

Tupper called to a nearby police officer: "Arrest this man," he said. "I will hold somebody responsible for the dog if he is dead." The officer apprehended the lineman and hauled him off to the station house. Tupper then went upstairs to tell his wife, who fell on the bed and wept. "She became prostrated on account of thinking so much of the dog," Tupper said softly.

Dash lay on the wire for fifteen minutes before someone lassoed him with a rope and dragged him off. Tupper put Dash on the floor of the restaurant, not ready to give him up for dead. A man from a lighting company told Tupper to scrape a hole in the earth and put Dash in it, because the earth "would draw the electricity from him." He did so. That was at about four in the afternoon. Six hours later, Dash began twitching his hindquarters and moving his head and eyes. Tupper and his wife offered him milk and brandy, which he refused, so they rubbed his body with alcohol and carried him to their bedroom, where he stayed for five days, gathering strength.

After Tupper related this tale, Dash himself made an appearance in the hearing room. The lawyers and witnesses examined the deep burns on his left foreleg and right shoulder. "He appeared to be well, but not frisky," the *World* observed.[5]

"This dog may do more to save Kemmler's life than all his lawyers," Cockran said.[6]

DASH WAS the sentimental favorite of the Kemmler hearings. Amid all the testimony regarding dog-killing experiments, the spectators were relieved to hear the story of a dog whose owners loved him and cared whether he lived or died. The dog also raised a scientific question. Kemmler's lawyers suspected that many of the dogs experimented upon at the Edison laboratory had merely been stunned, like Dash, and that the true agent of death had been the dissecting knife.

These matters had a peculiar resonance in the summer of 1889. A few months earlier a famous mind reader named Washington Irving Bishop had fallen unconscious during a Manhattan performance. Later that night doctors pronounced him dead and conducted an autopsy. They overlooked the note that Bishop always carried explaining that he was prone to deathlike, cataleptic trances and warning doctors to be cautious in pronouncing him dead. Bishop's wife charged that her husband had been murdered by dissection, and a prolonged scandal ensued.[7]

The Bishop episode revealed lingering uncertainties in the definition of death. At the end of the nineteenth century the fear of premature burial was still common, and doctors constantly created new ways to check for signs of life. Pulse and breath were still thought to be the most unequivocal evidence, but determining whether these functions had ceased proved difficult. Suspecting that breath might not be detected by the old method of holding a mirror to the nose, physicians submerged corpses in water to check for bubbles. Notwithstanding the availability of the stethoscope, some thought opening an artery was a better way to check for motion of the blood. Even in the 1880s some physicians checked for life by burning the skin, jabbing a needle into the heart, or applying a "nipple pincher." A few still insisted that the only sure sign of death was putrefaction.[8]

If death eluded definition even in normal circumstances, the added element of electricity—the force long linked to the spark of life—confused matters further. Victims of accidental shocks, like those of drowning or strangulation, had been known to revive. Many feared that electrical execution would simply stun its victims and fail to kill.[9]

Dr. Frederick Peterson, the eminent neurologist who had assisted at the Edison laboratory experiments and who had done more research on electrical killing than any other physician, was called to testify. He explained that after he cut open one of the dogs killed by electricity, its heart continued to beat for thirty-six minutes.

"You considered this dog dead?" one of Kemmler's lawyers asked with surprise.

"Certainly," Dr. Peterson replied. "That is not an uncommon thing."

"So, if you cut a man, or human being, open, and found his heart beating, you would continue the examination, regardless of the fact that his heart was beating?"

"Yes, if I thought he was dead."

"And his heart could be beating, and he be dead?"

"Yes, I think so."[10]

The referee pursued the line of questioning.

"How does death result from electricity?" he asked Dr. Peterson.

"That is not known."

"How does it result from hanging?"

"By cutting—either breaking the spinal cord or cutting the vertebrae, or by injuring the spinal cord, or by suffocation."

"And in the application of electricity you do not know how it is caused?"

"We know that breathing ceases, and that the heart will stop beating also," Dr. Peterson explained. "We say death results from that—but why those things stop we do not know."[11]

Witnesses expressed many opinions as to how electricity killed. Brown explained that men receiving shocks died either from "blows delivered to the nerves," "paralysis of the heart," or decomposition of the blood through the process of electrolysis.[12]

Elbridge Gerry said electricity "would cause death by an instantaneous paralysis of the whole system."

Cockran, incredulous, repeated the phrase: "The instantaneous paralysis of the whole system?"

"A general paralysis of the entire nervous system," Gerry said.

"Don't you think that is a good deal of gibberish?" Cockran asked.

"No," Gerry replied grimly.[13]

The explanations of the physicians offered much of the same. Dr. George Fell explained that electrical death was "caused by the coagulation of the protoplasm of the body through the force of the electrical current." Dr. A. D. Rockwell, the expert in electrical medicine who had assisted the state in its killing tests, said, "Well, such a current would, I suppose, by its mechanical effects, produce a rupture of tissues in the interior of the body, the tender tissues; and it would immediately stop the action of the heart through the interior—produce paralysis of the nerve centers." He paused, then admitted, "No ones knows exactly the details of the effects, but that is about it."[14]

The testimony exposed the shaky scientific foundations of electrocution. As many observers recognized, in the absence of a precise definition of death or a clear understanding of how electricity killed, the claim that electricity caused death "instantly" and "painlessly" was entirely vacuous.

WILLIAM KEMMLER'S legal team had a two-point strategy: to prove that electricity could not be counted on to kill, and to hint that those who supported the law had unsavory motivations. Dash the dog and other surviving victims of accidental shocks had supported the first point. Subsequent witnesses gave Cockran an opportunity to work on the second.

Harold Brown sat at the elbow of Deputy Attorney General William Poste throughout the hearings, making notes and whispering advice regarding technical points. Cockran had the first opportunity to examine Brown on the witness stand, and he asked him about how he became involved in the animal-killing experiments.

Brown replied, "In June and July of last year I began a series of experiments, not having any reference to electrical execution, but to determine as nearly as possible the comparative danger between the two different classes of current."[15]

Referee Becker interrupted with a few questions of his own for

Brown: "Are you connected in any way with any of the electric lighting companies?"

"No, sir."

"Or have you any connections with the Edison company?"

"No, sir."

"Or Mr. Edison?"

"No, sir, except a personal acquaintance. I have received a great many favors at his hands; it is entirely a personal friendship."

Cockran resumed his questioning: "Your motive was a purely philanthropic one?"

"And to defend my own reputation."[16]

Cockran then asked about the bitter patent battles being fought in the courts: "There is a contest between the Westinghouse Electric Light Company and the Edison Electric Light Company as to the use of these incandescent burners?"

"I understand so," Brown said.

"And there is considerable feeling between the two corporations?"

"Of that I cannot say."

"Don't you know anything about it at all?"

"Not from actual knowledge."[17]

At Cockran's request, Brown submitted to the court an illustration of the execution apparatus he designed, which showed a reclining chair with footrest and straps, a generator, a switchboard, and other electrical apparatus. Even here Brown had not been able to resist the opportunity for propaganda. On the generator, in the tiniest script possible, he had added a label: "Westinghouse Electric, Pittsburgh."[18]

Brown's counterpart on the opposing side was Franklin Pope, a Westinghouse employee. Pope used the pages of the *Electrical Engineer*, which he coedited, to attack both Edison and the electrical execution law. Cockran's positions on technical matters—even the particular examples he used—so closely matched Pope's that it is obvious the two developed their case together.

When Cockran asked him whether there was a difference in safety between the two types of current, Pope did admit that a "continuous

Harold Brown submitted this plan for electrical execution as evidence during the hearings phase of Kemmler's appeal. On the generator he added a label reading "Westinghouse Electric, Pittsburgh."

current is probably less dangerous than an intermittent or alternating current." Although alternating current could kill a man, Pope said, it would be "extremely painful."*

When attorneys for New York State had a chance to cross-examine Pope, they attacked his credibility. One attorney asked about Pope's employer, Westinghouse Electric: "They object to the use of their machines in the contemplated execution of criminals?"

"I believe they do."

"For what reason, do you know?"

"I suppose for the reason the public would naturally suppose that a machine that was used for the express purpose of killing or ending human life would be an unsafe one to put into commercial use."[19]

"Has the company engaged counsel to urge this objection?" (At this question, the *Times* reported, "Mr. Cockran suddenly became intensely interested in the architecture of the ceiling.")[20]

*Pope himself died from an alternating-current accident in 1895.

"I don't know of anything of the kind," the witness replied.

The state's lawyer, however, would not let the matter rest. He said that Pope's testimony could not be trusted: "He is a representative of the Westinghouse people."

"That does not disqualify him," Cockran shouted. "He is not a charlatan." Cockran glared at Harold Brown.[21]

MUCH OF THE TESTIMONY of both Brown and Pope focused on one technical matter: the electrical resistance of human beings. In order to kill, the current had to be strong enough to overcome the resistance of the victim's body. There were ways to lower a person's resistance—such as clipping the victim's hair and covering the electrodes with wet sponges—but the executioners had to have a rough idea of the victim's resistance in order to be sure that the current used would be strong enough to overcome it. Brown claimed that 1,500 volts would overcome the resistance of any man. But according to Cockran and Pope, the resistance of living creatures varied enormously, and there was no good way to measure it. As a result, they claimed, the state could never know how high a voltage would be needed to kill.

The testimony on resistance became so technical and so confused that Referee Becker asked that more experiments be made. As with earlier questions regarding electrical execution, it seemed only Thomas Edison's laboratory could provide the answers. On Friday morning, July 12, Becker, the lawyers, and a knot of reporters journeyed to New Jersey. Edison himself, along with Arthur Kennelly, was off in Pennsylvania experimenting with his latest project, a magnetic iron ore separator, so the experiments were conducted by Brown and another Edison employee. Each test subject placed his hands in two jars containing a zinc sulfate solution, and a low-voltage current was passed through his body between the electrodes. A device called a Wheatstone bridge—named for Charles Wheatstone, one of the inventors of the electric telegraph—was used to measure his resistance. The measurements of the men ranged from 1,000 to 10,000 ohms. According to Brown, this was within the expected range and proved that resistance would prove no roadblock to efficient killing. Cockran remained unconvinced.[22]

REFEREE BECKER encountered trouble securing the testimony of the death penalty commissioners. Matthew Hale telegraphed from Albany to say that he "could throw no light on the subject" of why the commission recommended electricity, a surprising admission that indicated he had held little power in the final decision. Alfred Southwick was vacationing in Europe. That left Elbridge Gerry, who at first declined to testify on the grounds that he was in Newport preparing for a cruise with the New York Yacht Club. Becker telegraphed to Newport to insist that Gerry postpone his cruise and return to New York. Gerry appeared at the hearings fresh from his yacht, his commodore's cap in his hand.[23]

Under questioning by Cockran, Gerry testified, "My individual idea is that the dose of morphia would be most efficacious."

"But you surrendered your opinion on account of your associates?" Cockran asked.

"Yes." Alfred Southwick pushed electricity from the start, Gerry explained, but he had resisted that method until the committee received advice from a man that he considered the "greatest electrician of modern times."

"That was Mr. Edison?" Cockran asked.

"Thomas A. Edison."

"Did you consider he was the greatest medical man?"

"No, the greatest electrician."

"Even then did you consider he was a good authority to tell you what would destroy human life?"

"Yes."

Cockran continued: "You have a very great confidence in electricity?"

"I have more confidence in the ability of certain electricians–Thomas A. Edison."

"And you think Edison somewhat of an oracle, do you not?" Cockran said.

"Yes."

"You think he knows more about it than anyone in the United States?"

"Yes, and I haven't seen any foreigner that knows as much about the subject."

"And it was in that way that you came to report that bill?"

"Yes, sir; that was one of the main reasons."

"You finally decided that where Edison spoke there was no room for doubt, and you recommended the bill?"

"I certainly had no doubt after hearing his statement."[24]

A FEW DAYS AFTER the hearings began, Harold Brown wrote to Samuel Insull, the vice president of Edison General Electric, urging him to persuade Edison to testify. "There has been so much absurdity in the testimony of Mr. Westinghouse's witnesses, that Mr. Edison could dispose of by a word," Brown wrote. "Mr. Johnson [Edison General's president] thinks this important." Referee Becker was so anxious to hear Edison's views that he offered to take the hearings to Edison if Edison could not come to the hearings. The inventor made things easier by appearing in the hearing room on the morning of July 23. Despite the crowded hearing room and the sultry weather, the inventor appeared cool in his black suit and wore an easy smile, with an unlighted cigar between his teeth. He offered $100 to Carpenter Smith, the Westinghouse man who claimed to have survived several 1,000-volt shocks, on the condition that he take an alternating-current shock of just 100 volts, with the money "paid to his heirs or assigns if necessary." Smith did not respond to the challenge.[25]

Edison was introduced to Cockran and shook hands with him cordially, then took the witness chair. Attorney General Poste, compensating for Edison's deafness, shouted his questions so loudly they could be heard out on Broadway.[26]

"Will you explain generally the difference between a continuous and alternating current of electricity?" the attorney general asked.

"A continuous current is one that flows like water through a pipe," Edison said. "The intermittent [alternating] current is the same as if the same body of water was allowed to flow for a given time through the pipe, and then the direction was reversed and it flowed in the opposite direction for a given time."

"Now, what kind of current is produced by the Westinghouse dynamo?"

"The reverse current—the alternating."[27]

Edison explained that although his company owned the U.S. patent rights for an alternating system used widely in Europe, it had decided not to sell it in the United States.

"Why don't you use it in this country, Mr. Edison?"

"I don't like it."

"Do you think it is dangerous?"

"Yes."

Edison told Poste that alternating current of 1,000 volts would produce rapid death without suffering. That opinion, he said, was based on experiments at his laboratory.

"The experiments made by Mr. Brown and Mr. Kennelly?"

"Some of them," Edison said. "I only saw one or two."

"Has Harold P. Brown any business connection with yourself or the Edison company?"

"Not that I know of," Edison replied.[28]

Poste concluded his examination and turned the witness over to Cockran. Edison was sitting in a corner of the room, and Cockran paced the room's center, speaking rapidly. The inventor had trouble hearing, so he dragged his chair into the middle of the room, and the lawyer poured out his questions into the inventor's good ear.[29]

"Are you a pathologist, Mr. Edison?"

"No, sir."

"Do you understand anything about anatomy?"

"No, sir."

"You do not claim to understand anything about the structure of the human body?"

"No, sir, only generally."

"That is," Cockran said, "you know we all have got arms and legs–you mean you have got an idea?"

"Got a good idea of what is inside of us."

"Now, you do not even know what the component parts of the human blood are, do you?"

"Pretty near."

"What are they?"

"Carbon, hydrogen, oxygen, and nitrogen."

"Now you have mentioned pretty much every gas, haven't you?"

"I cannot help it; that answers the question."

Cockran turned to the question of whether electrocution would mutilate its victims, and Edison readily admitted that applying a really powerful current might cause burning.

"You would burn him up quick?"

"Carbonize him," Edison said.

"Right away?"

"Instantly."

"Suppose you took this *wicked Westinghouse current*"–Cockran's voice was heavy with sarcasm–"the one that is going to be used on Kemmler . . . suppose you kept it up for five or six minutes with 1,500 volts–with the extra exciting dynamo working, and all other appliances that would get it in its most wicked and aggravating form–how long would it take before you would get him to feel the heat?"[30]

"I think his temperature might rise up about four or five degrees above normal, and continue there until he was mummified, until the water had all been evaporated out of him."

"That is a new term," Cockran observed. "You think Kemmler would not be carbonized, but mummified?"

"Mummified," Edison said.[31]

. . .

HAVING FAILED to ruffle him on technical questions, Cockran pressed Edison on the rumors of conspiracy.

"Now, Mr. Edison, do you know Mr. Brown pretty well?"

"Fairly well. I have seen him about a dozen times."

"When did you first become aware of his existence?"

"I think the first time I saw him, he came out there about this very business, to Orange."

"And he was a stranger to you then?"

"I think so. I do not think I had ever seen him before."

"Did he come up there and ask you to let him have your laboratory for the purpose of killing dogs?"

"He wanted to try some experiments."

"Are you in the habit of giving your laboratory to everybody that asks you?"

"Yes, sometimes I let them experiment there."

"Might I entertain the hope that I might be allowed myself to go there?"

"Yes, sir," Edison replied with a smile. "You can come at any time. I will be glad to see you."

"Mr. Brown evidently commended himself to your approval during these experiments."

"He seemed to be a pretty nice kind of fellow, and it was no trouble to me, and I let him do it."

Cockran decided to try a more direct approach: "Now, Mr. Edison, there is a great degree of feeling between you and Mr. Westinghouse?"

Edison paused for several beats. "I do not dislike Mr. Westinghouse," he said finally.[32]

"There is a contest between you in the courts, isn't there?"

"Yes, sir."

"And in relation to some of these electric-light inventions?"

"Yes, sir."

Surprisingly, Cockran let the matter drop there. As the lawyer paused, Referee Becker asked Edison whether an electrician could operate the execution machinery without hurting himself. "Yes, sir," Edison said.

Cockran jumped in: "That is, you believe he can. You do not know anything about it."

"Of course I have got to testify to my belief," Edison retorted. "I have not killed anybody yet."[33]

Cockran had had enough. He lit Edison's cigar and dismissed him from the stand.[34]

THE KEMMLER HEARINGS exposed profound disagreement among electricians concerning whether electricity could be counted on to kill, and physicians betrayed their ignorance about the effects of electricity on the body. Elbridge Gerry, the chairman of the death penalty commission, admitted that it was Edison who had persuaded him to back electricity. Yet Edison admitted that he knew almost nothing about human physiology or the details of electrical killing. Despite all of this the press, much like Gerry, remained in thrall to Edison's views. The *World* contrasted "the confused, heterogeneous and imbecile testimony" of the law's opponents with "the curt and unequivocal testimony of Thos. A Edison." "The Kemmler case at last has an expert that knows something concerning electricity," the *Albany Journal* wrote. In that newspaper's view, Edison was "a man who had nothing to sell and no purpose to buy."[35]

CHAPTER 15

The Unmasking of Harold Brown

On August 25, just a few weeks after the hearings ended, the *New York Sun* published stunning revelations regarding Edison's interest in the electrocution law and his relationship with Harold Brown. "FOR SHAME, BROWN!" the newspaper's headline read. "Disgraceful Facts About the Electric Killing Scheme; Queer Work for a State's Expert; Paid by One Electric Company to Injure Another."

The *Sun* had gained possession of nearly four dozen letters—stolen from a locked desk in Brown's office, it was later revealed—between Brown and the officers of two electrical manufacturers, Edison Electric and Thomson-Houston. The newspaper first ran a brief excerpt from the Kemmler hearings in which Referee Becker had asked Brown if he was "connected in any way with any of the electric lighting companies," and in particular if he had "any connection with the Edison Company . . . or Thomas Edison." Brown's reply had been "No, sir." The *Sun* then reproduced a letter Brown subsequently received from his father, who wrote, "I was a little surprised at your statement that you were not connected with any electric company. I thought you were." The rest of the letters—reprinted in full in the *Sun*—proved that Brown's father was right, although the arrangements were more complicated than the critics of Brown had imagined.[1]

The *Sun* printed a letter Brown received in February 1889 from Frank Hastings, the secretary-treasurer of the Edison Electric Light Company, asking Brown to send a copy of his pamphlet *The Comparative Danger of the Alternating and Direct Currents* to all of the "legislators and officers of the State of Missouri," who were considering a bill that would limit the voltage of alternating current. The cost of printing and mailing the pamphlet to every legislator in Missouri must have been enormous, and Brown was not a wealthy man. The tone of Hastings's letter made it clear that he was giving Brown instructions in the matter; Edison Electric probably paid for the printing and mailing costs as well.[2]

A month later, in March 1889, Brown sent a letter to Thomas Edison, explaining that he was trying to persuade the city of Scranton, Pennsylvania, to place restrictions on alternating current. Brown said that if Edison would publicly endorse his efforts, he would be able "to add Scranton to the list of cities which have shut out the high-tension alternating current." Edison told Brown that he would be happy to oblige.

> My Dear Sir:
> I have your letter of 17th instant., and take much pleasure in enclosing herewith a testimonial signed by myself, which I hope will answer your purpose.
>
> Yours very truly,
> Thomas A. Edison.[3]*

*The testimonial itself did not appear in the *Sun*; in all likelihood it was not in Brown's desk because it had been sent on to the officials in Scranton.

A few days later, after winning the state contract to supply execution equipment, Brown wrote to Edison with another request. The generators would cost thousands of dollars, Brown explained, but the state would not repay him until after the first prisoner was electrocuted. He could undertake the plan only "if $5000 is made available" for him to use, and he needed Edison's help in acquiring that money.

> In view of the approaching consolidation, the people at 16 Broad street do not feel like undertaking the matter unless you approve of it. Do you not think that it is worth doing, as it will enable me, through the Board of Health, to shut off the overhead alternating current circuits in the State, and will, by showing the lack of efficiency of the Westinghouse apparatus, head off investors, and prick the bubble, thus helping all legitimate electrical enterprises? A word from you will carry it through, and without it the chance will be lost. Is it not worth while to say the word?

"16 Broad street" was the Manhattan address of the Edison Electric headquarters. The "approaching consolidation" referred to the creation of Edison General Electric through the merger of the Edison Electric Light Company with the separate Edison manufacturing enterprises, a deal that was finalized later that spring. Brown's letter indicated that Edison Electric's managers had already agreed to give him the money, but that they also wanted the approval of Edison, who controlled the manufacturing companies.[4]

Edison's reply never surfaced, but he must have agreed to give Brown the $5,000, because in May Brown wrote to Edison again: "Thanks to your note to Mr. Johnson [president of Edison Electric] I have been able to arrange the matter satisfactorily; have supplied the State with Westinghouse execution dynamos."[5]

Because George Westinghouse had refused to sell his generators for the purposes of execution, Brown secured them through a second partner, the Thomson-Houston Electric Company. In April and May Charles Coffin, the president of Thomson-Houston, located several utility companies that owned Westinghouse generators and made arrangements to buy those machines and replace them with Thomson-Houston alternating generators. Brown used the money provided by Edison Electric to buy the Westinghouse dynamos from Thomson-Houston. On May 23 Thomas Edison, whom Brown had kept apprised of these arrangements, made an enigmatic statement to a *Pittsburgh Post* reporter: "The Westinghouse people deny the report, I know, but their machines were certainly purchased, although they may not have been obtained direct from the company." Only with the publication of the *Sun* letters three months later did the public understand what he meant.[6]

Charles Coffin helped Brown because he hoped to hurt Westinghouse—not on the safety issue (as Edison did) but through a comparative test of dynamo efficiency. Westinghouse's advertisements claimed that his generators were 50 percent more efficient than those of other companies. In an open letter published in several newspapers, Brown challenged Westinghouse to send one of his alternating generators to Johns Hopkins University, where a professor would compare its efficiency with a competitor's machine. When Westinghouse refused to cooperate, Brown arranged to loan one of the Westinghouse execution dynamos to Johns Hopkins. The *Sun* letters showed that Brown was working as an agent for Thomson-Houston in arranging this test. Brown told Coffin that he planned to mail copies of the efficiency challenge all over North America in an effort to undermine Westinghouse, and Coffin agreed to pay Brown $1,000 for "expense attending the Baltimore test." Coffin was so pleased with Brown's work that he promised him $500 more for "future expert services." Coffin expected that the Johns Hopkins tests would prove the Westinghouse generators much less efficient than advertised, thereby boosting the reputation of Thomson-Houston's own machines.[7]

Coffin had another reason to help Brown: If he failed to supply Brown with Westinghouse generators, Coffin ran the risk that Thomson-Houston generators—which could be just as lethal as Westinghouse's—would be used for executions. Brown warned Coffin of this possibility in a May 13 letter: "If anything happens to that [Westinghouse] machine another make of dynamo will have to be used." Later that month, after some delays in delivery of the equipment, Brown told Coffin that the Westinghouse people were "bringing tremendous pressure to bear to prevent the test and to get other apparatus than theirs used." By securing the Westinghouse dynamos for Brown, Coffin protected the interests of his own firm.[8]

According to Brown, George Westinghouse was so opposed to the use of his generators for executions that he had sent agents to Auburn prison with orders to sabotage the equipment. The agents, Brown wrote to Coffin, "will cripple it if the liberal use of money will do it. I know of their offering money to some of the prison officials some time ago. . . . I can assure you that there is a desperate attempt being made to have the use of the Westinghouse dynamo for this purpose a failure."[9]

Most of the letters reproduced in the *Sun* related to Brown's negotiations with Thomson-Houston for acquisition of the dynamos. After Brown received the $5,000 he requested from Edison Electric, that company mostly disappeared from the correspondence. Thomas Edison and Arthur Kennelly reappeared in the *Sun* letters on the eve of the Kemmler hearings to give Brown unsolicited advice about his testimony. "At Mr. Edison's instance, I beg to bring the following consideration before your notice," Kennelly wrote to Brown. He said that the only possible argument against electrocution was that it "may burn the flesh of the criminal." Kennelly and Edison proposed that Brown experiment by sending high-voltage current through a dead body. If the test did not produce "any external injury," Kennelly wrote, it would "effectually silence the mutilation argument." (Apparently, this test was never conducted.)[10]

 BROWN REPORTED the theft of his letters to the police in early September, after they were printed in the *Sun*, and he offered a $1,000 reward for the return of the originals. The letters were stolen more than a month before they appeared in the *Sun*, explaining why the last letter in the batch was dated July 18. Franklin Pope, a Westinghouse employee and the editor of the *Electrical Engineer*, printed information derived from the letters long before they appeared in the *Sun*. The theft from Brown's desk was a clear case of industrial espionage, carried out by someone who almost certainly was working for George Westinghouse.[11]

Some of the *Sun* letters had been faked, Brown claimed, but the only letter he specifically labeled a forgery—the one from Kennelly—is indisputably authentic; the original copy of it can be found in the archives of the Edison laboratory. Original copies of at least two other letters that appeared in the *Sun* can be found in the same archives.[12]

The reality of the plot described in the *Sun* letters is also supported by an internal document from the Thomson-Houston Company. In May, as Thomson-Houston was locating the Westinghouse dynamos for Harold Brown, Elihu Thomson, the firm's cofounder and chief inventor, wrote a letter to Charles Coffin objecting to the alliance with Brown. "Whether the matter ever becomes publicly known or not, although I think it must, I dislike to see steps taken to confer a reputation on Westinghouse apparatus which our own similar apparatus must share," Thomson wrote. Coffin ignored the advice.[13]

THE *SUN* ATTACKED Brown for his "crookedness" and said he could not "be trusted as an expert for the State." According to the *Electrical World*, the letters proved Brown had "been 'on the make' throughout." The *Electrician* denounced the "conspiracy against the Westinghouse company" and predicted that the revelation of Brown's letters would provoke "public disgust leading to the repeal of the law."[14]

Although the press heaped scorn on Brown, it had little to say about the roles of the Edison and Thomson-Houston companies. Typical was the response of the *Electrical World*, which described Brown as "a man who, by playing off one company against another in various ways and by different subterfuges, has succeeded in making a neat little sum of money for himself." Brown was presented as the mastermind, the electric companies his dupes. This interpretation ignored the evidence of the letters themselves, which showed that both Edison and Thomson-Houston eagerly provided Brown with money, expert advice, references, and equipment to further his schemes.[15]

Following the publication of the letters, Brown betrayed little concern, explaining, "I am opposing the Westinghouse system as any right-spirited man would expose . . . the grocer who sells poison where he pretends to sell sugar." He even tried to use the letters as a defense against the charge that he was working for Edison. "My fight is against the Westinghouse company, but not in favor of the Edison Company. If anything, the letters published show leaning toward the Thomson-Houston system, a rival of the Edison system."[16]

Brown chose to ignore the letter in which he asked Thomas Edison for $5,000 to purchase the Westinghouse dynamos, and he must have felt fortunate that another batch of documents did not become public at the time. Those records, preserved in the archives of the Edison laboratory, show that Brown's participation in the plot against Westinghouse was even more extensive than the *Sun* letters indicated.

Whereas the first of the *Sun* letters was dated January 1889, Brown's involvement in the conspiracy dated back at least as far as the dog experiments of the previous summer. At the Kemmler hearings, Edison, Kennelly, and Brown claimed that the tests in July 1888 had been entirely Brown's idea, but Edison's laboratory records tell a different story. Both Brown and Kennelly testified that the first dog-killing experiments at the Edison lab took place on July 9, 1888, but neither mentioned that Brown was not present for this experiment. Kennelly also did not acknowledge that he conducted tests on a fox terrier on July 6, also in the absence of Harold Brown. It is not until July 12 that Kennelly's notebook first notes that the experiments took place "before Mr. H.P. Brown." Brown assisted again on July 14, but on the seventeenth—when two more dogs died—Kennelly wrote in his notebook, "Mr. Brown not present at this experiment." More tests were performed in August and September, when Brown was occupied elsewhere. Rather than instigating the tests, Brown simply assisted in a series of experiments by Kennelly and Edison that were in progress before he appeared at the laboratory and that continued in his absence.[17]

Private correspondence from the laboratory also shows that Brown's role in the tests was more minor—and Edison and Kennelly's much greater—than any of the men later admitted. In a letter written to the *Electrical Review* in September 1888, Kennelly explained that although Brown had "taken part in" the tests, "all the experiments made and published on the dogs except at Columbia College have been carried out by your humble servant." In another letter Kennelly made it clear that he performed those tests on orders from Edison, who supervised and directed the labors of all of his assistants: "We are continuing the experiments for Mr. Edison, and can kill a dog now more swiftly than a rifle bullet when desired," Kennelly wrote to Edison Electric official Frank Hastings. "I have lately been trying various experiments on dogs," Edison himself wrote in a letter dated July 13, 1888, acknowledging that the physiological tests were not Brown's but his own.[18]

The evidence in the Edison archives indicates that, starting in the summer of 1888, Brown worked closely with Edison Electric while publicly maintaining that his only interest was promoting public safety. Although there is no direct proof that he was on the company's payroll, it is likely that he was paid for his services. The campaign against Westinghouse was coordinated by two Edison Electric officials: President Edward Johnson and Secretary-Treasurer Frank Hastings. As Brown prepared for the public dog-killing demonstrations at Columbia College in late July 1888, Hastings arranged for him to borrow equipment from the Edison laboratory. "Mr. Johnson and I are both very much interested in these experiments and are very anxious that everything should be there in time," Hastings wrote to Charles Batchelor, Edison's lab superintendent. "It is of course understood that these materials are to be loaned to Mr. Brown in our interests and at as little expense to us as possible." Hastings acknowledged that Brown's work was promoting the "interests" of Edison Electric.[19]

Six weeks after the Columbia experiments, in September 1888, Edward Johnson tried to persuade Manhattan authorities to replace drowning with electricity as the method of exterminating stray dogs. "I don't want to *buy* a Westinghouse machine just now," Johnson wrote to Edison. "Have you an alternating dynamo that will give us 1000 volts which you can loan us for a week or so?" If the demonstration worked, Johnson told Edison, the mayor "says he will have an appropriation made to buy a Westinghouse plant for this purpose." Edison loaned him the dynamo.[20]

Johnson's pointed mention of Westinghouse's dynamo revealed the motive behind the request: If the city used an alternating generator to kill dogs, it would bolster Edison Electric's claim that alternating current was too dangerous for use in America's homes.

Using Westinghouse machines to kill humans would make the same point even more strongly. By the fall of 1888, the state had still not decided whether to use alternating or direct current for executions. Working with Harold Brown, Edison Electric arranged to kill two calves and a horse before Elbridge Gerry and the Medico-Legal Society in December 1888. According to a letter Brown wrote to Kennelly the day after the experiments, Edison had engaged in some discreet lobbying in favor of execution by alternating current: "The results [of the experiments] were very satisfactory, especially so since Mr. Edison's talk with Mr. Gerry and the members of the committee of course carried great weight." Within a week of the experiments, the medical society advocated alternating current for electrocutions.[21]

Edison Electric paid for those December experiments. The Edison archives contain an invoice, prepared by Kennelly and submitted to Edison Electric, of "expenses incurred by Mr. Edison in connection with the experiments of electricity on animals [for] the New York committee." The costs included:

purchase of horse	15.00
leading of horse from Newark	1.00
car expense of leader	.30
food for calves	.50
removal of carcass of horse to Newark	5.00

Edison Electric financed all of the physiological experiments. The Edison laboratory billbooks contain careful records of all expenses—both materials and labor—connected with the tests and submitted the bills to Edison Electric. Between August 1888 and April 1889, the laboratory submitted six bills totaling $200.29.[22]

In May 1889, two months after the last of those experiments, Harold Brown secured the state contract to supply execution dynamos, and he used his official position to buy Westinghouse machines. The charges that critics had been making since the first dog experiments in July 1888 turned out to be true: Edison Electric was orchestrating a plot to discredit Westinghouse. Although no single document spells out the plan in detail, the evidence detailed in the *Sun* letters and in the Edison archives—Hastings's close alliance with Brown, Edison and Kennelly's experiments in dog killing that did not involve Brown, Johnson's promotion of Westinghouse machines for use at the dog pound, and Edison Electric's sponsorship of the experiments—clearly indicate that such a plan was being carried out. Brown's pose as an independent, public-spirited researcher provided cover for his role as attack dog for the Edison interests.

CHAPTER 16

Pride and Reputation

WHEN THE Harold Brown scandal broke in New York in August 1889, Edison was an ocean away. Not long after he testified at the Kemmler hearings, he and his wife, Mina, boarded the steamship *La Bourgogne*, bound for the Universal Exposition in Paris.

Edison was grandly received at the world's fair. French newspapers hailed him as "His Majesty Edison" and "Edison the Great," and when he ventured onto the street, crowds gathered to cheer and stare. In their rooms at the Hôtel du Rhin, floral offerings to Mina crowded every piece of furniture, and assorted European royalty dropped in to record their voices on Edison's personal phonograph. Edison also found himself besieged by visitors of lower rank; many of them, he complained, were crackpot inventors who begged him to "give the last touches to some lunatical invention of theirs." Mina found all the attention wearying. "We never get out as somebody is after him all the time," she wrote to her mother. Despite the complaint, Edison and his wife went out often, attending an incessant round of ceremonies and banquets in his honor. A special envoy of King Humbert of Italy named Edison a grand officer of the crown of Italy, and the French government raised him to the highest rank of the Legion of Honor.[1]

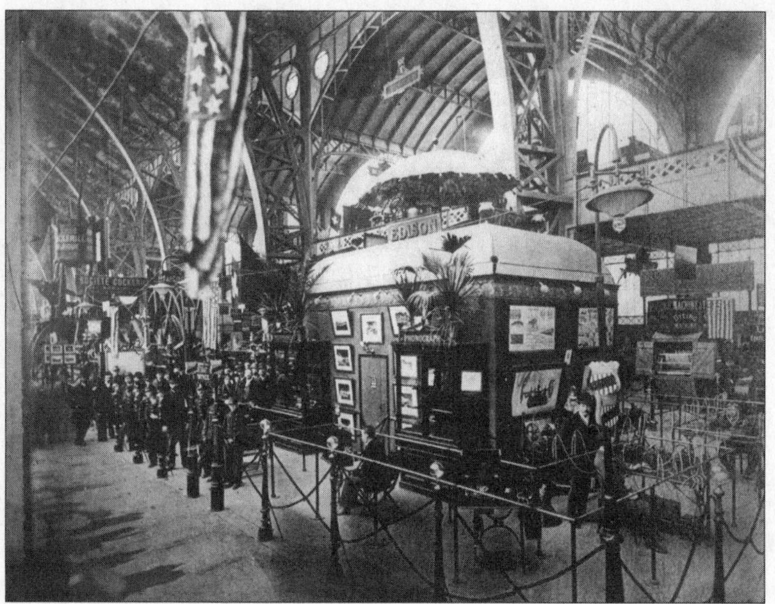

Part of the Edison exhibit at the Paris Exposition of 1889.

Edison was not always gracious enough to return the praise. "What has struck me so far chiefly is the absolute laziness of everybody over here," he told an American reporter in Paris. "When do these people work?"[2]

The fruits of Edison's own labor were obvious. To advertise his products, he had purchased the single largest exhibit space at the fair, covering a third of the area allotted to American companies. The display featured many of Edison's inventions: telegraphs, motors, electric railways, telephones, batteries, electric pens, typewriters, phonographs, and a complete central power station that included dynamos, underground conductors, and meters. To guide visitors to his display, Edison lit a beacon: a forty-foot-tall model of an incandescent lamp, lit from within by 13,000 standard incandescent bulbs and sitting atop a twenty-foot pedestal. Twelve steps of multicolored bulbs led up to the top of the pedestal, which contained a niche with a bust of the inventor. Above the display, *Edison* was spelled out in lights.[3]

The inventor paused in his self-promotion to advocate another new electrical device: He spoke so glowingly about execution by alternating current that the French Academy of Sciences decided to investigate the matter (but ultimately retained the guillotine).[4]

Just outside Edison's Paris hotel stood a statue of Napoleon. As he passed by it each day, he might have recalled the time when the explorer Henry M. Stanley dropped by the Orange laboratory to hear the phonograph. After listening for a while, Stanley asked Edison whom he would choose if he could "hear the voice of any man whose name is known in the history of the world."

"Napoleon's," Edison answered briskly.

The pious Stanley replied, "I should like to hear the voice of our Saviour."

"Well, you know," said Edison, not a bit flustered. *"I like a hustler!"*[5]

As he basked in the adoration of Paris, the unmasking of Harold Brown revealed that Edison played a starring role in one of the strangest hustles in the history of American business.

IN 1888, when controversies over electrical execution first arose, Edison was asked his opinion about capital punishment in general. "There are wonderful possibilities in each human soul, and I cannot endorse a method of punishment which destroys the last chance of usefulness," he said. "I think that the killing of a human being is an act of foolish barbarity. It is childish—unworthy of a developed intelligence."[6]

Edison, however, had no qualms about helping to create a better way to kill. Arthur Kennelly became a true believer in the humanity of electrical execution, which he called "the signal of a rising civilization." Edison's support for the new method was more measured but nonetheless genuine. As he told the *Sun*, "I am not in favor of executions, but if they are to take place electricity will do the work, and it is more certain and perhaps a little more civilized than the rope."[7]

This assertion would have been equally true for any make of alternating dynamo, but Edison helped to ensure that Westinghouse machines would be used. Although he never directly explained why he participated in the conspiracy, his rivals claimed to know the reasons. George Westinghouse believed Edison's actions were economically motivated, an attempt to neutralize Westinghouse's technological advantage by imposing government regulations on alternating current. Bourke Cockran thought it was personal; at the Kemmler hearings he hinted that Edison's pique over Westinghouse's infringement of his incandescent lamp patents drove him to seek revenge. Both believed that Edison's relentless focus on the question of safety was merely a dodge for selfish motives.[8]

If Edison's only goal had been to regain his hold on the electricity market, he had more conventional avenues open to him. In 1886 Edison Electric had purchased American rights to one of Europe's best alternating-current systems, and Edison conducted experiments on similar equipment. Despite his vested interest in direct current, Edison could have manufactured both systems, gradually shifting from direct to alternating current. Given the power of the Edison name and his deep experience in the electricity business, he would have had a good chance of routing Westinghouse in head-to-head competition for the alternating-current market.[9]

Many voices within the Edison fold wished he would do just that. At a convention of local Edison companies in the summer of 1889, Edison affiliates voiced a desperate need for a new system that could compete with alternating current, something using "higher pressures and consequently less outlay of copper than that involved by the three-wire method. We earnestly appeal to the parent organization to supply these deficiencies." Instead of building a rival alternating system, Edison tried to develop a five-wire circuit, which proved too complex. A high-pressure direct-current system was equally balky and, Edison feared, too dangerous.[10]

Later in 1889 Edison consented to build an alternating generator—but he did not agree to sell it. He thought that if he built a system exactly like Westinghouse's, he could use it to show the inefficiency and danger of alternating current. "Our condemnation of our apparatus would carry with it condemnation of theirs," Edison said. "We are, of course, agreed that the Edison Company has no desire, and no intention of actually selling alternating apparatus for electric lighting, if they can possibly avoid it."[11]

In his opposition to alternating current, Edison was fully supported by Edward Johnson, his old friend and the president of Edison General Electric. Johnson explained that selling alternating equipment would "destroy the reputation of the Edison Company, which has, in a large measure, been built up on the safety and economy of the Edison apparatus."[12]

IN 1882, when a humane society representative had inquired whether electricity might offer a better way to kill livestock, Thomas Edison told her that "the alternating machine of Gramme would kill instantly."* At that point Edison had no possible economic motive for saying this; the Gramme machine was used exclusively for arc lights, which Edison did not make, and an alternating incandescent lighting system to rival his own direct-current system was still four years in the future. Edison had a genuine fear of alternating current based on the theory that its rapid back-and-forth motion did more damage to the body than the steady flow of direct current.[13]

The tests performed at Edison's laboratory in 1888 had confirmed that alternating current was more dangerous. In the fall of 1889, those results were further buttressed by Jacques Arsène d'Arsonval, one of the most eminent physiologists in Paris, whose reports on electrical safety were translated and published in American electrical journals. Edison kept up with d'Arsonval's research and shared it with the pub-

*Zénobe-Théophile Gramme was a Belgian who built some of the most advanced generators of the 1870s.

lic. "Let me read you part of an article by Mr. d'Arsonval, the greatest authority in France upon electrophysiology," Edison told a newspaper reporter in October. "He says: '. . . at a mean equal pressure alternating currents are much more dangerous than continuous currents.'"[14]

Safety had long been Edison's preeminent concern. When he first introduced electric light, he argued that it was safer than illuminating gas because it was less likely to cause fires and would not asphyxiate people in their sleep. When in 1882 Brush Electric introduced its storage battery electric lighting system—which sent high-voltage direct current into batteries in people's homes—the Edison company warned that introducing high voltage into homes was unwise. "If deaths happen from such contact with electric wires," Edison Electric's president wrote to the *New York Times*, "a storm of undiscriminating public indignation will attack all methods of domestic lighting by electricity."[15]

Even in 1889, Edison believed that the electrical industry was in a precarious state. Only a tiny fraction of American homes were wired for electricity, and electrical utilities were burdened with enormous debt. The powerful illuminating gas companies had improved their technology and cut prices in an effort to bury their upstart competitors. Electrical utilities had yet to prove that they could compete successfully with gas lighting. In Edison's view, accidental deaths might be a nail in the coffin of the electric lighting industry. As he explained, "If we ever kill a customer it would be a bad blow to the business."[16]

Edison's statement—in its concern for the business rather than the customer—may appear to neglect the human costs of electrical accidents, but business meant more than money to Edison. In 1886 his lawyers had proposed merging the English branch of the Edison company with the firm of Joseph Swan, with the new company to bear both men's names. Edison was outraged: "The company shall be called the Edison Electric Light Company, Ltd., or at least shall be distinguished by my name without the name of any other inventor in its title," Edison wrote. "I am bound by pride of reputation, by pride and interest in my work. You will hardly expect me to remain interested, to continue working to build up my new inventions and improvements for a business in which my identity has been lost."[17]

Edison ultimately agreed that the new company would be known as Edison & Swan United Electric, but his attitude toward the merger reveals what drove him. Although he would make the same amount of money regardless of the company's name, it was not simply a question of economics; it was one of "pride" and "identity." After years of passionate labor, Edison "tamed the thunderbolt"—as the reporters phrased it—and created a system that safely brought electric light into people's homes. Thanks to that achievement, the terms *Edison* and *electricity* became virtually interchangeable in the American imagination, as well as in Edison's own mind. He vowed to protect his own good name and the reputation of the light he had invented.[18]

WHEN EDISON RETURNED from Paris on October 6, 1889, he received word that the U.S. Circuit Court had invalidated the paper-filament lamp patent—once owned by Consolidated Electric but by then controlled by Westinghouse—and dismissed Westinghouse's suit against Edison. (The other major lawsuit, in which Edison Electric was suing Westinghouse for violation of its basic lamp patent, was still unresolved.) Westinghouse planned to appeal this latest ruling to the U.S. Supreme Court, but the lower-court decision nonetheless constituted an important victory for Edison.[19]

"Westinghouse simply grabbed fifty-four of my patents and started into business, saying that he could sell his manufactures cheaper because he did not have to pay out money experimenting," Edison told the *New York Herald* on October 7. He added, "Westinghouse used to be a pretty solid fellow, but he has lately taken to shystering."

The day the article ran, a top Edison lieutenant ran into Westinghouse over lunch and reported the encounter to his boss: "Mr. Westinghouse remarked to me that he felt very much hurt by your calling him a 'shyster' in this morning's 'Herald.' He was really cut up about it."[20]

Rather than apologizing to his rival, Edison went on the attack, instructing his agents in Pittsburgh to gather all available information about Westinghouse's railroad air brake business. Since Westinghouse had invaded his territory in incandescent lighting, Edison considered turning the tables by striking at the heart of the Westinghouse empire. Although this plan was abandoned rather quickly, Edison soon found fresh grounds for assailing his rival.[21]

Three days after the decision in the patent case, a lineman for an electric company received a fatal shock atop a pole on Grand Street. In September two others had died in electrical accidents: A worker at East River Electric had lost his balance, grabbed a switchboard, and fell dead, and an Italian fruit vender slipped while cleaning the roof of his shed, fell against a wire owned by the United States Illuminating Company, and died. The deaths served as a reminder that the city was still festooned with dangerous overhead wires.

By the time of these deaths, the two firms with the most wires in the worst condition–United States Illuminating and Brush Electric Illuminating–were no longer independent companies; both had been acquired by Westinghouse Electric. This meant that most of the high-voltage overhead wires in New York City were under the control of George Westinghouse, who continued to resist efforts to remove them, make them safer, or place them in underground conduits.[22]

After the latest accidental death, Thomas Edison told the *Evening Sun*, "They say I am prejudiced, but if I had anything to say I would abolish the alternating current."[23]

SOME PEOPLE distrusted Edison's opinions on this matter because his money and his passion were tied up with direct current, but the same objection could hardly have been urged against alternating-current pioneer Elihu Thomson. Thomson feared alternating current nearly as much as Edison did, and he was just as outraged over the accidental deaths in New York and elsewhere. As the inventive mind behind the Thomson-Houston company, Thomson had sketched designs for alternating systems in 1885, but he delayed the work because he considered high-voltage alternating current too dangerous. Only after inventing and patenting a number of safety devices—such as fuses and "lightning arresters" to prevent dangerous shocks from reaching indoor wiring—did he consider the system safe enough to sell to the public. Unhampered by such concerns, Westinghouse had opened a large lead before Thomson-Houston entered the market. Despite the delay, Thomson believed his company had one clear advantage: It held the patents on all the important safety devices.[24]

As it turned out, those patents offered little leverage. Like the Edison and Westinghouse companies, Thomson-Houston made most of its money selling its dynamos, bulbs, and other equipment to utility companies, which then installed them and supplied light to customers. The manufacturing companies had little control over how carefully their equipment was installed. Most of the local utilities that bought the Thomson-Houston system did not purchase the safety equipment, and there were few governmental regulations that required them to do so. "In the general scramble for business there has been a neglect of proper precautions," Thomson explained.[25]

Thomson's private correspondence revealed his dismay. "The man-

Elihu Thomson

ner of installation in New York City is simply abominable," he wrote. The accidental deaths resulted from "gross carelessness and recklessness" on the part of the local lighting companies, and "it is only a wonder to me that fatal accidents are not more frequent." Even properly installed, however, such wires could not be safe, because "no insulation that has as yet been found is any too good." As Thomson saw it, the solution was to bury all of the wires used in heavily populated cities such as New York.[26]

When the overhead wire battle heated up in the fall of 1889, some Thomson-Houston managers asked Thomson to pen an article reassuring the public about the safety of the firm's equipment. He refused: "I certainly shall not put myself in a position to be criticized as Mr. Edison has been criticized in what he has said about wiring, as only said in self-interest." Many of the Thomson-Houston systems in service–including many plants in New York–were nothing short of lethal, and Thomson would not defend them.[27]

When Thomson did issue a public defense of alternating current, it was couched in the most cautious of language. High voltages were necessary for the affordable transmission of energy, Thomson wrote in *Electrical World*, but the public needed to be protected through safety devices, better insulation, and the placement of wires in underground conduits. Thomson did not downplay the dangers of the current he sold: "Alternating current is much less safe than . . . continuous currents of equivalent potential."[28]

George Westinghouse never made a similar admission. At the end of 1888 he publicly stated that "pressures exceeding 1,000 volts can be withstood by persons of ordinary health." By late 1889 he admitted that alternating current could kill under some circumstances, but he still insisted that "there have been hundreds of cases in which momentary contact with an alternating current of 1,000 volts and over . . . has resulted only in painful shocks, unaccompanied by permanent injury."[29]

As much as the lost business and patent dispute, it was Westinghouse's continued intransigence on the safety issue that outraged

Edison. By late 1889 Edison's long-standing views about the greater dangers of alternating current had been confirmed by his own tests and independent experiments in France. His rival, he felt, was destroying his business by stealing his patents, installing slapdash systems carrying lethal levels of electricity, and denying the dangers. Overhead wires continued to kill, and Edison feared the deaths might turn the public against all forms of electric lighting. Edison fought back, using the most dramatic means at his disposal–his public support of the electrocution law–to demonstrate the dangers of his competitor's current.[30]

ON OCTOBER 9, 1889, three days after Edison returned from France, that law survived its first legal challenge: Judge S. Edwin Day of the Cayuga County Court issued a ruling rejecting William Kemmler's appeal. One of the first reasons the judge cited was that, since hanging had been abolished, New York would have no capital punishment law if electrocution were declared unconstitutional. As a result, convictions of all murderers sentenced under the new law might be overturned. Many attorneys argued, to the contrary, that the state would simply revert to the law previously in effect, so that condemned prisoners could be sent to the gallows. But Judge Day raised doubt on the issue, and Kemmler's appeal became shadowed by the specter of prison doors clanging open and murderers strolling free.[31]

Judge Day had reviewed the transcript of the Kemmler hearings–two bound volumes running to more than 1,000 pages–but he had little to say about it, because his ruling rested upon a more basic question of the separation of governmental powers. As the judge saw it, electrical execution "became law after much more than ordinary consideration and deliberation," and he refused to contradict the legislature and the governor. "Every statute is presumed to be constitutional," he wrote, and the burden of proof rested with the party challenging the law. Although the hearing testimony was "conflicting," the judge ruled, Kemmler had not proven that electrical execution was beyond doubt cruel and unusual.[32]

"It's a victory for us," Harold Brown told the *World*. "By the way, there have been five deaths in this city from alternating currents since September 1."[33]

CHAPTER 17

The Electric Wire Panic

ALTHOUGH BROWN was pleased with the judge's ruling, the true goal of the Edison forces—banning alternating current or restricting its voltage—seemed no closer to reality. Westinghouse Electric and other lighting companies continued to resist New York law requiring burial of all wires, and accidents claimed more lives. Yet most people ignored Edison's warnings—until October 11, two days after the rejection of Kemmler's appeal, when one spectacular accident awakened the city to the terrors of electricity.[1]

John Feeks, a lineman for Western Union, was a calm man of medium height, with a sunburned face and a bristly red mustache. At noon on October 11, his job took him to the intersection of Centre and Chambers Streets in downtown Manhattan, an area with one of the densest networks of wires in the city. The pole Feeks stood under bore fifteen crossbars—nine running north and south, six east and west—carrying more than 250 wires. Some of those wires were dead, and the lineman's task that day was to cut them away so that they would not interfere with those still in use. Because he would be working only with low-pressure telegraph wires, he chose not to wear rubber boots or gloves. He had a wire cutter on his belt and metal spikes on the insteps of his leather shoes. He grasped the pole with his hands, planted a spike in the soft wood, and stepped up the pole as easily as climbing stairs.

It was the lunch hour in Manhattan's busiest district, and the lineman's ascent drew a crowd. About twenty-five feet up, Feeks slowed to ease past the first cross-arm, which bore two thick cables carrying the pole's only electric light current. After that, it was a clear shot of twenty feet up to the pole's top, where the other fourteen crossbars bore their heavy burden of telegraph wires. Feeks stopped again at the lowest crossbar and stared straight up, plotting a route through the dozens of interlaced wires. After shifting to the north side of the pole, he poked his head through a small opening in the wires and drew his shoulders together as he squeezed through. Then he looked for the next gap in the weave. When none offered itself, he yanked the wires apart with his hand and hoisted himself through. In this way Feeks passed the first, second, third, and fourth crossbars, before arriving at the place where he had work to do. He wrapped his left arm around a crossbar and braced his left foot against a wire on the crossbar below. His right leg dangling free, he drew the pliers from his belt and reached out to snip a wire. As he did so, he lost his balance and grasped a wire with his right hand to steady himself.

As far as Feeks knew, the only danger was falling, because the telegraph wires normally carried too little current to pose a risk. But somewhere, blocks away, an alternating-current light wire had crossed the same telegraph wire Feeks held in his hand. As the wires blew in the wind, the dead telegraph wire cut through the insulation on the light wire. High-voltage current diverted from its charted course and surged down the telegraph wire, into Feeks's right hand, and out his left foot. His body went tense, his right arm quivered, his mouth opened but emitted no sound. Feeks's head reared up, and his throat came to rest on the live wire. A tiny blue flame played around his right hand and his left foot, and small puffs of smoke drifted away on the wind.

The small group of men who had watched Feeks's ascent screamed, and immediately every face on the street turned and looked up to see

The death of lineman John Feeks, as illustrated in the *New York World*.

a man trapped like a fly in a web of wires. Before long, thousands of people filled the rooftops and blocked the streets and the approach to the Brooklyn Bridge. *World* reporter Nellie Bly, just a few days from setting off on her famous globe-girdling trip, pressed her way through the crowd in time to see the wires burn into the lineman's flesh. "The hand ceases to quiver," Bly wrote, "and a dark-red stream gushes from the wrist. Now it springs from the throat, spotting the pole and dripping down on the heads of the fleeing crowd."

After forty-five minutes another lineman, wearing rubber boots and gloves, climbed the pole, tied a rope around Feeks's waist, and tossed the other end over a crossbar and down to the street, where more Western Union men took up the slack. As the lineman clipped the wires that ensnared Feeks, they whipped to the street and sent the crowd running again. When the last wire was severed, Feeks swung free and was lowered slowly to the ground, doubled up, his hands touching his feet. "Killed first, cut afterwards, then roasted," Nellie Bly reported, "not by heathens, but by a monopoly. All at mid-day in the streets of New York."[2]

Electricity had killed other men in New York, but there had never been anything like the death of John Feeks. It was a shared trauma. Thousands witnessed the bloody spectacle in person, and hundreds of thousands more read about it and saw the illustrations in the newspapers. The *Tribune* wrote that it had been more than a decade since the city had experienced "so many unmistakable indications of popular agitation and anger." In shops and homes, in streetcars and on street corners, the death was the main topic of conversation for days afterward. "Until Feeks was killed it was popularly supposed that an ordinary telegraph or telephone wire was harmless under all circumstances," the *Sun* observed, but that was before "a multitude of thousands saw a poor fellow roasted upon a gridiron of fire-spitting threads that had never before shown a sign of danger."[3]

"Any moment may bring a similar horrible death to any man, woman or child in the city," the *World* warned.[4]

Panic-stricken building owners took the law into their own hands and chopped the wires running over their rooftops. According to the *World*, some people grew so terrified that they threw out their telephones, "as if the little wires which connect them went straight to the river of death."[5]

On the pole where Feeks died, a saloon keeper nailed a tin cracker box with a sign reading "Help the Victim's Widow," and within a few days the donations totaled more than $2,000.[6]

The day after Feeks died, Mayor Hugh Grant convened the Board of Electrical Control and ordered that all unsafe wires be cut down immediately. Before the removal of wires could commence, however, George Westinghouse set his lawyers to work and scored a quick legal victory, persuading Justice George Andrews of the state supreme court to issue an injunction barring the city from touching the wires of Westinghouse's subsidiaries, the Brush and U.S. Illuminating companies. The *World* hinted at the all-too-plausible possibility of bribery, charging the judge with "ignorance or worse" and asking, "What influence was it which induced Judge Andrews to render so absurd a decision?"[7]

George Westinghouse suddenly appeared to be the man that Edison and Brown had long described: a cold-blooded villain denying the dangers of alternating current and risking lives for the sake of profits. "How ignorant or how deceitful, and how contemptible in either case, appear the assurances of safety with which the community has been mocked for years past," a newspaper editorial charged.[8]

THE *TRIBUNE* took this occasion to bemoan the "extraordinary condition of impotence to which the complicated machinery of civilization has reduced this community." It was an unusual admission, for Americans rarely paused to acknowledge that economic growth was taking a heavy toll in lives. In fact, more Americans were killed and maimed in the peacetime economy of the 1870s and 1880s than in the Civil War, and the most dangerous industries–railroads and coal mining–were the same ones that fueled the boom. Every year 1 out of 100 railroad brakemen died, along with 6 percent of the workers in some Pennsylvania coal fields. The danger was not inevitable. In late Victorian England–hardly a model of the compassionate state–the rate of railroad worker deaths was much lower, thanks to more stringent regulation. The United States, on the other hand, did not begin to adopt effective safety regulations until after 1900. In the 1880s, the government and the courts showed little interest in protecting the lives of workers.[9]

Electricity accounted for fewer than 1 percent of the accidental deaths in New York in the late 1880s. If the raw numbers were the guide,

This illustration, which appeared in *Judge* magazine shortly after Feeks's death, captured the widespread fear that no one was safe on the streets of New York.

the city should have been in a panic over the many things—elevator shafts, street railways, illuminating gas—that killed far more people than overhead wires. But electricity was different. It was new and mysterious, a force of life as well as a cause of death. Electricity held people in thrall, and it was a short step from awe to dread.[10]

The death of John Feeks showed Americans that the danger was no longer restricted to workers in a few industries. Electric wires brought the terror home. "Death does not stop at the door," one expert said, "but comes right into the house, and perhaps as you are closing a door or turning on the gas you are killed. It is likely that many of the cases of sudden death we hear of from heart disease may come about this way." "There is no safety, and death lurks all around us," another expert warned. "A man ringing a door-bell or leaning up against a lamp post might be struck dead any instant."[11]

With word that judicial intervention once again blocked removal of dangerous wires, the public's impotence turned to rage. A population accustomed to violent death, meek before the corrupt alliance of government and business, suddenly found its voice. One marker of the depth of public anger was the unanimity of newspaper opinion. Joseph Pulitzer's crusading *World*, as expected, concluded that "men's lives are cheaper to this monopoly than insulated wires." More surprising were the reactions of the *Times* and *Tribune*, which usually sided with corporations and private property: Both papers urged that the dangerous light wires be cut down, and that the companies' officers be indicted for manslaughter. The *Times* claimed that the Board of Electrical Control was as responsible as the electric companies. If the board's members were convicted of murder in Feeks's death, the newspaper wrote, they would receive the death penalty and "the deadly current might be put to good use."[12]

Electrical execution was never far from people's minds during the wire panic. The grisly spectacle of Feeks's death accomplished in forty-five minutes what Edison and his allies had been trying to do for more than a year: It convinced the public that the Westinghouse current was terrible and deadly.

The *World* drafted Harold Brown as its in-house expert and sent a reporter to accompany him as he measured leakage from alternating-current wires. The *Times* quoted a lengthy tirade by Brown against alternating current, which included the reminder that "a certain electric light syndicate"—Westinghouse—"has recently acquired the Brush and United States Illuminating Company's stations," the two main offenders in the current string of deaths. The electric wire panic rehabilitated Brown's reputation. A magazine enlisted him to write an article on electrocution and the dangers of alternating current, and an electrical journal asserted that he "has done his utmost, either from pure philanthropy or the love of gain, to bring home to the public a possible danger." At this point few seemed to care that Brown had conducted secret deals and abused his contract with the state in order to malign the Westinghouse company. Brown slipped comfortably into the guise he had tried to wear since the summer of 1888: that of an altruist warning the public about a lethal threat.[13]

During the panic, Edison consulted with Brown several times, asking for statistics on accidental deaths and plotting strategy on how best to turn the publicity to their advantage. An investor in the Edison system took an optimistic view of Feeks's death, telling Edison that "there never has been such a grand occasion" to attack alternating current. "A communication from you to the principal newspapers" promoting direct current "would greatly benefit our companies . . . and boom the stocks," the man wrote.[14]

In this regard, Edison needed little prompting. When a reporter from New York appeared at his doorstep not long after Feeks's death, Edison greeted him with a question: "Have they killed anyone there today?" With his gray hair and gleaming gray eyes, Edison was half prophet of doom, half reassuring grandfather. He predicted that more innocents would die soon, and that the problem would not stop when the wires were buried. "When under ground," he warned, "the dangerous current will creep into your house, and will come up the manholes." Although eager to sow fear, Edison also offered a path to safety. "Is there not a law in New York against the manufacture of nitro-glycerine within the [city] limits?" he asked. "Well, there must be one against deadly currents. Let the Mayor keep the pressure reduced to 700 volts continuous current and to 200 alternating." These restrictions, Edison said, could be enforced "under police regulation, just as steam boiler pressure is."[15]

The inventor soon earned an even bigger forum. The *North American Review*, one of the nation's most influential opinion journals, asked Edison to contribute an essay, published in November as "The Dangers of Electric Lighting." It opened with an invocation of Feeks's death: "If the martyrdom of this poor victim results in the application of stringent measures for the protection of life," Edison wrote, "the sacrifice will not have been made in vain." He said the tragedy could have been avoided had authorities heeded his earlier warnings. Alluding to his dog-killing experiments, he wrote, "I have taken life—not human life—in the belief and full consciousness that the end justified the means."
These tests had shown that the passage of "alternating current through any living body means instantaneous death."

"Burying these wires," Edison believed, "will result only in the transfer of deaths to man-holes, houses, stores, and offices, through the agency of the telephone, the low-pressure systems, and the apparatus of the high-tension current itself."

"I have no intention, and I am sure none will accuse me, of being an alarmist," Edison said, having just raised the specter of people being shocked dead in their homes as they picked up the telephone. He said he was simply calling attention to an unseen danger and proposing a remedy. His own low-pressure system was commercially successful and perfectly safe. He therefore advocated "rigid rules for the restriction of electrical pressure," although he would have preferred to go a step farther: "My personal desire would be to prohibit entirely the use of alternating currents. They are as unnecessary as they are dangerous."[16]

IN LATE NOVEMBER and early December, three more Manhattan men died in electrical accidents. Two of the victims were light company employees, but it was the third death that further stoked the public's rage. While closing up shop for the night, a clerk in an Eighth Avenue dry-goods store picked up a tall metal display case to move it from the sidewalk into the store. The case touched a low-hanging Brush arc lamp, and the clerk fell dead from the shock. It was exactly the type of tragedy Edison warned about: a private citizen struck dead on the sidewalk while performing the routine tasks of life.[17]

"Mr. Edison has since declared that any metallic object—a doorknob, a railing, a gas fixture, the most common and necessary appliance of life—might at any moment become the medium of death," the *Tribune* warned. New Yorkers took heed. Some refused to have doorbell wires strung through their homes, fearful that the touch of a button might bring instant death. The *Evening Post* observed, "One scarcely ventures to put a latch key into his own door." An electrical journal branded such fears "lunatical" and "nonsensical," but the public was not reassured. The *Tribune* and other New York newspapers endorsed Edison's call for voltage limits, as did papers in other states. Referring to Edison as "the highest authority on the matter of electricity," a South Carolina paper called for limits on voltage. "Mr. Edison is right in his position that electric tension should be regulated by law," another paper said. "The only reasonable solution of the whole problem lies in making every electric wire safe, not because it is insulated but because in its nakedness it carries no death-dealing power."[18]

The movement gained ground in New York as well. The *World* pointed out that the city's Board of Health had the power to remove from the streets anything "dangerous to life or health." At the *World*'s request, Harold Brown filed a petition with the health board asking it to prohibit "any current liable to cause death or injury to human life." The city health department conducted tests at a local light company and found that even new wires leaked, exposing the public to grave dangers. The board passed a resolution calling for limits of 250 volts on alternating current, but it deferred enforcement to the electrical board. "In the estimation of Thomas A. Edison," the *World* said, such limits would "afford a guarantee of safety not promised by any other plan." Mayor Grant rejected the plan, fearful of violating Judge Andews's injunction against the removal of the wires and worried that such voltage limits would leave Edison as the sole light company standing. "I will not vote for such a monopoly," Mayor Grant said.[19]

The health board's plan failed, but it terrified the alternating-current forces. Facing removal of their wires or even voltage limitations, they embraced the remedy that they had fought bitterly only a few months before: the burial of their wires. George Westinghouse wrote "A Reply to Mr. Edison" for the *North American Review*, insisting that alternating current would be perfectly safe if tucked under the streets. This tardy concession did not appease the public. Feeks and a dozen other men were dead, the *World* charged, only because "the Electric-Light Companies scorn the law, defy its officers and twiddle their fingers at a helpless public."[20]

On December 13, two months after the death of John Feeks, the state supreme court dissolved the injunction against removal of the wires and issued a ruling harshly condemning the light companies: "When

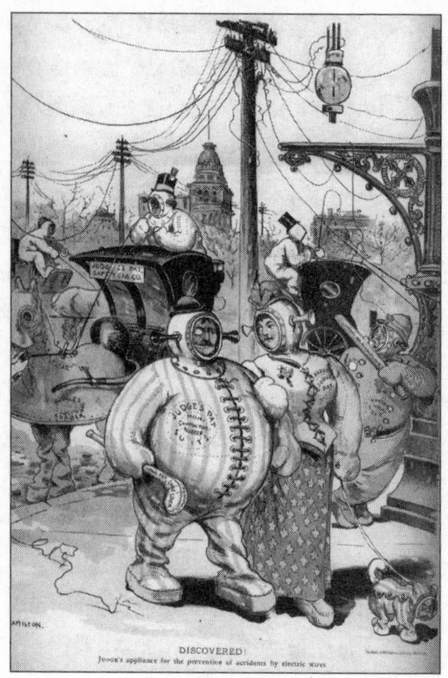

Judge magazine offered the tongue-in-cheek proposal that rubber suits offered the only means of protecting oneself (and one's animals) from the dangers of electric shock.

they claim that the destruction of these instruments of death, maintained by them in violation of every duty and obligation which they owe to the public, is an invasion of their rights of property, such claim seems to proceed upon the assumption that nothing has a right to exist except themselves."[21]

Time had run out for the Westinghouse lighting interests in New York. The electrical board ordered them to shut off their current by eight o'clock the next morning, and the removal of lines began at precisely half past nine. The Department of Public Works hired twenty-five men, equipped them with rubber gloves and insulated wire shears (known as "nippers"), and divided them into four gangs. Each gang started at a central station and followed the wires radiating from it, searching for violations: bad insulation, unauthorized poles, lamps hanging too low, wires affixed to telegraph poles or elevated train platforms. When they spotted problems, a lineman climbed the pole and snipped the wires, and then one of his fellows went at the pole with an ax. The crews toppled twenty-three poles and stripped nearly 50,000 feet of wire on the first day, a Saturday. Eager to continue, they started in again the next morning, although the *World* noted that "churchgoers expressed disapprobation of the Sunday work."[22]

New Yorkers took gleeful pleasure in this assault on private property. After the carnage of the previous months, the attack on the poles became cathartic. Crowds gathered to cheer and watch the chips fly as workers used axes to fell electric light poles. "Destruction of Property Goes Merrily On," read a headline in the staid *Times*. When the work continued on Christmas Eve, the newspaper described "the music of the axes and nippers" as "fitting accompaniment to the spirit of holiday merrymaking." The destructive frenzy tailed off on December 30, by which time more than 1 million feet of wire—a quarter of the city's total—had been stripped from the streets.[23]

Without wires, however, there was less light. The companies could lay new wires underground, but that would have to wait until the spring thaw unlocked the ground and the laying of conduits resumed. The only light company unaffected was Edison Illuminating, whose wires had always been underground. But Edison did not light the streets. A few months earlier, Brush, United States Illluminating, and a few other companies sent current to more than 1,000 arc lamps that threw a harsh white glare on the city's night life. On New Year's Eve 1889, all of those lights were off. Early in the evening the interior lights of stores and saloons threw a glow onto the sidewalks, but when businesses closed, the streets grew black. New Year's revelers lit candles and lanterns and picked their way along slushy streets, islands of light moving cautiously through the black night. A new decade had begun, but to many New Yorkers it appeared that the city had taken a step back into a gloomy past.[24]

CHAPTER 18

Designing the Electric Chair

As CITY WORKERS in Manhattan nipped wires and chopped poles, William Kemmler suffered another legal defeat. On December 30 a three-judge panel of the New York Supreme Court—the state's second-highest appellate level—affirmed the lower court's decision in Kemmler's case. If the legislature had prescribed a punishment that involved "torture and a lingering death," it would be the duty of the courts to intervene. The question was whether electrical execution was cruel, and there was no reason to believe that the court's judgment on this matter would be more competent than the legislature's.

The judges also declared that it was "within easy reach of electrical science" to use electricity "to produce instantaneous, and therefore painless, death." As evidence they pointed to Manhattan's recent electrical fatalities: "The frequency and publicity of death by accidental contact with electric wires during the last few years, and especially the last few months, has made the deadly power of the electric current shockingly familiar wherever the newspaper is read."[1]

. . .

THE DAY AFTER the New York Supreme Court issued its ruling, Harold Brown traveled to Auburn, where William Kemmler was starting his seventh month in prison. Brown had first worked on the Auburn electrical plant back in August, just after the Kemmler hearings. Now his machinery was to be tested by the state electrocution commission, which consisted of the physicians Carlos MacDonald and A. D. Rockwell, as well as Columbia College's Professor Louis Laudy, who had been Brown's host for the first public dog-killing experiments in the summer of 1888. For the Auburn tests the commission was joined by Dr. George Fell.[2]

Just as the tests began on December 31, 1889, one of the pulleys linking the steam engine to the dynamo snapped. While a workman repaired it, the prison warden, Charles Durston, took the visitors on a tour of his domain. Auburn prison was built in the 1820s as part of the movement to reform rather than punish criminals. Prisoners labored in prison workshops daily, but they were forbidden from talking to each other, and they could not receive visitors or correspond with friends or family. The "Auburn system" of prison discipline became internationally famous. Alexis de Tocqueville's *Democracy in America* grew from his travels in 1831 and 1832, but his official business on that trip was to study prisons, including Auburn. Enthusiasm for the penitentiary waned in the sixty years after Tocqueville's visit, but Auburn's program of strict discipline and manual labor survived. Warden Durston took his physician visitors to the prison workshops, where inmates—known as "stripeds," after their uniforms—earned their keep by making chairs, baskets, and other goods.[3]

By the time the tour ended, the pulley was repaired, and commissioners began their work. An old horse died from a thirty-second shock at 1,200 volts, and a four-week-old calf succumbed to a ten-second shock at the same voltage. As soon as the calf collapsed, Dr. Fell performed a tracheotomy and attached a "Fell Motor," an artificial respiration device he invented. The commission had invited Dr. Fell and his motor to Auburn in order to lay to rest the fears that electricity might stun rather than kill its victims. After thirty minutes of artificial respiration, the calf showed no signs of recovery and was pronounced dead.[4]

THE TESTS AT AUBURN showed that the Westinghouse generator had the power to kill a horse. Executing a human being required more complex considerations, such as the positioning of the victim. There was nothing inevitable about the choice of a chair. A writer in the *Medico-Legal Journal* suggested a tiny room, "something like a sentry-box or watchman's hut," with a metal-lined floor and an electrode descending like a shower head from the ceiling. In an illustration accompanying the article, a man—barefoot and with trouser legs rolled, as if ready for a walk on the beach—stands in the hut, ready to receive the current from the top of his head to the soles of his feet. If shocked while in a standing position, he would fall in a heap to the ground. This might do for horses and dogs, but it would rob a human execution of its proper solemnity. The writer therefore proposed that "spring locks" be attached to the hut and closed around the prisoner's neck and limbs to keep him upright. Another New York electrician advised standing the victim on a metal plate, with his hands stretched above his head and attached to wires, in the position of a scourging victim. The current would run from the floor plate to the hands, and "the body would be prevented from falling by the wires." Neither of these arrangements would have prevented the prisoner from foiling the execution by lifting his feet off the floor.[5]

The chair offered an easier solution. In an early proposal for electrical execution, published in *Scientific American* in 1873, the writer advised placing the prisoner in a chair, and a German writer offered the same idea a few years later. Alfred Southwick—perhaps inspired by his professional experience with dental chairs—advocated a chair from the start, although he changed his mind about what form it should take. Interviewed late in 1886, Southwick explained that a floor plate under the seated prisoner's feet would serve as one electrode, while the other,

One scientist suggested that criminals be executed in an electrical hut, with the current running between its metal-lined floor and an electrode descending from the ceiling to the victim's head.

a metal rod, would be "brought in contact with the back of the neck over the spinal column." Upon further consideration, however, Southwick came to believe that "an arm-chair, with metal arms, would be more convenient. The condemned would be seated in the chair and, at the proper moment, receive a full electric charge through the metallic arms." Another year passed, and Southwick changed his mind again. The death penalty commission report recommended "a chair, with a head and foot-rest, in which the condemned could be seated in a semi-reclining position; one electrode would be connected with the head rest, and the other with the foot-rest, which would consist of a metal plate." Others suggested a chair with metal plates adjustable to each side of the torso, or with two metal bands, one to fit around the neck, another around the chest.[6]

All of these plans involved applying metal electrodes to the skin. When current encountered skin, which had a high resistance, some of the electrical energy was transformed into heat, which caused burning.

At the Edison laboratory experiments, copper wire electrodes were wrapped around wet cotton to lower resistance, and Dr. Frederick Peterson adapted this idea for use on humans. He advised wrapping the electrodes in "a sponge or chamois skin, thoroughly wet in order to prevent burning."[7]

There was also the question of restraint. When Harold Brown was asked why he experimented upon dogs rather than cats, he replied, "Because the cat is very apt to wiggle around when you attempt to apply the electrode, and they also have claws." If the condemned man behaved less like a dog and more like a cat, the authorities would have trouble executing him. In his earliest public statements on execution, Thomas Edison proposed that manacles double as electrodes. "No matter what position the prisoner took, nor how much he twisted and turned, whether he stood or sat down, he could not escape the shock," Edison explained. In its 1888 report, the Medico-Legal Society noted the "unseemly struggles and contortions" that marked many executions and advised proper restraints: "A stout table covered with rubber cloth and having holes along its borders for binding, or a strong chair should be procured. The prisoner lying on his back, or sitting, should be firmly bound upon the table, or in the chair."[8]

This is the only point in the entire debate on electrical execution at which anyone raised the possibility of a table. A few lines later the report added, "We think a chair is preferable to a table." The reason for this preference was not explained, but it is likely that the chair was considered more dignified for the prisoner. On the gallows a prisoner stood tall and proud, a full participant in the ritual of retribution. Strapped to a table, he would be utterly helpless, resembling a bit too closely an experimental animal strapped to a laboratory bench for vivisection. The chair occupied a middle ground, allowing the necessary restraint and support but also paying at least minimal respect to the prisoner's humanity.[9]

The most controversial question involved the placement of electrodes. In Edison's view, the dozens of accidental deaths caused by grasping a wire with both hands proved the worth of a hand-to-hand

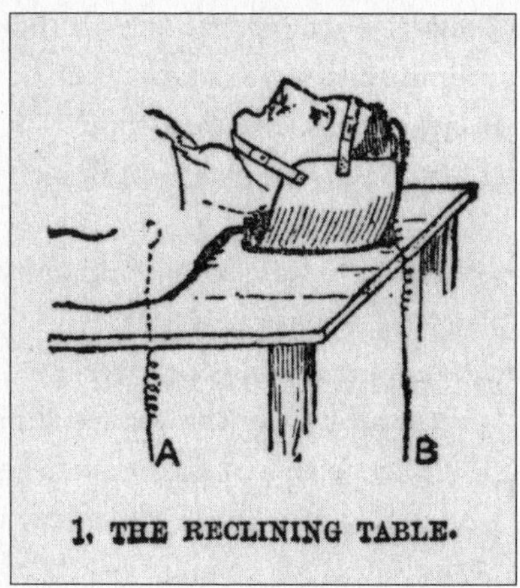

The Medico-Legal Society proposed an "electric table" before concluding that a chair would be preferable.

arrangement. The Medico-Legal Society's physicians tested this theory by running current between the forelegs of a horse, but they were not satisfied with the results. Wishing to attack "the seat of consciousness" and the areas that "exercise jurisdiction over the lungs and heart," the report advocated placing electrodes on the head and spine. The positive pole, the doctors said, should go on the head, because "the electric stream flows from the positive to the negative pole"; the physicians apparently were unaware that with alternating current each pole alternated between positive and negative. However tentative the theories, experimental results seemed to support these recommendations. The report endorsed fixing one electrode to the chair such that "it will impinge upon the spine between the shoulders," while the other should be attached to "a sort of helmet" attached to the back of the chair.[10]

The Medico-Legal Society enjoyed no official authority, but the state commission appointed early in 1889 adopted most of its recommendations, with one slight revision. In March 1889 the commissioners killed nine animals at the Edison lab with the electrodes placed at different areas of the animals' bodies. They concluded that a current passing from the head to leg—rather than head to spine—worked best.[11]

It was up to Harold Brown to translate these recommendations into an execution device. His design—which he unveiled to the press in May 1889—featured an oak reclining chair with a head electrode in a "metal cap" and feet electrodes encased in "electrical shoes." In June or July of that year, a carpenter at the Auburn prison built a chair to Brown's specifications.[12]

In December 1889, after Dr. Fell and the state commissioners killed a horse and a calf at Auburn to test the execution dynamo, they examined Brown's chair. Dr. Fell believed that it was poorly designed, and he explained his views to Austin Lathrop, the state prison superintendent. Dr. Fell agreed that one electrode should go on the head, but he thought the second should be affixed near the spine on the lower back, so that the current flow "would include the heart and produce the greatest density in the neck, including the region of the medulla oblongata." This opinion contradicted that of the official state commission, which opted for a foot electrode, but the superintendent brushed this matter aside and told Dr. Fell to get to work. Back in Buffalo, the doctor designed a new chair, hired a carpenter to build it,* and shipped it to Auburn prison on February 12, 1890.[13]

ON THAT SAME DAY in February, Thomas Edison was bound for a vacation. "I am pretty well broken down with overwork and am going down in the North Carolina Mountains to freshen up," he wrote in a letter to Henry Villard. On his way to the mountains Edison stopped off in Richmond, Virginia, to lobby for his favorite cause. A state senate committee was considering bill no. 238, "For the Prevention of Danger from Electric Currents," which called for limiting alternating current to 200 volts. Edison's appearance drew such a crowd—including "dozens

---

*Although it has been claimed that the great American furniture designer Gustav Stickley built Auburn's first electric chair, the credit belongs to the anonymous Buffalo carpenter hired by Fell. Stickley did serve as director of manufacturing operations at Auburn prison from 1892 to 1894, and he may have built the three-legged chair that replaced the original one in 1893.

of ladies," the *Richmond Times* noted—that the hearings were moved from a committee room to the legislature's main chambers. The crowd cheered and applauded when Edison was announced. He smiled and bowed, then launched into a familiar speech, explaining electricity in folksy terms and charging that alternating-current companies risked lives to save money. Harold Brown followed Edison to the examination stand and explained New York's new execution law and how he had come to support it.[14]

The bill died in committee not long after Edison boarded his train for North Carolina. To the Virginians, Edison's motives seemed transparent. "Though purporting to guard the interest of the people of Virginia," one opponent said, the bill "was in reality a continuation of the struggle for supremacy of two electrical companies." Richmond, like most Virginia cities, was lit by alternating current, but the allegedly lethal current had yet to kill a man there. "Why should the promoters of the bill come here from New York and New Jersey, where persons had lost their lives by pure carelessness?"[15]

Ohio proposed similar restrictions on alternating current a month later. On Edison's orders, Arthur Kennelly packed up an alternating generator and took it to Columbus. The legislative committee considering the bill gathered at the local Edison lighting station, where Kennelly and Harold Brown quickly dispatched a calf and a horse. "Committee evidently impressed," Kennelly reported to Edison. The Ohio bill failed nonetheless.[16]

When New York's legislative session started in Albany in January 1890, the state senate's Committee on Laws proposed an investigation into the matter of electrical safety. Considering the carnage in New York City the previous fall, it seemed a reasonable undertaking. The newspapers, however, familiar with the rampant corruption in the legislature, believed that bribery was the true motivating force. In an article headlined "Electricity and the Spoilsmen," the *Herald* called attention to "two very fat birds known as the Edison and Westinghouse roosters. Both are more willing to be plucked than to be slaughtered. It is well known that either of these birds is willing to give up a great deal of 'corn' rather than see the other fattened by legislative preference." The *Times* declared that the senators leading the investigation were quick "to grasp a matter that is agitating the public mind, and bend it to conform with their personal interests."[17]

The committee's hearings, which stretched through March and April, were a tedious and predictable affair, with both Westinghouse and Edison producing lawyers and experts testifying in precisely the way everyone expected that they would. The hearings could not even rely on the excitement of an appearance by Edison, who was still vacationing in North Carolina. One senator complained that the hearings without Edison were "like 'Hamlet' with Hamlet left out."[18] But the committee's investigation was more farce than tragedy. As the press saw it, the legislators were stuffing their pockets with bribes and neglecting the safety of the public. The lawmakers had hijacked an issue of great public concern, stalled any action through the expedient of endless hearings, and waited for public outrage to dissipate. Fortunately for them, no spectacular electrical accidents took place during the hearings to reawaken public fear. Edison's campaign to ban alternating current was dying with a whimper.[19]

AS THE STATE SENATE conducted its bumbling investigation into electricity regulation, one clause of the electrical execution act came under attack. Many saw the "gag law" provision–which barred newspapers from printing any execution details beyond that bare fact that it had taken place–as an unconstitutional violation of press freedom. The *Electrical Review*'s editor worried that "secrecy would enable the executioners to cover up any blunders," while the *New York Press* editor expressed the darker fear that scientists "might experiment upon the condemned men and indulge in any species of brutality if they knew they were to be screened by the law." With a nearly unanimous voice, newspapers vowed to break the law. One editor recognized the situation as an opportunity: "They can arrest me if they choose, and if they arrest all of the editors who publish the reports, that will make another sensation and be another good story for us to publish."[20]

In February 1890 a state assemblyman introduced a bill to rescind the gag law and allow newspapers to print full accounts of executions. Elbridge Gerry traveled to Albany to fight the new bill, arguing that printing details of executions was "an incentive to crime rather than a deterrent." Laws forbidding publication of lewd materials such as "advertisements of bawdy houses" had withstood constitutional challenge, Gerry explained, and he saw no reason why the obscenities of violence could not be restricted as well. His view carried the day, and the gag law stood.[21]

Kemmler's sentence stood as well. Defeated in the New York Supreme Court, the prisoner met the same fate at the state's highest appellate level, the Court of Appeals. In a unanimous decision issued March 21, the court declared the electrocution law constitutional. In creating the law, New York's legislature had acted "with care and caution and unusual deliberation," the justices wrote. "It would be a strange result indeed if it could now be held that its efforts to devise a more humane method" had produced precisely the opposite result.[22]

CHAPTER 19

The Conversion of William Kemmler

ON THE MORNING of March 31—a year and two days after he murdered Tillie Ziegler with a hatchet—William Kemmler put on a brown suit, a multicolored scarf, and an imitation diamond pin. Around his right wrist was a handcuff that bound him to the left wrist of Daniel McNaughton, his keeper. Joined by warden Durston, the pair boarded a train that carried them to Buffalo. The law required that Kemmler be sentenced again in the court where he was convicted, and he arrived in time for a hearing in the early afternoon. The courtroom was filled to overflowing, and spectators stood on tiptoes in the corridors, trying to catch a glimpse of the prisoner. Standing before the judge, Kemmler seemed numb and cold. Asked if he had anything to say, he replied, "No, sir."

"Then the order of the Court is that the former sentence in your case be carried into effect within the week beginning April 28," the judge said.[1]

There was some talk of an appeal to the U.S. Supreme Court, but the previous December Bourke Cockran had vowed that if he lost in the state Court of Appeals, he would carry the case no farther. "It is generally believed that nothing further will be done," the *World* now reported. William Kemmler, it seemed, had less than a month to live.[2]

Kemmler and his escorts boarded the return train to Auburn that same day, and before midnight the prisoner was back inside his cell. He was the first occupant of that particular cell. Not long after Kemmler arrived in Auburn, the warden built two solitary cages side by side in the prison's basement to house condemned men. Since all earlier executions had taken place at county jails, this was the state's first "death row." Steel plates lined the floor, ceiling, and walls, while the front had iron bars set two inches apart. The cell was furnished with an iron bedstead, a chair, a stool, and a small stand. Just outside the door was a chair occupied at all times by one of Kemmler's two keepers, who alternated in twelve-hour shifts. Neither had worked in the prison before being hired to keep what was known as the "death watch." Kemmler slept from ten to six. He ate breakfast at eight, dinner at noon, and supper at six. He was allowed to pace back and forth in front of his cell for an hour in the morning and an hour in the evening.[3]

Kemmler usually slept well, but after returning from the Buffalo sentencing he spent a restless night. The next day he could not speak without bursting into tears. At the midnight shift change, Keeper McNaughton told his relief, Bill Wemple, to keep an eye on the prisoner. A few minutes later Wemple heard a voice from the cell: "Who are you up there?" Kemmler said. "What do you want with me?" The keeper rushed to the cell and asked what was wrong. Kemmler pointed to the ceiling and said, "I saw a man up there who said that he was Jesus Christ. He had, oh, such a good face, and he said to me that he would forgive all my sins."[4]

When he had first arrived in Auburn the previous May, Kemmler was "no more than a wild beast," according to the newspapers. He boasted about murdering "the old she-devil" Tillie Ziegler and said he would do it again a hundred times. But he had changed, through the combined effects of solitary confinement, enforced sobriety, and the influence of a few visitors. Keeper McNaughton read to Kemmler from the Bible and told him about the saving power of Christ. These efforts were assisted by the warden's wife, "a sort of Florence Nightingale in the prison," in the *Herald*'s view. As was common practice at the time, Gertrude Durston lived with her husband on the prison grounds and worked alongside him. With a dramatic escort of two huge dogs—a Saint Bernard and an English mastiff—she went on daily rounds of the workshops, the mess, and the hospital, distributing religious tracts to the prisoners.[5]

Mrs. Durston made a special project of William Kemmler, the first condemned man to come within range of her ministry. She gave him gifts—a Bible, a pictorial Bible primer, and a writing slate—and set to work on the twin pillars of American uplift: religion and literacy. She helped him write on his slate and read him Bible stories. In the hours when she was absent, Keeper McNaughton took on the same tasks. After a few months of their quiet ministrations, Kemmler became more gentle and started to take a genuine interest in religion. In his quiet hours he practiced on his slate, or looked at his pictorial Bible, or even tried to read a few verses on his own. His vision of Christ was less a bolt from the blue than the culmination of a year of religious training.

The *Herald* broke the story of Kemmler's vision of Christ on April 7, and it was quickly picked up by newspapers across the country. Mrs. Durston received dozens of letters from New York, Boston, Philadelphia, even Texas, encouraging her to continue the work of the Lord. The prisoner also began to receive visits every other day from two spiritual advisers: Reverend Horatio Yates, the prison chaplain; and Reverend Dr. Oscar A. Houghton, the pastor of Auburn Methodist Episcopal Church. On each visit the ministers discussed a Bible verse with Kemmler, then offered a prayer. "The Bible says, 'If a man shed blood, by man shall his blood be shed,'" Kemmler told Reverend Houghton. "I love Jesus and I am not afraid."[6]

Some newspapers doubted the authenticity of Kemmler's conversion. The warden barred reporters from visiting the prisoner, so all reports of his behavior were filtered through the Durstons, the clergymen, or the keepers. Some reporters discovered that his favorite possession was not the pictorial Bible but a game called "Pigs in Clover," which had become a fad that year after it was learned that President Benjamin Harrison played it at the White House. Kemmler sat for hours with the handheld toy, guiding marbles (the pigs) through a circular maze to their resting point at the center (the clover). The *Buffalo Evening News* learned that Kemmler "chews tobacco . . . and hums the common songs of the day. Does he look over his pictorial Bible? Yes, but he would just as soon look over a pictorial Boccaccio."[7]

The *World* took an even harsher stance, decrying "the on-pouring of maudlin gush over Kemmler's beatific piety, his 'experience' of religion and his 'changed' heart," not to mention "the romancing of saintly feminine influences and their regenerating effect on the murderer's dull, sodden faculties."[8]

The *World* understood that the story being told about Kemmler's conversion was standard melodrama, familiar from the countless tear-jerkers of the Victorian stage. But the very conventionality of the tale was what made it compelling. One of the goals of the new method of killing was to rob execution of ritual, to turn it into a medical procedure and destroy the old hanging ceremony of procession, prayers, confession, and drop. Kemmler may have been locked away in a basement cell and denied contact with the public, but the execution ritual reasserted itself through the medium of the newspaper. Kemmler played a role not unlike that of Jesse Strang, the man hanged before a crowd of 30,000 in Albany in 1827. Both prisoners expressed sorrow for their sins, submitted to the will of God and the state, and accepted meekly the punishment meted out. Regardless of whether it was true, the story of Kemmler's conversion reassured Americans that their system of punishment was fair and good. They were killing someone, certainly, but they were also protecting society and saving a soul.

KEMMLER'S EXECUTION had been set for the week beginning April 28, with the warden deciding the exact day. Kemmler spent a great deal of time that week scribbling on cards for the benefit of those who wanted his signature for their autograph collections. Each day he wrote "William Kemmler, Auburn, N.Y., April 1890" on dozens of plain white cards. It was a laborious process, given his still-shaky writing skills. Sometimes he confused April with Auburn and wrote "Aprilburn" or "Aubril," which forced him to tear up the card and start again. Keeper McNaughton read him stories about his case from the New York newspapers, and Kemmler complained that the newspapers' illustrations did not look a bit like him. The window of Kemmler's cell afforded him a view of the comings and goings at the prison's front gates. On more than one occasion he was heard to say, "By God, there's another reporter."[9]

None of the reporters indicated that Kemmler was anything but calm in the face of death. According to some reports, Gertrude Durston told Kemmler to "go to the chair like a man," and the prisoner said that he would. He gave Mrs. Durston a message for her husband: "Tell Charley not to put on the current so strong that it will burn me." Later the prisoner offered reassurances to the warden: "You have been kind to me and I shall try to make your work easy."[10]

The law allowed Warden Durston to invite about two dozen witnesses. He kept his list a secret, but on April 28 the men began to appear in Auburn. Alfred Southwick and George Quinby, the district attorney who had convicted Kemmler, arrived from Buffalo and proceeded to the prison. As the warden walked them toward the death chamber, he said, "Hush! Talk low!" He nodded at a wall: "Kemmler is in that room." The men heard the steady tramp of a man's boots. "That is Kemmler marching up and down in the corridor in front of his cell," the warden explained.[11]

Durston had placed the electric chair in a room immediately adjacent to Kemmler's cell. Formerly the reception room where new prisoners were bathed and issued their stripes, the death chamber still contained a bathtub and a sink. The room measured about seventeen by twenty-five feet and was dimly lit by two heavily grated windows. The ceiling was low, the floor boards rough, the walls freshly painted in a dull white. Whereas Brown's electric chair had been a reclining model, the one Dr. Fell designed was a simple oak chair with a ladder back, wide arms, and a footrest that could be extended or stowed under the seat. Newspapers reported that it looked like "an ordinary barber's chair." "There is nothing uncomfortable about the chair save the deadly current which goes with it," the *Herald* observed.[12]

The deadly current would be applied through two electrodes, one at the lower back, the other at the head. Attached to the top of the chair's back was a figure-4-shaped frame that projected out over the seat and that could be raised or lowered, depending on the prisoner's height. Dangling from this frame was a bell-shaped piece of rubber containing the sponge-covered head electrode. A similar electrode was attached to the lower back of the chair so that it pressed against the victim's spine. The chair was bolted to—and insulated from—the floor, and it bristled with leather straps: two for each arm and leg, two for the torso, and a broad mask for the face. The death chair's

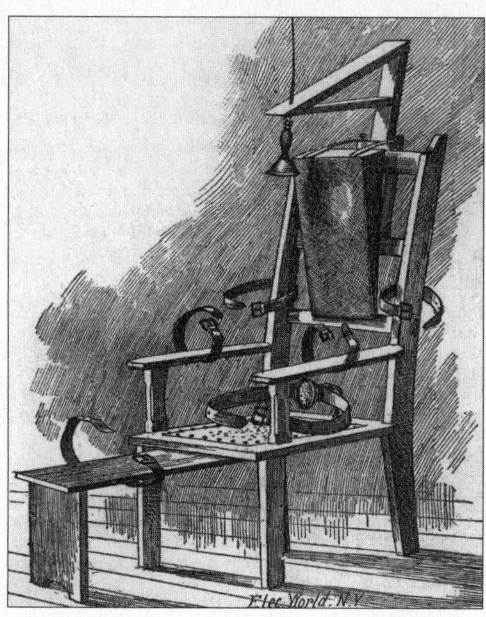

The first electric chair, designed by Dr. George Fell and built in Buffalo, featured a retractable wooden footrest.

seat was perforated with drainage holes, because death released the bladder and bowels.[13]

On Monday morning, April 28, Reverend Houghton persuaded Kemmler to be baptized in preparation for his death. The minister called for a bowl of water, and the prisoner knelt on the floor as the water was poured over his neatly combed hair. Gertrude Durston also visited Kemmler. She told him they would not meet again in this world. Normally stoic, Kemmler began to weep. He made her a gift of fifty autographed cards to distribute as she wished. Mrs. Durston told him to keep his spirits up and trust in the warden. She bid him a hasty good-bye, and within the hour she had boarded a train that took her out of town. She wanted to be far away when William Kemmler died.[14]

"ELECTROCUTION: Painless Death of Murderer Kemmler." So read the headline in the *Weekly Sentinel* of Port Arthur, Ontario. The article reported that after Kemmler was strapped into the chair, "his spiritual adviser repeated slowly the words of the Lord's prayer, the doomed man repeating it after him. When he came to the words 'For thine is the kingdom, the power and the glory,' the electrician in charge of the apparatus touched the electric button sending the charge of over 7,000 volts through the chair and its unfortunate victim."[15]

It was a dramatic account, but it did not have the virtue of being true. More accurate reports of events in Auburn on April 29 could be found in the New York and Buffalo papers, which ran headlines such as "He Still Lives" and "Kemmler's Doom Delayed by Law." About noon a lawyer named Roger M. Sherman arrived from New York and presented Warden Durston with a writ of habeas corpus delaying the execution while an appeal was made to the U.S. Supreme Court.[16]

Warden Durston brought Kemmler the good news. "Well, Kemmler, you've got a reprieve," the warden said. "You have two months and perhaps longer to live." "All right," the prisoner replied. "It makes me feel a little easier." Since the human execution was called off, Warden

Durston and about ten of the invited witnesses executed a calf in the death chamber, and Dr. Fell tried unsuccessfully to revive the animal with his Fell Motor. Harold Brown did not attend this calf killing. As he left town, he reminded reporters that his contract to supervise electrocution equipment for the state would expire on May 1, and he said he would have nothing to do with Kemmler's execution, should it ever take place. Brown said that he was "glad to be relieved of the responsibility," although he did not explain why. Perhaps, having ensured that Westinghouse dynamos would be used, he considered his true purposes to have been fulfilled.[17]

NEWSPAPERS FIXED THE BLAME for this latest legal maneuver upon George Westinghouse. On the day of the reprieve, one of the Pittsburgh tycoon's lawyers was spotted on a train near Auburn, but he claimed to be on unrelated business.

A reporter buttonholed Roger Sherman as he left Auburn by train: "Mr. Sherman, who is your client?"

"Why, Kemmler."

"Have you seen him today?"

"No, I didn't care to."

"Have you ever seen him?"

"Well, I don't care to answer that question. It is of no public interest."

"Are you in any way connected with the Westinghouse people? Have they retained you to save Kemmler?"

"Absolutely, no. I do not know any one connected with the Westinghouse Company, and I am not in their employ, directly or indirectly."

"You are not doing this for humanity's sake?"

Sherman laughed and said, "There are four men in Sing Sing Prison, three in Clinton and one in Auburn who will probably be executed by electricity." His true client, he hinted, was one of the other men. Sherman pointed out that the state Court of Appeals had not ruled on the question of cruelty and had instead simply deferred to the judgment of the legislature. Sherman said, "If Kemmler's lawyers were stupid enough to sit down supinely and accept such a decision when the issue could be settled in a higher court, and another man has the sense to come in and show them the proper thing to do, why shouldn't he do it?"[18]

Within forty-eight hours of Kemmler's reprieve, a piece of astounding news arrived on the wires from Albany: The lower house of the state legislature had voted to abolish capital punishment. Every year Representative N. M. Curtis, a principled foe of the death penalty, introduced a bill to abolish it, and every year he watched it die of mockery or neglect. This session, however, the bill passed, without debate, by an overwhelming majority of 74–30. The reason for the vote was not hard to divine. The backroom dealing was so brazen that one lawmaker told his colleagues on the floor of the Assembly, "Westinghouse money is passing this bill." Newspapers, which normally couched their corruption charges in hints and innuendo, came right out and charged the Westinghouse company with bribery. According to the *World*, "Such a bill would never have been passed except for the aid of Westinghouse's money." The bill was "purely and simply in the interest of the electric light company," the *Herald* wrote. The *Times* described passage of the bill as "probably the most disgraceful exhibition ever made of itself by a legislative body in a civilized country."[19]

When the bill was forwarded to the state senate, it died in a committee, but this outcome did little to stem the tide of invective sweeping over the Westinghouse company. George Westinghouse felt compelled to publish a denial in the New York newspapers: "Neither I nor the Westinghouse Electric Company, nor any person associated with me, has any connection, direct or indirect, with the habeas corpus proceedings instituted by Mr. Sherman in the Kemmler case or with the effort to abolish capital punishment in this State by legislative enactment. I make this denial without any reservation of any character."[20] Incredulity greeted Westinghouse's disavowals. "Poor

Mr. Westinghouse," the *Herald* wrote mockingly, claiming that the industrialist had only himself to blame. The danger of his alternating current had received "a thousandfold the advertising that it would have got had he remained quiet and never started the ball of investigation and publicity rolling."[21]*

THE WRIT OF habeas corpus acted as a stay, allowing time for Sherman to appeal to the U.S. Supreme Court. When he appeared before the Court in Washington on May 20, Sherman argued that electricity was likely to inflict a slow and lingering death. The justices of the nation's highest court, like their counterparts in New York, were unmoved. The court's opinion in *Ex parte Kemmler* provided a definition of cruelty that would become standard: "Punishments are cruel when they involve torture or a lingering death; but the punishment of death is not cruel, within the meaning of that word as used in the Constitution. It implies there something inhuman and barbarous, something more than the mere extinguishment of life." Given this standard, the Court held that electrical execution was not cruel. The New York State Legislature had acted within its proper authority in passing the electrical execution act, and the state courts had affirmed the law's validity. The Supreme Court had no reason to intervene.[22]

A few thin rays of hope remained. There were reports that Westinghouse Electric would try to repossess the execution dynamos on the grounds that Harold Brown had obtained them fraudulently, but the rumors proved false. Bourke Cockran jumped back into the game to file a new appeal with the state courts. On June 24 he argued before the Court of Appeals that the state constitution vested the power to

---

*It was later revealed that Westinghouse wrote a letter to Roger Sherman on May 7 explaining why he thought electrical execution would be unreliable. Although the letter does not explicitly state that Westinghouse had hired Sherman, many observers viewed it as further evidence that Westinghouse was financing Kemmler's appeal. The *New York Tribune* estimated that Westinghouse spent $100,000 on the case.

execute criminals with county sheriffs, and the law violated the constitution by granting that privilege to the prison warden. It was a desperation move, and the court swiftly rejected it.[23]

The long road of appeals had come to an end. When he learned that he would die during the week beginning August 4, Kemmler took the news stoically. "Well, the sooner it's over with now the better," he told Keeper McNaughton. "I'm tired of this monkeying."[24]

CHAPTER 20

The First Experiment

KEEPER McNAUGHTON entered William Kemmler's cell at five o'clock on the morning of August 6, 1890, turned a key on an iron lighting fixture, and struck a match to light the gas jet. By the flickering light, the prisoner ate a breakfast of dry toast and coffee, then combed his beard and wavy brown hair. He carefully stepped into his favorite yellow patterned trousers, then put on a white linen shirt, a dark gray coat and vest, and a black-and-white checked bow tie. The prisoner dressed, one reporter noted, "as carefully as if he were going to a ball."[1]

Kemmler collected his worldly possessions—Bible, pictorial Bible primer, writing slate, Pigs in Clover puzzle—and placed them on a small stand, along with his last will and testament. He left the items to his friends in the prison: The Bible primer would go to Mrs. Durston, the Bible to Keeper McNaughton, the slate to Chaplain Yates, the puzzle to Reverend Houghton. Just before six the chaplain, the minister, and the keeper crowded into his cell. They were joined by Joseph Velling, a deputy sheriff from Buffalo who had befriended Kemmler in the process of guarding him before and during his trial the previous year. Velling produced a pair of clippers, and the prisoner reluctantly allowed the hair at the crown of his head to be shorn very close. Then Velling split the seam on Kemmler's trousers just below the waistband. When these physical preparations were completed, the four men turned to the spiritual. They got down on their knees and began to pray.

AT THE GATES of the prison a crowd of 1,000 people gathered under a cloudless morning sky. Although the date of the execution was kept secret, all of the witnesses had arrived in town the day before, leading to speculation that Kemmler's execution was imminent. The platform of the New York Central Railroad Station across the street from the prison was filled to capacity. A few intrepid souls climbed trees, and the *World* reporter had constructed a special platform twenty feet up a telegraph pole. But these lofty perches only afforded a better vantage on the prison's fanciful turrets and ivy-covered walls, for the death chair was hidden in a basement cell. In a dimly lit freight room next to the train station, dozens of telegraph operators—sent up

Auburn prison in 1890. Kemmler's cell and the death chamber are in the basement to the left of the entrance. The woman in the foreground may be the warden's wife, Gertrude Durston, accompanied by her English mastiff.

from New York City by Western Union just for the occasion—sat at makeshift tables, ready to relay the news of William Kemmler's historic death.

Starting a little before six, well-dressed men in groups of two or three began to emerge from the Osborne House, the local hotel, and walked briskly toward the prison. The men forced passage through the throng at the gate and presented themselves to the guards, who escorted them to the warden's parlor. Durston was pacing the room, angry that the witnesses were late in arriving. He needed the execution to be over by seven, so that the steam engines being used to power the dynamos could begin their usual work of running machinery in the prison's factories. Finally, at twenty past six, the last witnesses appeared, and the warden led them down an iron spiral staircase to the prison basement and into the death chamber.

When the machinery was first installed in April, the electric chair and the switchboard had been in the same room, but at the beginning of August the warden decided to move the chair to an adjacent room, while leaving the switchboard in its original place. His purpose, apparently, was to shield the identity of the man who threw the switch. Formerly the keepers' messroom, the new death chamber was about eighteen by twenty-five feet. Two iron-grated windows, partially covered by Virginia creeper ivy, looked out toward the crowds at the main gate. The death chair sat at the center of one end of the room.[2]

Wooden chairs were arrayed in a half-circle around the electric chair, and the witnesses took their seats. Despite the provision of the law prohibiting newspapers from publishing details of executions, the warden had invited two wire service reporters, one from the Associated Press and another from the United Press. Alfred Southwick—now referred to as the "father of the electrical execution law"—was in the room, as was Dr. Fell. A few men were notable by their absence. Harold Brown had not been invited, and Elbridge Gerry turned down his invitation in favor of a cruise with the New York Yacht Club. Edison did not attend, nor did his chief electrician, Alfred Kennelly, who a year earlier had expressed a wish to do so.[3]

Of the twenty-five official witnesses, fourteen were physicians. They included the editor of the *Medical Record*; the head of the state's Board of Health; a deputy coroner from Manhattan (invited because he had autopsied many men killed in electrical accidents); and the coroner who autopsied Lemuel Smith after his 1881 death in the Buffalo arc lighting plant. Dr. Carlos MacDonald, who had been the state's electrocution expert for nearly a year, served as one of the official execution physicians. The other was Dr. Edward C. Spitzka, the current president of the American Neurological Association.[4]

When everyone was seated in the room, Durston called aside his two official physicians. Astonishingly, the warden had not yet decided how long the prisoner should be subjected to the current, so he asked the doctors for advice.

"Fifteen seconds," Spitzka told him.

"That's a long time," said the warden, who feared burning Kemmler.

"Well, say ten seconds at least," MacDonald offered.

It was agreed that Spitzka would decide when to turn on the current and when to turn it off. He asked if anyone had a stopwatch, and MacDonald pulled one from his coat.[5]

Warden Durston abruptly left the death chamber and walked down the hall to Kemmler's cell. He greeted the prisoner and the ministers, then read the death warrant. "All right, I am ready," Kemmler said. He bid good-bye to Keeper McNaughton, who declined to be present in the death chamber. The prisoner fell into step behind the warden, the ministers followed the prisoner, and Sheriff Velling brought up the rear. The execution procession was brief, requiring just a few steps down a hallway and into the death chamber.

The witnesses had been whispering nervously among themselves, but they fell silent as the warden and the prisoner entered the room. Kemmler walked toward the death chair, then paused, uncertain as to whether he should sit in it. He peered at the warden's face, like an actor uncertain of his cue.

"Will some gentleman give me a chair?" the warden asked. A witness pushed a common kitchen chair into the circle, and the warden placed it in front and a little to the right of the death chair, facing the witnesses. He pulled another chair next to it. He and Kemmler sat side by side facing the witnesses, the warden's arm over the prisoner's shoulder.

"Now, gentlemen, this is William Kemmler." The condemned man bowed slightly and looked around the arc of faces as if expecting greetings. No one spoke. The warden's hands and voice trembled. "I have just read the death warrant to him and have told him he has got to die." He turned toward Kemmler: "Have you anything to say?"

Kemmler's face brightened. He started to stand, then decided to remain seated. His feet were set wide on the stone floor, a hand on either knee, elbows akimbo. "Well, I wish everybody good luck in the world," he said in a deliberate voice. "I believe I am going to a good place."

"Amen," said the ministers.

At a nod from the warden, Kemmler stood. "Take off your coat, William," Durston said. Kemmler slipped off the coat and folded it neatly over his chair. The witnesses could see the slit that Velling had cut below the waistband of his trousers. The warden bent down and began drawing the tail of Kemmler's shirt through the hole and cutting it off with scissors, dropping the scraps to the floor. When he was finished, a patch of skin at the base of the prisoner's spine was exposed to the warm, damp air. Durston motioned to the prisoner. Kemmler turned and lowered himself into the electric chair.

ON THE SECOND FLOOR of the east wing of the prison—more than 1,000 feet from the death chamber—was the Westinghouse dynamo, which was under the care of a Rochester electrician named Charles Barnes. Attached to the dynamo were wires made by the Edison Electric Company. The wires, insulated in rubber and affixed to glass and porcelain insulators, ran out the window, over the roof, around the prison's ornamental dome, down the front wall, through a basement window, and into the switchboard room, which was under the direction of an electrician named Edwin F. Davis. The switchboard held two voltmeters, an ammeter (for measuring amperage), a bank of twenty incandescent lamps, and two jaw switches—metal bars eighteen inches long that swung in an arc of 180 degrees, from open to closed. The first switch allowed current to flow to the lamps, which were used, along with the meters, to gauge the strength and steadiness of the current. The second switch sent the current through wires that led to the adjoining room, where the electric chair was located.

Kemmler sat in the chair in a natural, easy posture. The warden had decided not to use the footrest, so Kemmler's feet rested on the floor. He lifted his arms high to allow the chest straps to be wrapped around him. The warden's hands trembled so much as he started to fasten the prisoner's arms that he could hardly thread the straps through the buckles. "It won't hurt you, Bill. It won't hurt you at all," said the warden, perhaps offering reassurance more to himself than to the prisoner, who did not appear to need it.

"Don't get excited, Joe," Kemmler said when Velling began to fumble with the straps. "I want you to make a good job of this."

When the arm, leg, and body straps were cinched tight, Kemmler was completely immobilized. Durston pushed the rubber cup of the lower electrode through the hole in the back of Kemmler's trousers, and the spring mechanism held it tight against his spine. Durston slid the other electrode down against the ragged tonsure on Kemmler's head.

The prisoner moved his head from side to side, to show that it was not snug. "I guess you'd better make that a little tighter, Mr. Durston," he said, and the warden granted the request.

Durston affixed the leather mask, which pulled Kemmler's head hard and tight against a rubber-covered cushion on the chair's back. It covered the prisoner's chin, forehead, and eyes and smashed down his nose, but it left his mouth exposed.

Dr. Fell stepped forward with a syringe and soaked the sponges with a saltwater solution to lower resistance and prevent burning. Dr. Spitzka said, "God bless you, Kemmler," then nodded to the warden.

The Kemmler execution, as pictured in the *New York Herald*. The switchboard room was behind the door.

Durston edged over toward the door leading to the switchboard room.

"Good-bye, William," he said.

"Good-bye," came the muffled response from the chair.

DURSTON RAPPED TWICE on the door, a prearranged signal. The dynamo had been humming smoothly for several minutes, sending more than 1,000 volts of electricity through the switchboard. At Durston's signal, someone in the switchboard room—his identity was never revealed, but it might have been one of Kemmler's fellow convicts—closed the switch, diverting the current into the electric chair.

Kemmler gave a quick, convulsive start. His mouth twisted into a ghastly grin. Every muscle in his body contracted, straining against the leather straps. His right index finger doubled under with such strength that the nail cut into the palm and blood trickled out onto the arm of the chair.

Dr. Spitzka tiptoed to the chair and stared intently at the face of the bound figure. After seventeen seconds, Spitzka cried, "That will do! Turn off the current. He is dead." Another voice echoed, "Oh, he's dead."

The warden rapped on the door to the switchboard room, and the current to the death chair was cut off. At the switchboard, Edwin Davis pressed a button that rang a bell in the dynamo room, and the dynamos were shut off. Kemmler's muscles relaxed, and he slumped against the straps. The head electrode was removed.

"Observe the lividity about the base of the nose," Dr. Spitzka said. "Note where the mask rests on the nose–the white appearance there." The other doctors gathered round and pressed their fingers against Kemmler's face, noting the play of white and red when the fingers were removed. The hue of Kemmler's skin, Spitzka said, was "unmistakable evidence of death."

Electrical execution had been quick, clean, and painless, just as its advocates had argued. "This is the culmination of ten years work and study," Southwick proclaimed. "We live in a higher civilization today."

HE SPOKE TOO SOON. A cut on Kemmler's hand was dripping rhythmic pulses of blood. One of the witnesses shouted, "Great God! He is alive!" Another said, "See, he breathes!"

And he did. Kemmler's chest heaved, and from his mouth came a rasping sound, growing quicker and harsher with each suck of breath. A purplish foam from his lips splattered onto the leather mask. Saliva dripped from his mouth and ran in three streams down his beard and onto his gray vest. The chest straps squeaked as he struggled for breath, and he groaned, an animal cry that witnesses found impossible to describe. His whole body shook and shivered.

"Turn on the current! Turn on the current!" Someone shoved the electrode back down against Kemmler's skull. The warden signaled to the switchboard room, and the switchboard room signaled to the dynamo room, where the operators struggled to restart the Westinghouse machine. More than two minutes after the electricity had been cut off, the witnesses heard the thunk of the switch from the next room, and the current flowed.

Once again Kemmler's muscles contracted, his body rising up, rigid as a statue. This time there would be no mistake. The current stayed on for between one and two minutes–in the confusion, no one remembered to keep time.

The sponge of the back electrode dried out and burned away, allowing the bare metal disk to press sizzling against Kemmler's skin. At the head electrode his hair began to singe. The stench of burning hair and flesh filled the room.[6]

One witness turned aside and vomited. The United Press reporter fainted, and another witness propped him on a bench and fanned him with a newspaper. District Attorney Quinby rushed from the room in horror. Some turned away and hid their faces in their hands; others were so repulsed that they could not avert their eyes from the spectacle.

"Cut off the current," Spitzka shouted, and once again the wires fell dead. The smell of urine and feces mingled with the acrid smoke in the air. Dr. Fell doused a small fire that had started on Kemmler's coat near the back electrode. Another doctor held a bright light to Kemmler's eyes, and the optic nerve showed no response.

This time, Kemmler was dead. Fell made a small incision at the temple and drew off a blood sample for later testing. The warden loosened both electrodes and unbuckled the mask and the straps. There were livid blue marks where the mask had pressed Kemmler's face, and purple spots began to mottle his arms and neck. The witnesses filed out of the room, leaving Kemmler slumped in the chair. One of the witnesses–the sheriff of Erie County–was crying as he walked through the throng of reporters and townspeople at the gates of the prison.[7]

CHAPTER 21

After Kemmler

NEWSPAPERS ignored the ban on printing details of electrocutions. In an article headlined "Far Worse Than Hanging," the *Times* asserted that "no convicted murderer of modern times has been made to suffer as Kemmler suffered." Recalling the electric wire panic of the previous fall, the *Herald* noted that Kemmler suffered the same fate as John Feeks, "the lineman who was slowly roasted to death in the sight of thousands." In London the *Standard* described the execution as "a disgrace to civilization." "Kemmler's Death was Disgusting," the *Buffalo Express* stated plainly.[1]

The suspicions that electricity might stun rather than kill had never entirely been laid to rest. The warden had forbidden Dr. Fell from trying to resuscitate Kemmler with his Fell Motor; to make sure Kemmler would not awaken spontaneously, the doctors chose to wait three hours after the execution before performing the autopsy. As Southwick explained, they wanted to explode the notion "that if the electric shock did not kill him the surgeon's knives would."[2]

The postmortem revealed that a large part of the brain had been "carbonized"—burned to a crisp black—while the skin of the lower back exhibited a burn four inches in diameter. One doctor reported

The only known photograph of the first electric chair. Note the brackets beneath the seat that held the footrest, which was never used.
The chair was replaced by a new design in 1893.

that the spinal muscles under the burn were "cooked, like 'overdone beef,' throughout their entire thickness." When the autopsy was concluded, the physicians took samples of blood, brain, and spinal cord to study at leisure in their own laboratories. Kemmler's corpse was put inside a pine coffin and driven at midnight to the convict burying ground that adjoined Auburn's Fort Hill Cemetery. There the coffin was reopened and, in accordance with the law, a barrel of quicklime poured over the body. After nailing the coffin shut again, the sextons lowered it into the ground and covered it with earth.[3]

Bourke Cockran described Kemmler's death as "a sort of ghastly triumph" for him, because it seemed to confirm his arguments against the method. Some believed that Cockran or his Westinghouse employers had sabotaged Kemmler's execution. "Yes, there might have been corrupt reasons for this," Dr. Spitzka said. "The interests of the company who manufactured the dynamos would certainly be advanced . . . if this execution was a botch." Warden Durston also suspected "crooked work," and the *Herald* proposed "either that the dynamos were faulty or that the interested company had bribed some one to make them seem so."[4]

No one ever proved sabotage. The charges simply distracted attention from the more likely cause of the problems: shoddy preparations. Durston's decision to put the chair and the switchboard in different rooms meant that the men who decided when to turn the current off and on had no idea how strong the current was running. Moreover, the switchboard and the dynamo were located more than 1,000 feet apart, and the switchboard room could communicate with the dynamo room only through a crude electric bell signal system. The men in the dynamo room had no outgoing communication system at all, so they could not inform those in the death chamber about problems with the machinery.

And there had been problems. The dynamo rested on a wooden floor that vibrated up and down more than an inch when the dynamo ran at full power. The leather belts linking steam engine to dynamo were brand-new and had not been used enough to get the stretch out. When Kemmler was thrown into the circuit, the resistance of his body put such a strain on the dynamo that the belts began to slip badly. Charles Barnes, who supervised the dynamo operation, later described the chaotic scene as he and three convict assistants tried to keep the dynamo running. One convict listened for the switchboard room's bell signals, another was "busily oiling the dynamos and putting rosin on the belt to try to stop the slipping, while the third was busy holding a board against the pulley to keep the belt on." Barnes estimated that during the execution the current was running at 700 volts–about half what the physicians thought they were using.[5]

WITNESSES DIFFERED WIDELY in their assessments of the execution. Alfred Southwick flatly denied the reported horrors of the death chamber: "A party of ladies could have been in that room and not known what was going on, so silent was the process—not a cry from the subject, not a sound." Other witnesses claimed that it had been a gruesome spectacle, but the question of whether Kemmler had suffered remained in dispute. "I will see that bound figure and hear those sounds until my dying day," said the electrician Charles Huntley, who came to the reasonable conclusion that Kemmler's moaning was evidence of great pain. However, most of the physicians present believed that Kemmler had suffered no pain whatsoever, because he was knocked unconscious at the start of the first shock. They did not explain how they came to this conclusion.[6]

Some were comforted by this claim and saw in it hope for the future of electrocution. "The failure was due not to the system but to the bungling, inefficient way in which the execution was managed," the *Herald* claimed. In the view of Thomas Edison, the electric chair was like any other new device, requiring a few trials to work out the bugs. The next electrocution, Edison predicted, "will be accomplished instantly and without the scene at Auburn today."[7]

George Westinghouse disagreed. "It has been a brutal affair," he told a reporter. "They could have done better with an axe."[8]

In the weeks following the execution, more people agreed with Westinghouse than with Edison. Many believed that no one else would ever die in an electric chair. Dr. Spitzka, despite his assertions that Kemmler died quickly, nonetheless predicted, "There will never be another electrocution." According to the *World*, "The first experiment in electric execution should be the last." Newspapers all across the country—the *Philadelphia Times*, *Terre Haute Express*, *Indianapolis Standard*, and *Boston Globe*—agreed with this judgment. The *Sun* urged the state legislature to repeal the law, then acidly added, "Civilization will find other lines on which to manifest its progress."[9]

In the official execution report, filed in October, Dr. MacDonald described electrocution as "the most efficient and least painful method that has yet been devised." He proposed boosting the execution voltage to 2,000 volts and placing the voltmeter in the same room as the chair. MacDonald also urged the state to build a special dynamo for the purpose in order to avoid doing "injustice to any electrical lighting company."[10]

Also in the fall of 1890, other physicians present at Kemmler's death published their views on the matter in medical journals. According to Dr. Spitzka, Kemmler was "dead, in the usual sense of the word, after the first passage of the current." The blood flowed from Kemmler's cut not because he had a pulse but because the electricity broke down its structure and inhibited normal postmortem coagulation; the spattering mucus and saliva were caused by postmortem muscle contractions; and the apparent breathing was simply a release of air that had been trapped in Kemmler's lungs by current-induced muscle contractions. Spitzka's claim about the breathing contradicted the reports of every other witness, who reported that Kemmler had inhaled as well as exhaled. Dr. Fell offered an ingenious explanation that played on the uncertainties concerning the definition of death. In Kemmler's execution, Dr. Fell said, "effective respiration survived the final heart arrest." In other words, Kemmler was dead despite the fact that he was breathing. These opinions, however dubious, were published by prominent physicians in respected medical journals. As the horror of Kemmler's death faded, many newspapers tempered their criticism and began to suggest that electrocution deserved another chance.[11]

HAVING FAILED IN HIS BID to save Kemmler, Roger Sherman reappeared as the attorney for Shibuya Jugiro, a Japanese immigrant convicted of murder and sentenced to death. In November 1890 Sherman presented Jugiro's case before the U.S. Supreme Court, offering Kemmler's death as a "practical illustration" of the cruelty of electrocution. Without issuing an opinion, the Supreme Court rejected Jugiro's appeal on the authority of its decision in *Ex parte Kemmler*.[12]

During the oral arguments in the Jugiro case, one of the Supreme Court justices provided a clue to his reasoning. When Roger Sherman claimed that electricity could not be counted on to destroy life, the justice pointed out that "in New-York City persons have been killed in a short time by accidental contact with electric wires."[13]

The Supreme Court's decision ended the hopes of the men awaiting death in New York penitentiaries. Sing Sing's warden decided that all four murderers on his death row–Jugiro, Harris Smiler, James Slocum, and Joseph Wood–would die on the morning of July 7, 1891. When tests showed that the Sing Sing ammeter was faulty, it was sent to Arthur Kennelly at the Edison laboratory, who recalibrated the device and sent it back. Dr. MacDonald also made some changes in the apparatus. Rather than electrodes at the crown of the head and base of the spine, as in Kemmler's execution, the Sing Sing chair featured a larger head electrode that covered the forehead and temples and a second electrode to be affixed to the calf.

The details of how these new arrangements worked were kept secret, because the warden prohibited witnesses from speaking to the press after the executions. "I cannot give you any minute particulars," Dr. A. D. Rockwell told reporters upon emerging from the prison, "but they were all highly successful." Although newspapers grumbled about the gag law, most accepted the official line that the executions went off flawlessly. "A proud day for the Empire State!" the *Herald* crowed.[14]

A more complete picture emerged months later, in Dr. MacDonald's official report. All of the prisoners received shocks ranging from 1,400 to 1,700 volts. James Slocum, the first in the chair, survived an initial shock of twenty-seven seconds, with a strong pulse and "noisy respiration." A second shock of twenty-six seconds was required to kill him. MacDonald and the other doctors present conferred and decided that several shorter shocks, rather than one long one, might kill more quickly. Harris Smiler, the next victim, received three contacts of ten seconds each, with a pause between shocks just long enough to rewet the sponges. Even after these shocks, however, "the pulse was beating so firmly and regularly" that a further contact of nineteen seconds was required to kill him. The doctors consulted again, deciding that the first two experiments had shown that "the duration of the current was quite as important an item as the making and breaking of the contact." In the next execution, therefore, Joseph Wood took three contacts of twenty seconds each. Afterward doctors detected a faint heartbeat, but it disappeared within twenty seconds. The final prisoner, Shibuya Jugiro, received three contacts of fifteen seconds, after which "a very slight fluttering was felt at the wrist." This pulse disappeared within fifteen minutes, and Jugiro was pronounced dead.[15]

MacDonald's report highlighted the experimental nature of electrocution. The press gag law gave the physicians and prison officials the secrecy they required to improve the killing process free from the threat of public criticism.[16]

ALTHOUGH THE LAST of these four victims died rather quickly, his body nonetheless had been scorched and burned. Hoping to prevent this unsightly effect, electrocution's boosters again sought advice from Thomas Edison.[17]

In his testimony at the Kemmler hearings, the inventor had advised running the current from hand to hand and using jars of water as electrodes to prevent burning. In the wake of the four Sing Sing executions, the idea was revived. At Alfred Southwick's urging, Edison sent Arthur Kennelly back to the laboratory. In September 1891 Kennelly filled two jars with salt water and placed a zinc electrode in the bottom of each. He then took a two-foot-long piece of raw beef, placed one end in each jar, and sent 780 volts through it. Although the meat caught fire at a spot between the jars, Kennelly emphasized another observation: "The beef was not scarred or altered in appearance over the part that had been immersed in water."[18]

Kennelly warned Southwick in a letter that the skin of the prisoner's hands would offer more resistance than raw beef, but he nonetheless concluded that it would be "almost impossible to burn the skin" if liquid electrodes were used. Southwick replied, "Allow me to thank you personally and also Mr. Edison through you." The new electrode arrangement, Southwick wrote, "will fully settle the question of electrocution. It will remove all the unsightly part of an execution (scalding)."[19]

Southwick thought the new method would be used in the execution of convicted murderer Martin Loppy, who exhausted his appeals in October and was scheduled to die in December. For unknown reasons, however, the authorities chose not to adopt the liquid electrode plan. When Loppy was strapped into the Sing Sing chair on December 7, the electrodes were affixed to his forehead and calf, as they had been for the last four victims. As at the last executions, the warden swore all witnesses to secrecy, but this time a few broke the silence and revealed that Loppy's execution had been horribly bungled.[20]

Loppy's death created another sensation in the press and provoked calls for a repeal of the electrocution law. At the very least, the editors wrote, the press needed access to the death chamber, because only with clear information could the public judge the method on its merits. Press opposition to the gag law had intensified the previous summer, when Manhattan's district attorney indicted the *Herald* for violating the law by printing details of the four executions at Sing Sing. The other newspapers charged that the district attorney was playing favorites; they wanted indictments, too, because self-righteous defense of the First Amendment made good copy. The district attorney obliged by indicting all but one of New York's morning papers. As it turned out, the indictments were moot. Governor Hill—the man who signed the electrical execution act into law—left office at the end of 1891, and the new governor came out in opposition to the gag law. An obliging legislature repealed it in early February.[21]

The press thus gained access to the death chamber just in time for the execution of Charles McElvaine, which was set for the morning of February 8, 1892. Given the outcry at Loppy's fate, state authorities decided to try Edison's electrode arrangement. Arthur Kennelly was on hand as a witness. Before the prisoner entered the room, Dr. MacDonald explained the motive behind changing the position of the electrodes: "Eminent electricians—Mr. Edison in particular—have publicly

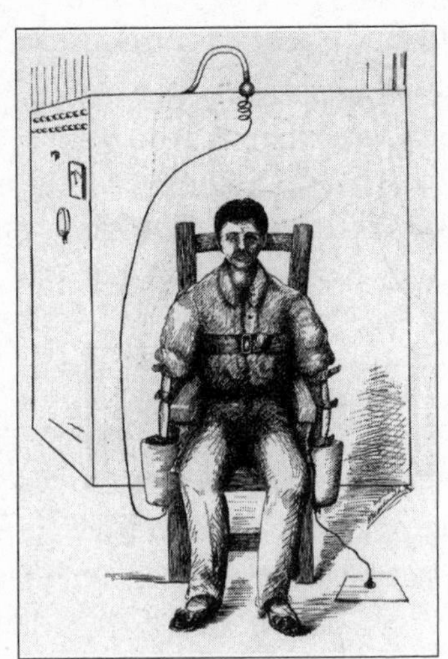

Upon Edison's suggestion, New York officials attempted to use liquid hand electrodes in the 1892 execution of Charles McElvaine at Sing Sing. This illustration from the *New York Medical Journal* is inaccurate: As a backup, the executioners also attached electrodes to McElvaine's head and calf and used these to kill McElvaine after Edison's hand-to-hand arrangement failed.

expressed their opinion that contact should be made through the hands." McElvaine entered the room and sat in the newly designed chair, which had a wooden basin of salt water suspended from the end of each arm. McElvaine's arms and wrists were strapped down such that his hands were fully immersed in the water. MacDonald hedged his bets by attaching a backup pair of electrodes in the usual spots on the head and calf.[22]

The preparations complete, the anxious prisoner shouted "Let her go!" MacDonald obliged by signaling to the switchboard. After about fifty seconds at 1,600 volts, MacDonald cut off the current, but McElvaine began to sputter and moan. "Turn it on!" MacDonald yelled. "The other way." The electrician threw a switch that sent the current into the leg and head electrodes. This time the current flowed for forty-three seconds. McElvaine was dead, his corpse a frightful sight, covered in burns and blisters.[23]

The criticism was directed at the new electrode arrangement. "Mr. Edison was entirely mistaken in recommending that the current be applied through the hands," the *Times* wrote.[24]

Electrocutions continued with the old electrode arrangement. In May 1892 authorities executed a man at Auburn prison—the first to die there since Kemmler—and reports indicated that this one went more smoothly. There were ten electrocutions in 1893, two in 1894, six in 1895, five in 1896, eight in 1897. By the turn of the century, more than fifty men had died in the chair. The killings started to become routine, rating only short articles in the daily newspapers.[25]

Not all of the later executions went smoothly. During his 1893 execution in Auburn, William Taylor's legs contracted so strongly that the front leg of the chair was torn free, sending him pitching forward.* Officials tried to turn the current on again but discovered that the dynamo had burned out. Taylor revived and began to groan, so the

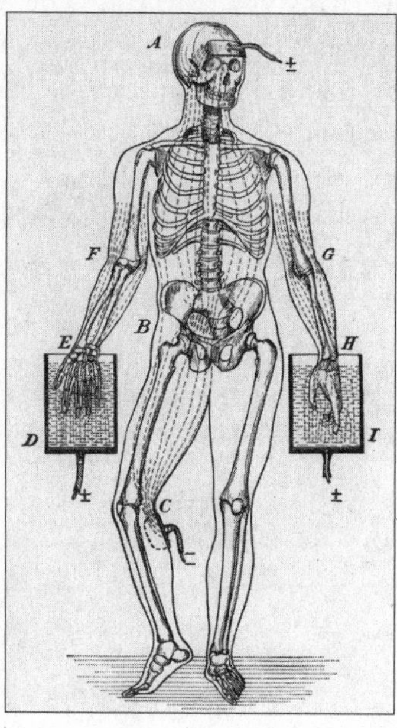

An illustration from the *New York Medical Journal* showing the flow of electricity through the human body in both the head-to-calf and hand-to-hand electrode arrangements.

*By the time of Taylor's execution, the original chair had been replaced by one with only three legs—two in back, one in front.

physicians gave him morphine and chloroform to ease his suffering while carpenters repaired the chair and electricians tapped into the city's electrical supply for more current. An hour later, Taylor's unconscious body was carried back to the chair for a second jolt. Some witnesses claimed that the drugs had killed him before the second shock was administered.[26]

Taylor's case was extreme, but botched electrocutions were not infrequent. As a result, there were few outright enthusiasts of electrocution, but alternatives were few. Many physicians proposed different methods of scientific killing, usually involving poison—an injection of morphine, chloroform, a sealed chamber filled with carbon monoxide or illuminating gas—but no one carried through on these ideas. A New York State assemblyman suggested a return to hanging, but his bill went nowhere, probably because bungled hangings remained common in states that retained the gallows. The *Herald* expressed the general mood of resignation: "Electricity seems, on the whole, to answer the purpose better than anything else." The *Journal of the American Medical Association* observed, "All public clamor against the method may be said to have been effectively stilled, for the present at least. Now it remains to be seen if other States will adopt the measure."[27]

CHAPTER 22

The End of the Battle of the Currents

IN THE WEEKS before and after Kemmler's execution, Thomas Edison and Arthur Kennelly were busily engaged in new experiments on alternating current. They were spurred to action by an angry letter from Samuel Insull, the vice president of Edison General Electric. At a meeting of Edison utility companies in the summer of 1889, managers from many cities had complained that they were being crushed by competition from Westinghouse. Insull had promised the local companies that within six months Edison General would have its own alternating system available. Because of Thomas Edison's intransigence, that promise was not kept. As another meeting of the local Edison companies approached in the summer of 1890, Insull reminded Edison of the earlier promise. "A year has now expired since that meeting was held, and we are no further ahead to-day than we were at that time." Insull had instructed Edison General's engineers to begin work on alternating generators, and he now urged Edison to get started on a transformer. "It is of the utmost possible importance that we should be able to go ahead on alternating apparatus," Insull told Edison. "I think that it is particularly important that in our new organization any promises I make to our people should be absolutely adhered to, and I shall be glad if you will bear this in mind when you reply to this letter."[1]

Although he was under contract to develop electrical equipment for Edison General Electric, Thomas Edison was not used to receiving such peremptory marching orders. He nonetheless complied. He and Kennelly worked on a transformer in August, and during the fall they began designing a generator as well. Before long they had a working alternating system.[2]

In the fall of 1889, during the frenzied months of the electric wire panic, Edison's goal of having alternating current banned had looked surprisingly attainable, but one year later all of his lobbying efforts had ended in failure. Although he still had doubts about the safety and efficiency of the technology—as late as 1891 he would say that "the use of the alternating current is unworthy of practical men"—he conceded that his firm needed an alternating system. He also tempered his rhetoric. "The death rate [from high-voltage alternating current] will be reduced to a minimum, or rather there will be none at all," the inventor told a reporter in November 1890. "The electric wires—I mean the small, cheap, high tension wires—ought to come under strict inspection laws." A year after having advocated a complete ban, Edison claimed that the safety of alternating current could be assured simply through proper regulation.[3]

Although Edison's statements indicated that the battle of the currents was inching toward a conclusion, competition among the leading electric light firms remained fierce. Edison, Westinghouse, and Thomson-Houston were fighting city-by-city skirmishes for lighting and electric streetcar contracts, and patent litigation grew feverish. Competition took a heavy toll on the firms, which struggled to remain profitable. The situation turned darker in the fall of 1890, when the failure of a major London brokerage house roiled world money markets. Westinghouse was forced to reorganize and find new investors, and Edison General struggled to remain solvent.[4]

The crisis confirmed Edison General director Henry Villard's view that competition—with its expensive takeovers, patent litigation, and price slashing—was killing the electrical industry. Villard and Charles Coffin, the head of Thomson-Houston, had discussed a merger as early as 1889, but Thomas Edison refused to go along. "If you make the coalition, my usefulness as an inventor is gone," he told Villard. "My service wouldn't be worth a penny. I can only invent under powerful incentive. No competition means no invention." Edison knew how much money his business demanded, but he clung defiantly to an earlier model of entrepreneurial culture, one based on pride in craft, in which the inventor's identity and the company's identity were one and the same. Edison managed to hold off Villard's merger plans in 1889 and again early in 1890.[5]

After the financial crisis of late 1890, Villard again explored consolidation with Thomson-Houston. Again, the talks went nowhere, but this time the major obstacle was not Thomas Edison's pride but the lamp patent infringement suit against Westinghouse, which in July 1891 was decided in Edison's favor. The presiding judge ruled that Edison had invented the first commercially viable incandescent lamp by placing a high-resistance carbon filament in a one-piece glass globe exhausted of air. The decision, which gave Edison interests a monopoly over a central element of all electrical systems, convinced them they could outcompete their rivals and therefore did not need to merge. That confidence proved misplaced. Both Thomson-Houston and Westinghouse found ways to skirt Edison's lamp patent—for instance, by using a bulb with a stopper rather than Edison's one-piece globe—and the legal decision had little effect on competition between the three major firms.[6]

The failure to translate legal victory into industrial dominance convinced Edison General's investors that they could no longer go it alone. The company had potent resources—major factories, a national marketing structure, the Edison name, the Edison laboratory—but Thomson-Houston had built a more efficient organization. In 1891 Thomson-Houston and Edison General had roughly comparable sales figures—$10.3 million and $10.9 million, respectively—but Edison General earned profits of only $1.4 million, whereas its rival cleared nearly twice as much. Thomson-Houston managers saw benefits in consolidation with Edison General: It would create a diverse product line, boost profits by quelling price competition, and attract new investors. In February 1892 a committee of six financiers—including Edison investor J. P. Morgan and Henry Higginson, Thomson-Houston's primary broker—hammered out the details of a merger and formed a new company. Edison's named was dropped, and the new firm was known as General Electric. Charles Coffin became its president, and four of his five top officers were from Thomson-Houston. As the directors saw it, the Thomson-Houston men—and not their Edison General counterparts—were best equipped to manage the company efficiently.[7]

Thomas Edison owned about 14 percent of the new company's stock, and he professed to be thrilled with the new arrangement. As rivals, the inventor said, Edison General and Thomson-Houston had cut prices so low that they were selling equipment at a loss and losing their shirts. Asked about rumors that he opposed the merger, Edison only laughed. "I haven't been either frozen out, euchred, or turned down," he said. "My stock is worth more now than it was before."[8]

But there was a great deal that he did not say. The formation of General Electric marked the Edison company's full surrender in the battle of the currents, because it lost its identity as a company primarily committed to direct current. According to Alfred Tate, Edison's personal secretary, the inventor had been unaware of these merger plans until the day they were publicly announced. Samuel Insull acknowledged in his memoir that "Mr. Edison was not in real sympathy with the consolidation," and that the merger led to a falling-out between the two men. Reports at the time claimed that Edison blamed Insull for the turn of events, believing that Insull's mismanagement of Edison General made the merger necessary. If he was upset with his own managers, he was outraged by the men chosen to lead General Electric. Unlike his earlier enterprises, it would not be run by his close friends, the craftsmen and machinists who had worked with him for years. Edison's inventions—a decade's worth of sleepless nights—fell under the control of the professional managers from Thomson-Houston, the very men he had accused a few years earlier of "having boldly appropriated and infringed every patent we use."[9]

Some reports held that the new directors summarily dropped the "Edison" from the new company's name, but according to Insull it was the inventor himself who insisted that General Electric do without his name. Edison's creation had outgrown him. He was trapped between a desire for personal control of a company and the need for the capital required to spread his inventions around the globe. General Electric would be an enormous company run by professional managers, pooling the resources of thousands of investors, employing the patents of dozens of inventors. Since it was no longer his company, it should no longer bear his name.

Edison's emotions were best captured in a memorandum he wrote to Henry Villard regarding an earlier merger proposal. "You will see how impossible it is for me to spur on my mind under the shadow of probably future affiliations with competitors, to be entered into for financial reasons," Edison wrote, then continued in a mournful vein: "I think up to date I have performed every duty asked of me, made every concession, and I would now ask you not to oppose my gradual retirement from the lighting business, which will enable me to enter fresh and more congenial fields of work."[10]

Alfred Tate reported that only once did Edison reveal his bitterness about the merger. "I've come to the conclusion that I never did know anything about [electricity]," Edison told Tate. "I'm going to do something now so different and so much bigger than anything I've ever done before people will forget that my name ever was connected with anything electrical."[11]

Such forgetting was unlikely, but Edison did turn to other ventures.

In 1888 he had prepared a patent application for a "kinetoscope," which he described as "an instrument which does for the Eye what the phonograph does for the Ear, which is the recording and reproduction of things in motion." In 1892 Edison and his assistant William Dickson created prototypes for the kinetograph (a camera) and kinetoscope (a peephole motion picture viewer). In 1894 "Edison Kinetoscopic Record of a Sneeze" became the first motion picture to receive copyright protection, and in the following years Edison emerged as the pioneer of the American motion picture business.[12]

Another project proved less successful. Forsaking the delicate consumer goods that had made his reputation and his fortune—the phonograph, the lightbulb—Edison began to devote his time and his passion to the crude work of hard rock mining. Newspaper men dutifully trekked to Edison's sprawling mine complex in the mountains of northern New Jersey, where the wizard told them about his grand plans to revolutionize the iron ore industry. The scheme was doomed. In the coming years he sold off most of his General Electric stock and poured the money down a mine.[13]

He never expressed any regret over the failure. Perhaps he saw the loss of his electricity fortune as a ritual purging, the severing of ties with the industry that he had created, only to have it escape his control.[14]

THE TWO COMPANIES that formed General Electric complemented each other perfectly, with Edison General's strengths in direct current balancing Thomson-Houston's experience in alternating current and arc lighting. Both firms had lucrative lines in electric railways. Boasting solid patents and expertise in the full range of electrical equipment, General Electric was well positioned to slug it out with its lone rival, Westinghouse Electric.

The biggest battleground was the market for alternating equipment, both lighting systems and motors. Despite his strong patents, Nikola Tesla failed to produce a practical alternating-current motor while on the Westinghouse payroll in 1888 and 1889. The idea, though, was too important to be abandoned. Inspired by Tesla's patents, other inventors continued to experiment. Their results were unveiled in spectacular fashion at an 1891 electrical exhibition in Frankfurt, Germany. A German and a Swiss company pulled off the greatest feat of electrical transmission to date, sending 15,000 volts of alternating current along more than 100 miles of cable, from a hydroelectric plant in the Alps to the exhibition building in Frankfurt. As impressive as the transmission itself was, it was matched by the machinery it drove in Frankfurt: a 100-horsepower alternating motor. Both the transmission system and the motors were far more efficient than anyone expected them to be. At long last, direct current had a rival in the field of electrical power.[15]

The Frankfurt display caught the attention of electricians worldwide, but none were more interested than those working on a project in upstate New York. Niagara Falls was the Holy Grail of power engineering. Draining one Great Lake into another, the falls offered one of the heaviest, steadiest sources of hydraulic power in the world, but no one knew how to harness it. Financiers, J. P. Morgan among them, created the Cataract Construction Company in 1889. To make the enormous project pay, the company needed to supply power to factories fairly close to the site, as well as to Buffalo, more than twenty miles away.

In the fall of 1889 the Cataract company had sought advice from Thomas Edison, who at the time was in Paris collecting honors and reviewing his displays at the Universal Exposition. The inventor wired his response: "No difficulty transferring unlimited power. Will assist. Sailing today." Back in the United States, he admitted that there were in fact considerable difficulties. Direct current was unsuitable for high-voltage, long-distance transmission. Alternating current was perfect for transmission, but the lack of a good motor meant that the current could be used only for light and not for the industrial applications needed to make the project pay. The Niagara plans stalled for lack of attractive options. Then came the Frankfurt exposition in 1891, which demonstrated alternating current's clear advantages over direct. Throughout 1892 both General Electric and Westinghouse scrambled to create alternating-current proposals for Niagara.[16]

They also battled for another prize: the contract to light the world's fair planned for Chicago in 1893. Officially known as the World's Columbian Exposition, the fair would commemorate (a year late) the quincentennial of Columbus's voyage to the New World. Like all world's fairs, its real purpose was to showcase industrial progress. This would be the first great American fair of the electrical age, and the contract to supply incandescent lighting carried great prestige. After submitting a bid of $1 million, General Electric was shocked to learn that Westinghouse had secured the contract with a bid of $399,000.[17]

The Chicago fair centered on the "White City," a collection of elaborate classical buildings surrounding a central lagoon. There were buildings devoted to Fine Arts, Agriculture, Mining, Transportation, and Manufacturing, but electricity captured the spirit of the times. At the last great American fair, in Philadelphia in 1876, Americans saw a few crude dynamos and arc lamps. Less than two decades later, a new industry was in full flower. Westinghouse lit the grounds and buildings with 100,000 incandescent lamps, while other companies added 5,000 arc lamps. Spotlights with colored filters turned iridescent the jets of water in the fair's fountains. Visitors could board an electric train or putter about the lagoon in an electric launch. The massive Ferris wheel–invented for the fair–was studded with 1,500 bulbs.

At the entrance to the Electricity Building was a heroic fifteen-foot-tall statue of Benjamin Franklin, kite in hand, and displays inside traced the history of electricity over the previous centuries. Thrusting up at the building's center was the Edison Tower, a tall shaft encircled with thousands of colored lamps and topped by a huge incandescent bulb. Many of the fair's 27 million visitors were experiencing the wonders of electricity for the first time, and the exposition "dissolved much of the mystery that had pervaded its domain," one official claimed. "It brought electricity to the people in the light of a servant not as an awful master." Fairgoers admired the latest phonographs, telephones (including an underwater version for divers), railway signals, pickpocket detectors, seismographs, clocks, tabulating machines, cash registers, and egg incubators. The military-minded could see an electric torpedo, while industrialists enjoyed electric fans, hoists, riveters, welders, pumps, drills, conveyor belts, and air compressors. Medical electricity was not neglected. In addition to devices for giving therapeutic shocks, there were dental mallets and drills, surgical lamps, cautery devices, and a lamp-tipped catheter for illuminating kidney stones. The Edison Manufacturing Company displayed a physician's kit in a black walnut case that included a battery, sponge electrodes, and an "interchangeable fast and slow vibrator." The average citizen was most interested in devices for the home, including the "model electric kitchen" that featured ovens, kettles, frying pans, saucepans, griddles, coffeepots, dishwashers, coffee mills, cigar lighters, hot-water heaters, irons, hair curlers, and boot polishers. The myriad uses of electricity struck one fairgoer as "little short of miracle or witchcraft."[18]

Although the exhibits showcased the electrical future, the real revolution at the fair was the system providing the power. In 1893 electrical equipment ran on many types of current, with the division between alternating and direct current being only the most basic. Alternating generators produced current of different types—known as one-, two-, and three-phase—that were not interchangeable. Arc lamps required high pressures (1,000 to 2,000 volts), incandescent lamps low (50 to 100 volts). Power needs were even more numerous. The new alternating motors worked well for devices that ran at a constant speed, such as dentist's drills and fans, but electric streetcars and elevators—the two major uses for electrical power in the 1890s—required direct-current motors, which were better equipped to run at varying speeds under heavy loads. Individuals, companies, and municipalities had invested in many different kinds of electrical equipment. The Pearl Street district in New York could not afford to scrap its direct-current lighting system and start all over with alternating, just as an early Westinghouse lighting company that was heavily invested in single-phase alternating current could not convert to two- or three-phase. And any city with electric railways needed direct current to run them. It was like having a dozen railroads, each running on a different-gauge track. The current, ad hoc solution to the problem of incompatibility was to generate different types of electricity for each application.

Together with Nikola Tesla, George Westinghouse created a universal system of electrical supply. At the core of the Tesla Polyphase System used at Chicago was a two-phase alternating generator (soon three-phase would become the industry standard). Flexibility was provided by new coupling devices, such as phase-converters that changed two- or three-phase current into single-phase. Even more important was a device called a rotary converter that was capable of changing alternating current into direct. The universal system could supply, from one generator, electricity to serve any need: direct-current motors, single-phase or polyphase alternating motors, electrochemical processes (which required direct current), and incandescent and arc lamps requiring direct current, single-phase, or polyphase alternating current.[19]

George Westinghouse's bid for the Chicago fair contract was so low that he lost money on it, but the true purpose of his Chicago installation was to stake a claim to Niagara. Not accidentally, the Westinghouse univeral system was perfectly adapted to meet the needs of the Niagara power station. In the fall of 1893, not long after the fair closed, Westinghouse won the Niagara contract. Two years later Niagara delivered power to its first customer, the Pittsburgh Reduction Company, which built near the falls to exploit the cheap electricity. In 1896 Buffalo received its first Niagara power, and before long growing demand forced an expansion of transmission lines. Although Westinghouse won the initial contract, Niagara was very much a cooperative affair. General Electric built transformers, transmission lines, substations, and rotary converters. When generating capacity was expanded, the company supplied some of the new generators. By 1900 Westinghouse and General Electric together controlled virtually all of the electrical

The Westinghouse Electric and General Electric exhibits at the World's Columbian Exposition in Chicago, 1893.

manufacturing market, and both firms prospered throughout the next century.[20]

The Niagara project marked the symbolic victory of alternating over direct current, but Thomas Edison's beloved current did not disappear. Some of Niagara's heaviest customers used direct current for streetcars, industrial motors, and electrochemistry, and Edison's old three-wire direct-current systems in urban centers remained viable—and profitable—long into the twentieth century. For decades, most electrical systems entailed both alternating- and direct-current applications; alternating generators, however, were at the core of those systems.

The End of the Battle of the Currents

Alternating current triumphed in the battle of the currents for the same reasons that attracted George Westinghouse to it in the first place. It could be produced at relatively low voltages, stepped up to higher voltages with transformers for economical transmission over long distances, and then stepped down to any voltage required. As electricity expanded into every aspect of life, alternating current offered a flexibility that direct current could not hope to match. By 1917 more than 95 percent of the electricity generated in the United States would be alternating current.[21]

CHAPTER 23

The Age of the Electric Chair

GERMAN chancellor Otto von Bismarck wrote to his consul-general in New York in 1888 seeking information on electrocution, but German officials decided against adopting the method after learning of the problems with Kemmler's execution. American states were less cautious. Ohio adopted electrocution in 1896, and two men died in the chair the next year. In 1898 Massachusetts became the third state to adopt electrocution, and convicted murderer Luigi Storti became the state's first victim three years later.[1]

With electrocution protected by the courts, prison officials set out to refine execution techniques. Their task was made easier by rapid advances in the field of electrical engineering. The 1890s saw the first golden age of modern electrical theory, and the improved understanding of current also transformed the study of its effects on the human body. At the turn of the century two researchers at the University of Geneva conducted classic experiments in electrophysiology that confirmed the views of Edison, Kennelly, and other researchers:

Alternating current was, in fact, more dangerous than direct. The researchers also distinguished between the physiological effects of high- and low-voltage currents. Low voltages (less than 120 volts) of alternating current killed by causing ventricular fibrillation. (This fact provided the clue that, half a century later, led to the invention of the cardiac defibrillator.) At higher voltages, alternating current damaged the central nervous system, causing loss of consciousness and respiratory failure. When the heavy current met the resistance of the body, electrical energy was transformed into heat, which destroyed the brain by cooking it.[2]

Arthur Kennelly left the Edison laboratory in 1894 to become an independent consultant, but his support for electrocution never wavered. In 1895 the eminent French physiologist Jacques d'Arsonval—who five years earlier had confirmed Edison's views of the greater dangers of alternating current—created a sensation by reviving the charge that the electric chair merely stunned its victims, who were then killed by autopsy. To refute the claim, Kennelly observed a Sing Sing execution in January 1895 and conducted more experiments on dogs. His report, published in the *Journal of the American Medical Association*, concluded that "where electrocution is properly carried out, there is not even a remote possibility of subsequent resuscitation of the criminal."[3]

Edison backed up his former employee. D'Arsonval's statements were "nonsense," Edison told a reporter; "electricity in the death chair kills instantly beyond possibility of resuscitation."[4]

Another outspoken electrocution proponent was Dr. Edward A. Spitzka. A professor at Jefferson Medical College in Philadelphia and a famous brain anatomist, Spitzka was the son of Dr. Edward C. Spitzka, who witnessed Kemmler's execution. The younger Spitzka served as official physician at several executions and had the honor of penning the "electrocution" article for the classic eleventh edition (1910) of the *Encyclopedia Britannica*, in which he assured general readers, "When properly performed the effect is painless and instantaneous death."

Spitzka's view became the conventional wisdom. Medical journals—including *JAMA*—greeted electrocution with quiet acceptance or hearty approval.[5]

THE SAME MESSAGE spread through more popular venues. In 1901 Thomas Edison's motion picture company released *Execution of Czolgosz,* a film reenactment of the electrocution of President William McKinley's assassin. In the film, an actor playing Leon Czolgosz was strapped into an electric chair. The current was turned on and off three times in rapid succession, and each time he reared back and then relaxed. After the third shock, a doctor checked his heart with a stethoscope, found him dead, and gave a curt nod of satisfaction. The whole process took less than ten seconds. The film, which enjoyed enormous success, brought alive the fantasy of the quick, clean death that supporters of the electric chair had long promoted, while omitting the gruesome details that marked real electrocutions.[6]

Two years later, Edison's film company filmed an actual electrocution—but not of a human being. Topsy, a rogue Coney Island circus elephant, was condemned to death in 1903. Her owners proposed hanging her from a scaffold with a huge rope and pulley, but the ASPCA objected. Instead, Topsy was fitted with copper-lined wooden sandals on her right front and back left legs, and the sandals were wired to a dynamo at the local Edison lighting plant. As a Sunday crowd of 1,500 people looked on, 6,000 volts of electricity were sent through Topsy. She stiffened, then crashed over on her right side, dead. Cameramen from the Edison Manufacturing Company caught the execution on film. As with the Czolgosz film, *Electrocuting an Elephant* was distributed across the country and watched by thousands of viewers eager to see the killing power of electricity.[7]

With widespread popular and scientific support, electrocution spread to more states. In 1906 New Jersey abolished hanging and a year later killed a murderer in the electric chair. Present at this execution was Edward A. Spitzka, as well as two technicians of electrical death, Edwin F. Davis and Carl F. Adams. Davis, who controlled the switchboard room at Kemmler's execution, had become New York's official electrocutioner. In 1897 he was awarded a U.S. patent for an "electrocution-chair," and he supervised the first such deaths in Ohio and Massachusetts, charging $150, plus expenses, for each execution.[8]

Not long after New Jersey introduced its death chair, Davis retired from the business, and Carl Adams replaced him as the entrepreneur of electrical death. As the chief electrician at the New Jersey state prison in Trenton, Adams constructed the state's electric chair, which proved so successful that Adams Electric Company had a new line of work. When Virginia became the fifth state to adopt electrocution in 1908, Adams submitted a bid to install "one complete Electrocuting Plant." The bid—$3,200—included an assurance of success: "We fully guarantee our machine to give perfect satisfaction if operated according to our instructions." Six months after the contract was signed, Adams informed Virginia officials that he would give them free of charge a newly designed component of the chair that was not specified in the contract, because "the more efficient we make the outfit the better it will be for our future business with other States." In 1912 Adams installed an electric chair at South Carolina's new death house.[9]

The first four states to adopt electrocution—New York, Ohio, Massachusetts, and New Jersey—were in the North and had long been at the center of America's humanitarian reform movements. The next four states to make the switch were Virginia, North Carolina, Kentucky, and South Carolina. The electric chair assumed a different identity in the South, where the alleviation of physical suffering had never been a top priority. In the years after Reconstruction, the white population brutally enforced the disenfranchisement and social subjugation of African-Americans, about 2,000 of whom were lynched in the South between 1890 and 1910. Accused of some infraction against racial codes, the victims were seized by white mobs and hanged or burned, often before a cheering crowd.[10*]

After its founding in 1909, the National Association for the Advancement of Colored People began a campaign against lynching. For the first time southern leaders became fearful that the killings and publicity surrounding them might damage the region's economic prospects, and they began to condemn lynching. The number of lynchings began to decline after 1910, largely because of the rise of "legal lynching": quick judicial trials that offered black defendants only marginally more constitutional protection than did a lynch mob. As the number of lynchings fell, the number of state-imposed executions rose just as quickly. In a half-dozen southern states, those executions took place in the electric chair, that latest symbol of humanitarian progress. Between 1908 and 1930, Carl Adams's chair in Virginia killed 148 people, all but 17 of them black.[11]

IN 1913 ARKANSAS, Indiana, Nebraska, Oklahoma, Pennsylvania, Tennessee, and Vermont adopted the electric chair, and by 1930 those states had been joined by Texas, Alabama, Florida, Georgia, the District of Columbia, Illinois, and New Mexico. In 1924 Nevada became the first state to adopt an even newer method: placing prisoners in an airtight chamber and filling it with hydrocyanic gas. For a decade Nevada remained the only state to use gas. Critics believed the new method—with its claustrophobic chamber and invisible agent of death—was worse than electricity, and the recent experience of poison gas deaths in World War I offered no reassurance. In 1933, however, Colorado and Arizona built gas chambers, and by 1955 eight more states followed suit, convinced that the gas chamber marked another

*In 1896 a white boy at a southern fair paid a nickel to hear the Edison phonograph and was treated to a recording of two black men being burned alive.

step in the march of progress. The matter remained open to dispute, because bungled gas executions sometimes caused prisoners to gasp and choke as they died. As a result, some states continued to opt for electrocution. Between 1935 and 1949 five states adopted the chair: Connecticut, South Dakota, Louisiana, Mississippi, and West Virginia. In all, twenty-five states plus the District of Columbia passed electrocution laws.[12]

The controversy surrounding electrocution never disappeared. There was evidence that, in some cases, electrocution killed fairly quickly. But throughout the century electrocutions often went horribly wrong, causing prisoners to suffer the extreme pain that accompanies cardiac arrest, tetanic muscle contraction, asphyxiation, boiling body fluids, and severe burns.[13]

In the nineteenth century there was strong support for capital punishment at all levels of society. Over the course of the twentieth century, doubts grew about the morality of the death penalty and its effectiveness as a deterrent. Nine states abolished capital punishment between 1907 and 1917; however, in the wake of the race riots and Red Scare that followed World War I, five of those nine states reinstated it. By the 1920s, all but five states imposed capital punishment. The number of executions peaked during the Great Depression in 1935, when 199 criminals were put to death.[14]

For the next four decades, the execution rate declined steadily. Many people came to believe that crime resulted from mental illness or corrupting childhood influences; if criminal acts did not result from free choice, the argument for deterrence lost all force. Notorious cases in which the guilt of the condemned was widely doubted—such as those involving Sacco and Vanzetti or the Scottsboro boys—also produced doubts about the wisdom of this irrevocable sentence. Gallup polls showed that between the 1930s and the 1960s, support for capital punishment declined in the United States, until opponents slightly outnumbered supporters, even in the South. By 1965 fourteen states had abolished capital punishment, the most ever. Even in states that retained the death penalty, juries were reluctant to impose it. Judges, affected by the same social trends, became more receptive to appeals from death-row inmates. By the late 1960s, the U.S. legal machinery of execution had ground to a complete halt.[15]

Much of the concern focused on racial disparities: African-Americans were sentenced to death far more frequently than whites convicted on similar charges. The legal battle was headed by the NAACP's Legal Defense Fund, which scored a victory in 1972 with the Supreme Court decision in *Furman v. Georgia*. By a 5–4 margin, the Court invalidated the capital punishment statutes of every state. According to the justices who cast the pivotal votes, the problem was that some defendants were executed for crimes for which others received only prison sentences. Capital punishment was "so wantonly and so freakishly imposed" that it was an arbitrary and therefore unconstitutional form of punishment.[16]

The justices clearly implied that the death penalty would be constitutional if it were imposed in a less capricious manner. States set about drafting new laws that detailed in advance which particular types of murder cases—such as those committed during the commission of another felony—warranted a capital sentence. The Supreme Court upheld such laws in 1976, and by the late 1970s a new era of American executions had begun.[17]

The new laws did not end the arbitrariness of capital punishment. The decision as to whether to seek a death penalty in a particular case often had less to do with the "aggravating circumstances" of the crime than with whether the prosecutor was up for reelection soon. Although African-Americans were no longer sentenced to death at higher rates than whites, a different racial disparity emerged: A murderer was much more likely to be sentenced to death if his victim was white rather than black. Despite many instances of prisoners being released from death row after having been proven innocent, a majority of Americans continued to support capital punishment. By 2000, more than 3,000 convicted criminals were sentenced to death each year, the most since the Justice Department began keeping statistics in the 1950s.[18]

The situation in the United States stood in stark contrast to that in Great Britain, West Germany, Austria, Italy, Australia, Denmark, Belgium, the Netherlands, Ireland, and Mexico, which by the end of the 1960s had stopped executing criminals. In 2001 China led the world in executions, followed by Iran, Saudi Arabia, and the United States.[19]

The United States followed a path so different from its peers for a number of reasons, not the least of which was that the homicide rate in America was higher than that in any other Western nation. Another fact distinguished the U.S. case: It was the only Western nation to experiment with scientific methods of capital punishment. In the early 1950s a British royal commission conducted a systematic study of execution methods and concluded that hanging should be retained. In the commission's view, there was no solid evidence that either electrocution or the gas chamber was more humane than the gallows. Germany also had considered scientific methods but decided to keep using the guillotine and the ax, a decision that resulted not from a dispassionate assessment of available options but from an "ingrained, almost mystical prejudice" in favor of the traditional methods. The nations of western Europe stayed with nineteenth-century methods of execution until they abolished the death penalty after World War II. In the United States—where a mystical prejudice favored the embrace of the new—scientific methods made executions more palatable. Americans reassured themselves that killing was a legitimate state function, so long as it was done gently.[20]

When executions resumed in the late 1970s, states began to abandon their electric chairs and gas chambers and to execute criminals with hypodermic injections of drugs. First adopted by Texas and Oklahoma in 1977, lethal injection swept the country far more quickly than either electrocution or the gas chamber had. State after state made the switch, with a few offering condemned prisoners a choice between the needle and the chair. Although problems with lethal injection did occur, it came closer than any other method to fulfilling the ideal often mentioned in the capital punishment debates of a century before. Quick, clean, and unspectacular, lethal injection turned execution into a medical procedure. "It's extremely sanitary," a Missouri prison chaplain reported. "The guy just goes to sleep."[21]

In 2001 the Georgia Supreme Court deemed electrocution unconstitutionally cruel and replaced it with lethal injection. In a court opinion, the justices quoted with approval from the 1885 speech in which New York governor David Hill said that "the science of the present day" might provide a "less barbarous" method of execution. Those words had led to the creation of the death penalty commission and to New York's electrical execution law. Just as hanging was deemed unacceptable in the 1880s, the Georgia justices wrote, so now, at the start of the twenty-first century, "electrocution offends the evolving standards of decency that characterize a mature, civilized, society."[22]

As of April 2003, 4,432 men and women had died in the electric chair in the United States. Nebraska was the only state with electrocution as its sole method of execution. A few more prisoners may be electrocuted, but lethal injection has emerged as the method of the future. The era of the electric chair has come to an end.[23]

EPILOGUE

The New Spectacle of Death

SHORTLY AFTER World War I, when he was about seventy years old, Thomas Edison began taking annual camping trips with Henry Ford, Harvey Firestone, and the naturalist John Burroughs. A story was told of a time the party's vehicle broke down in West Virginia. A local mechanic took a look under the hood and pronounced the problem to be mechanical. "I am Henry Ford," the automaker said, "and I say the motor itself is in perfect order." The mechanic then suggested a short in the electrical system. "I am Thomas A. Edison," came the response, "and I say the wiring is all right." The mechanic then peered skeptically at the white-bearded Burroughs and said, "And I suppose that must be Santa Claus."[1]

Although still very much alive, Edison had entered the realm of the mythical, occupying an outsized place in the American imagination. Until his death at age eighty-four in 1931, he continued to experiment; his most successful later ventures involved the production of cement (8,000 tons of which helped build Yankee Stadium) and the manufacture of electrical storage batteries. In 1913 readers of *Independent* magazine named him the "most useful" man in the United States. There could be few higher compliments in a nation so earnestly devoted to the practical, but the award did Edison something of a disservice. His greatest feats lay in coupling the useful with the magical—a machine that talked, light without fire—in ways that transformed the lives of everyone in the modern world.[2]

In his later years Edison gave a great deal of thought to his place in history. He collaborated on an official biography, published in 1910, and in an endless succession of newspaper interviews he reminisced—with great enthusiasm and intermittent accuracy—about his life of invention. There was one topic, however, that Edison omitted from the biography and almost never mentioned to reporters: his role in the creation of the electric chair. During a 1905 interview, he broke his silence on the matter and revealed that his views had changed little since 1889: He still believed capital punishment was a "barbarity," and he still considered electrocution to be the quickest and therefore most humane of

Edison in 1895.

methods. When asked whether he had invented the electric chair, Edison grew indignant: "I did not invent such an instrument."[3]

New York's introduction of electrical execution was a momentous change that required the political, medical, and technological skills of many men, including Alfred Southwick, Elbridge Gerry, Governor David Hill, Harold Brown, George Fell, Arthur Kennelly, and Carlos MacDonald. But because of his immense fame and powerful reputation, Thomas Edison's opinions carried the greatest weight. Through the experiments at his laboratory and his tireless promotion of the new law, Edison—more than any other single person—ensured that electrocution would be the state's method of capital punishment.[4]

Edison genuinely believed that the electric chair was more humane than hanging, but he went to such lengths to defend it only because he saw it as a useful weapon in his battle against alternating current. Although inspired in part by an intense dislike for George Westinghouse, Edison's most powerful motivation in the battle of the currents was his genuine fear of alternating current.

In one limited sense Edison succeeded in his campaign. It was only in the midst of the electric wire panic of 1889, when he was pressing for severe restrictions on alternating current, that Westinghouse and the other recalcitrant lighting firms began to cooperate with efforts to place their wires underground, thereby lowering the risk to the public. Ultimately, however, Edison lost the battle of the currents, as alternating current became the industry standard. He certainly had overestimated its risks, probably because he generalized too broadly from the case of Manhattan, where high population density, sloppy installations, and lax regulation combined to make high-voltage current especially dangerous. In other cities—Paris, Berlin, Chicago—underground wires and better regulation made it safer. A ban on alternating current also would have handed much of the electricity market to the Edison interests, and the public recognized the danger of this. As the *World* explained, "We do not want to be grilled by electricity or oppressed by monopoly."[5]

The question is not so much why Edison's campaign failed as why

The placement of Manhattan's electrical wires in underground conduits beautified the city and lowered the risk of accidental shock. The illustration on the left appeared in *Harper's Weekly* in 1889; the photograph on the right was taken about 1910.

he thought it might succeed. "It is impossible now that any man, or body of men, should resist the course of alternating-current development," an electrical journal stated in 1889. "Joshua may command the sun to stand still, but Mr. Edison is not Joshua." Edison's nearest approach to triumph—during the wire panic—was also the moment of his defeat. When the city work crews chopped down the offending electric light wires, the resulting gloom served as an advertisement for electric light; only when the light disappeared did people realize how dependent upon it they had become. Alternating current could satisfy the demand much more easily than direct. If Edison had succeeded in banning alternating current, he would have saved many people from accidental death. Yet he also would have stalled the spread of electricity and stymied industrial growth.[6]

On one occasion when Edison called for strict limits on alternating current, a reporter challenged his motives. "Your statements are shaped very much by your business connection," the reporter charged. "Assume that they are, which they are not," Edison replied. "That does not alter the truth of my statements. Would I be likely to make statements on a scientific matter which could be proved as wrong?"[7]

But this was not a matter of scientific truth or falsity. Edison and his opponents were engaged in the assessment of risk: How much danger would people tolerate in exchange for the convenience of electric light? After the wire panic, the public's fears about electrical safety quieted. Part of this was due to better insulation and better regulation, but people continued to die. Medical journal articles from the early twentieth century—such as "A Case of Death from the Electric Current While Handling the Telephone"—seemed to confirm Edison's worst fears about the domestic dangers of the current. By the 1920s—the decade when electricity finally became common in American homes—about 1,000 Americans died annually from electric shock. The panic over electrical accidents waned not because the deaths stopped but because they became routine. Accidental deaths involving street railways—and, later, automobiles—also initially created great public alarm. But, as with electricity, people eventually grew accustomed to the carnage.[8]

Edison had worried that if accidents led to deaths, people would decide to stick with cheaper sources of illumination such as gas and kerosene lamps. His greatest mistake was an uncharacteristic lack of faith in his own inventions. He failed to understand that electricity could kill people and still retain the public's enthusiasm.

Because of the Edison conspiracy against Westinghouse's alternating current, two distinct issues—whether the current was safe enough to use for lighting purposes, and whether it could kill criminals painlessly—became hopelessly confused, and neither received the attention it deserved. George Westinghouse was forced both to defend the safety record of his current and to attack its use for electrocution. With a stubbornness that matched Edison's, Westinghouse continued to insist that alternating current was harmless long after the claim lost plausibility. The many deaths from electrical accidents convinced the public, as well

as the courts considering William Kemmler's appeal, that alternating current was lethal and that Westinghouse was less than honest. This loss of credibility damaged efforts to fight the electrocution law. The

Kemmler hearings—the only serious inquiry into the legitimacy of electrocution ever undertaken—revealed that execution equipment was liable to failure and that medical knowledge of electrical death was woefully inadequate. But most of those speaking out against the new method were Westinghouse men. The opposition to electrocution never got a fair hearing, because any objections could be dismissed out of hand as the cynical machinations of George Westinghouse. Few appeared troubled that Thomas Edison's motives for defending electrocution were equally suspect.[9]

BY DESTROYING THE GALLOWS, the builders of the electric chair hoped to take the spectacle out of executions. They promised a death that would be private, simple, and painless, leaving no mark on the corpse. As Edison told reporters in 1888, "Touch a button, close the circuit, it is over." The execution protocol that finally emerged was very different, requiring a hooded victim, elaborate machinery, an unseen and mysterious force, an attending priesthood of engineers speaking an arcane language, and a burnt offering. Deftly linking high technology and ancient sacrifice, the electric chair created a thoroughly modern spectacle of death.[10]

The public responded with enthusiasm. Prison wardens received thousands of letters from people who wanted to witness an electrocution. Almost all were turned down, but they found other ways to satisfy their curiosity. A New York dime museum featured a waxworks diorama titled "Execution of Criminals by Electricity" alongside such displays as "Beheading in Morocco," and an electric chair was among the many electrical wonders on display at the 1893 world's fair in Chicago. In the twentieth century electric chair acts became a staple of carnival sideshows; one performance featured a girl, said to be "immune" to electricity, taking shocks of 20,000 volts. Edison's *Execution of Czolgosz* and *Electrocuting an Elephant* were only the first of dozens of films to feature the new method of execution, ranging from the Clark Gable classic *Manhattan Melodrama* to John Waters's *Female*

Hamilton's drugstore in Auburn, New York, sold this postcard in the early twentieth century.

Trouble. Andy Warhol, a connoisseur of American icons, produced a series of prints titled *Electric Chair* in the 1960s and 1970s. Decommissioned electric chairs are now displayed at museums and draw large crowds. "It's one of the most popular things we've got," said an employee of a New Jersey museum. "Everyone who sees it goes 'Wow! The electric chair!' "[11]

AT THE TIME the electric chair was invented, Americans were in thrall to the fantasy of the push button: the belief that, in the future, machines would do all of their work for them. Many tried to let machines do their thinking for them as well, and no one exemplified this engineering mentality better than Thomas Edison, the opponent of capital punishment who helped invent a killing machine. Like many death penalty foes, Edison believed that making killing more humane was a sign of progress, a step down the road to complete abolition. The strategy backfired. By making executions appear painless, Edison helped the death penalty survive. The electric chair—and the later scientific methods it inspired—masked the barbarity of killing in the civilization of the machine.

Mark Twain explored the new technologies of death in *A Connecticut Yankee in King Arthur's Court*, first published in the midst of the electric wire panic of 1889. Twain tells the story of a nineteenth-century mechanic who is knocked cold in a fight and wakes up in England in the year 528. The Boss—as the Yankee is known—grows powerful by introducing to Camelot such nineteenth-century wonders as newspapers, telegraphs, electricity, and gunpowder, but he runs afoul of church and royalty by trying to improve the lot of the oppressed peasantry. A plainspoken mechanical genius who tirelessly promotes the wonders of industrial society, the Boss bears a striking resemblance to Thomas Edison.

Written in the spring of 1889—at the time when electrocution experiments were taking place at the Edison laboratory and wires were killing people on the streets of New York—the ending of *Connecticut Yankee* offers Twain's dark views on the battle of the electric currents. In the final chapters, the Boss and a few allies find themselves trapped in a cave, with all the knights of England massed for attack outside. The Boss has a dynamo with him, and he fortifies the cave entrance with a fence of live electric wires—"naked, not insulated." At dawn one morning, the Boss inspects the fence and discovers a dead knight, "a dim great figure in armor, standing erect, with both hands on the upper wire—and of course there was a smell of burning flesh." Others followed and died so quickly they had no chance to warn their fellows. The electric fence killed 11,000 knights, so that the cave "was enclosed with a solid wall of the dead—a bulwark, a breastwork, of corpses, you may say."

Although nineteenth-century technology destroyed the best the sixth century had to offer, the Boss and his men remained pinned in the cave, with thousands of rotting corpses just outside and, beyond them, a considerable number of surviving enemy forces. "We were in a trap, you see—a trap of our own making," the Boss's top lieutenant explained. "If we stayed where we were, our dead would kill us; if we moved out of our defences, we should no longer be invincible. We had conquered; in turn we were conquered."[12]

The novelist and critic William Dean Howells, a good friend of Twain's, also saw the trap of scientific killing. On Christmas Day 1887, just after news had leaked of the death penalty commission's report, Howells wrote a satiric letter to the editor of *Harper's Weekly* that perfectly captured the era's rhetoric of blithe technological optimism.

> I understand that the death-spark can be applied with a minimum of official intervention, and without even arousing the victim, or say patient, from his sleep on the morning fixed for the execution of the sentence. . . . I have fancied the executions throughout the State taking place from the Governor's office, where his private secretary, or the Governor himself, might touch a little annunciation-button, and dismiss a murderer to the presence of his Maker with the lightest pressure of the finger. In cases of unusual interest, the Executive might invite a company of distinguished persons to be present, and might ask some lady of the party to touch the button. Or, as when torpedoes are exploded or mining blasts fired in the completion of a great public work, a little child might be allowed to discharge the exemplary office.

Howells worried not that electrocutions would be bungled but that they would proceed precisely as planned. In the image of a young child—with "the lightest pressure of the finger"—pushing a button to end a man's life, he captured the true terror of the electric chair: When killing is made scientific, when it is made easier, it becomes not less but more horrifying.[13]

ACKNOWLEDGMENTS

This book is the issue of many long, largely happy days in the air-conditioned comfort of research libraries. I offer my profound gratitude to the staffs of the New York Public Library, New York Academy of Medicine, New-York Historical Society, Buffalo and Erie County Public Library, and the libraries of New York University, Cornell University, Columbia University, Washington University in St. Louis, the University of Texas at Austin, and Rutgers University in New Brunswick. My thanks as well to the staffs of the Death Penalty Information Center, New York State Archives, the Prints Division of the Metropolitan Museum of Art, Cayuga County Historian, Chicago Historical Society, Clements Library at the University of Michigan, Harvard University Archives, Historical Society of Western Pennsylvania, Westinghouse Museum, the Smithsonian Institution's National Museum of American History, Painted Post Historical Society, Ohio Historical Society, and the Barber Museum in Canal Winchester, Ohio. Elizabeth Ihrig of the Bakken Library and Museum photocopied and mailed important materials. Barbara Carr, the executive director of the Erie County SPCA, shared her desk, her thoughts, and the archives of her organization during a fascinating afternoon of research. Eileen McHugh, the executive director of the Cayuga Museum in Auburn, New York, answered questions about the prison and provided three photographs, including an extremely rare one of the first electric chair. Rob Cox, the manuscripts librarian of the American Philosophical Society, guided me through the Elihu Thomson Papers and located a portrait of Thomson therein. Judy Folkenberg of the National Library of Medicine helped identify and secure images. Dr. Scott Swank of the Dr. Samuel D. Harris National Museum of Dentistry photocopied dozens of images of nineteenth-century dental chairs to satisfy my curiosity as to whether they might have inspired electric chair design (they did not).

Leonard DeGraaf, the archivist for the Edison National Historic Site, tracked down crucial documents and images and made beautiful high-resolution scans for me. Some of my research was rendered embarrassingly easy by the prodigious labors of the Thomas A. Edison Papers staff, who have made tens of thousands of documents accessible in books, on microfilm, and over the Internet (http://edison.rutgers.edu).

There is more to writing than research materials. The New York Public Library's glorious Rose Main Reading Room and its eccentric habitués provided, respectively, inspiration for and necessary distraction from my work. Bagel Bob's on University Place sustained me during mornings of writing; Thai Cafe and the Garden in Greenpoint fed me at night. For their warmth, sharp conversation, and willingness to go out for a drink or three, my thanks to Matt Abramovitz, Katherine Biers, C. A. Carlson, Andrew Cocke, Eddie Del Rosario, Trace Farrell, Cori Hayden, Juliet Hooker, Katherine Lieber, Karen Lillis, Elisa Slattery, Peter Tarr, and Howard Yoon.

Mark Baumann and Leah Shafer masterfully handled a great deal of scanning. C. A. Carlson and Peter Tarr read multiple drafts, serving as sounding boards and unpaid but extravagantly appreciated editors. Peter Dear, John Essig, David Essig-Beatty, Mary Essig-Beatty, Juliet Hooker, Jenna Land, Larry Moore, and Lauris Moore read the manuscript (or parts thereof) and saved me from errors of fact, infelicities of style, and narrative deviations. I hold them entirely responsible for what blunders may remain.

Paul Israel, director and editor of the Thomas A. Edison Papers and author of the definitive Edison biography, showed great generosity to a stranger by reading and suggesting changes to a near-final draft. Joan Jacobs Brumberg, Peter Dear, and Michael Kammen trained me in the ways of history, as did Larry Moore, who continues to provide a model of graceful scholarship and gracious living. Jackie Johnson, editor, read the manuscript nearly as many times as I without once losing her equanimity or her sharp eye for detail. George Gibson, publisher, took a chance on a wild card and displayed unflagging enthusiasm throughout. My agent, Gail Ross, deftly guided me through the thickets of trade publication. Howard Yoon, excellent friend and savvy adviser, nurtured the project from conception to birth; without him the book would not exist.

Melissa Cole read not a prepublication word but made me immoderately happy during the final weeks of revision, and beyond. My siblings, nieces, and nephews—Len, Mary, Dan, Heather, David, Emily, Katherine, Jacob, Joseph, and Ella—provided entertainment and boundless affection. I dedicate this book to my parents, Dorothy and John Essig, whose love has made it, and all else, not only possible but joyful.

NOTES

ABBREVIATIONS USED IN NOTES

COMMISSION REPORT: *Report of the Commission to Investigate and Report the Most Humane and Practical Method of Carrying into Effect the Sentence of Death in Capital Cases* (Albany: Troy Press, 1888).

ENHS: Edison National Historic Site Archives, West Orange, New Jersey.

KEMMLER HEARINGS: *People of the State of New York, ex rel. William Kemmler Against Charles F. Durston, as Warden of the State Prison at Auburn, N.Y.* (Buffalo: J. D. Warren's Sons, 1889).

TAEB: *Papers of Thomas A. Edison* (book edition), ed. Reese V. Jenkins et al., 4 vols. (Baltimore: Johns Hopkins University Press, 1989–). Example: TAEB 4:858 refers to volume 4, page 858.*

TAEM: *Thomas A. Edison Papers: A Selective Microfilm Edition*, ed. Thomas E. Jeffrey (Frederick, Md.: University Publications of America, 1985–). Example: TAEM 138:355 refers to microfilm reel 138, frame 355.*

THOMSON PAPERS: Elihu Thomson Papers, American Philosophical Society, Philadelphia, Pennsylvania.

PROLOGUE. EDISON ON THE WITNESS STAND

1. *New York Telegraph*, July 23, 1889 (TAEM 146:463).

2. Testimony quotations are from Kemmler Hearings, 623, 629–30, 636. Also see *New York Daily Graphic*, July 23, 1889; *New York Sun*, *New York Tribune*, July 24, 1889.

3. *London Standard*, as quoted in *New York Times*, August 17, 1890.

4. Edison to Alfred Southwick, December 19, 1887 (TAEM 138:355).

5. The two most important academic articles on this topic are Thomas P. Hughes, "Harold P. Brown and the Executioner's Current: An Incident in the AC-DC Controversy," *Business History Review* 32 (1958): 143–65; and Terry S. Reynolds and Theodore Bernstein, "Edison and 'The Chair,'" *IEEE Technology and Society Magazine* 8 (March 1989): 19–28. Also see Terry S. Reynolds and Theodore

---

*Most of the documents in TAEM and TAEB also can be found on the Edison Papers Web site: http://edison.rutgers.edu.

Bernstein, "The Damnable Alternating Current," *Proceedings of the IEEE* 64 (1976): 1339–43; Theodore Bernstein, "'A Grand Success': The First Legal Electrocution Was Fraught with Controversy Which Flared Between Edison and Westinghouse," *IEEE Spectrum* 10 (February 1973): 54–58; Thomas Metzger, *Blood and Volts: Edison, Tesla, and the Electric Chair* (Brooklyn: Autonomedia, 1996); Craig Brandon, *The Electric Chair: An Unnatural American History* (Jefferson, N.C.: McFarland, 1999); Arnold Beichman, "The First Electrocution," *Commentary* 35 (1963): 410–19; Roger Neustadter, "The 'Deadly Current': The Death Penalty in the Industrial Age," *Journal of American Culture* 12 (1989): 79–87; Paul A. David, "Heroes, Herds and Hysteresis in Technological History: Thomas Edison and 'The Battle of the Systems' Reconsidered," *Industrial and Corporate Change* 1 (1992): 129–80. Richard Moran, *Executioner's Current: Thomas Edison, George Westinghouse, and the Invention of the Electric Chair* (New York: Knopf, 2002), was published too late for consideration here.

6. *Brooklyn Citizen*, November 4, 1888 (TAEM 25:580).

CHAPTER 1. EARLY SPARKS

1. J. L. Heilbron, *Elements of Early Modern Physics* (Berkeley: University of California Press, 1982), 160–61.

2. Quotation from ibid., 172. Also see Geoffrey V. Sutton, *Science for a Polite Society: Gender, Culture, and the Demonstration of Enlightenment* (Boulder, Col.: Westview Press, 1995), 290–96.

3. Sutton, *Science for a Polite Society*, 296–314; Heilbron, *Elements of Early Modern Physics*, 173–80.

4. Quotation from Heilbron, *Elements of Early Modern Physics*, 185. Also see Sutton, *Science for a Polite Society*, 314–19; Park Benjamin, *The Intellectual Rise in Electricity: A History* (New York: Appleton, 1895), 522.

5. Quotation from Benjamin Franklin to Peter Collinson, July 29, 1750, *Papers of Benjamin Franklin*, ed. Leonard W. Labaree (New Haven: Yale University Press, 1959–), 3:19–20. Also see I. Bernard Cohen, *Benjamin Franklin's Science* (Cambridge, Mass.: Harvard University Press, 1990), 28–29.

6. Cohen, *Benjamin Franklin's Science*, 66–109; Heilbron, *Elements of Early Modern Physics*, 196–200; Sutton, *Science for a Polite Society*, 325–31.

7. Quotations from Franklin to Peter Collinson, February 4, 1851, *Papers of Benjamin Franklin*, 4:111–13; Franklin to Barbeau Dubourg and Thomas Francois Dalibard, May 1873, *The Ingenious Dr. Franklin*, ed. Nathan G. Goodman (Philadelphia: University of Pennsylvania Press, 1931), 71–73.

8. Margaret Rowbottom and Charles Susskind, *Electricity and Medicine: History of Their Interaction* (San Francisco: San Francisco Press, 1984), 15–30; Samuel J. Rogal, "Electricity: John Wesley's 'Curious and Important Subject,'" *Eighteenth-Century Life* 13 (November 1989): 79–90.

9. Marcello Pera, *The Ambiguous Frog: The Galvani-Volta Controversy on Animal Electricity*, trans. Jonathan Mandelbaum (Princeton, N.J.: Princeton University Press, 1992), 19–24, 53–86; Rowbottom and Susskind, *Electricity and Medicine*, 37–42.

10. Quoted in Joost Mertens, "Shocks and Sparks: The Voltaic Pile as a Demonstration Device," *Isis* 89 (1998): 303; Rowbottom and Susskind, *Electricity and Medicine*, 42–43.

11. Jan Golinski, *Science as Public Culture: Chemistry and Enlightenment in Britain, 1760–1820* (Cambridge: Cambridge University Press, 1992), 188–235.

12. W. James King, "The Development of Electrical Technology in the 19th Century, Part 3: The Early Arc Light and Generator," *United States National Museum Bulletin* 228, paper 30 (Washington, D.C.: Smithsonian Institution, 1962), 345.

13. Paul Israel, *From Machine Shop to Industrial Laboratory: Telegraphy and the Changing Context of American Invention, 1830–1920* (Baltimore: Johns Hopkins University Press, 1992), 24–51; Herbert Ohlman, "Information," in *An Encyclopaedia of the History of Technology*, ed. Ian McNeil (New York: Routledge, 1990), 710–17; Iwan Rhys Morus, *Frankenstein's Children: Electricity, Exhibition, and Experiment in Early-Nineteenth-Century London* (Princeton, N.J.: Princeton University Press, 1998), 194–230; G. R. M. Garratt, "Telegraphy," in *A History of Technology*, 5 vols., ed. Charles Singer et al. (Oxford: Clarendon Press, 1958), 4:644–62.

14. Quotations from Edwin G. Burrows and Mike Wallace, *Gotham: A History of New York City to 1898* (New York: Oxford University Press, 1999), 675, 677. Also see Israel, *From Machine Shop to Industrial Laboratory*, 37–56.

15. Israel, *From Machine Shop to Industrial Laboratory*, 55, 103–16; Ohlman, "Information," 710–17; Garratt, "Telegraphy," 644–62.

CHAPTER 2. THE INVENTOR

1. Edison's Autobiographical Notes, September 11, 1908 (TAEB 1:630); Paul Israel, *Edison: A Life of Invention* (New York: Wiley, 1998), 6–19; Neil Baldwin, *Edison: Inventing the Century* (New York: Hyperion, 1995), 28–32.

2. Quotations from Robert Conot, *A Streak of Luck* (New York: Seaview, 1979), 8–9. Also see Israel, *Edison*, 6–11.

3. Quotation from Israel, *Edison*, 17.

4. "Edison's Autobiographical Notes," September 11, 1908 (TAEB 1:630); Israel, *Edison*, 6–19; Baldwin, *Edison*, 28–37.

5. Quotation from TAEB 1:662–63. Also see Israel, *Edison*, 22.

6. Quotation from Israel, *Edison*, 22. Also see Conot, *Streak of Luck*, 27–28.

7. Quotations from Baldwin, *Edison*, 47; *Telegrapher* 5 (January 30, 1889): 183 (TAEB 1:111). Also see Israel, *Edison*, 20–47.

8. Edison's Autobiographical Notes, September 11, 1908 (TAEB 1:633–41); Israel, *Edison*, 40–55.

9. Quotation from Edison to Samuel and Nancy Edison, October 30, 1870 (TAEB 1:212).

10. Israel, *Edison*, 55–101.

11. Quotation from ibid., 66.

12. Quotations from Thomas P. Hughes, *American Genesis: A Century of Invention and Technological Enthusiasm* (New York: Penguin, 1989), 28; Israel, *Edison*, 120.

13. TAEB 2:444.

14. Quotation from *New York Sun*, April 29, 1878 (TAEM 94:186). Also see "Edison's Autobiographical Notes" (TAEB 4:859, 863); Israel, *Edison*, 130–46; Matthew Josephson, *Edison* (New York: McGraw-Hill, 1959), 165–74; Conot, *Streak of Luck*, 108–9.

15. Quotations from Andre Millard, *Edison and the Business of Innovation* (Baltimore: Johns Hopkins University Press, 1990), 22; *New York Daily Graphic*, April 10, 1878 (TAEM 94:158).

16. Wyn Wachhorst, *Thomas Alva Edison: An American Myth* (Cambridge, Mass.: MIT Press, 1981), 21–22.

17. Quotation from "Edison's Autobiographical Notes," September 11, 1908 (TAEB 4:858). Also see Israel, *Edison*, 162–63. For a contemporaneous account of a trip into the West, see W. G. Marshall, *Through America; or, Nine Months in the United States* (London: Sampson Low, Marston, Searle, & Rivington, 1881), 122–35.

18. Quotation from *New York Daily Graphic*, April 2, 1878 (TAEM 27:776). Also see "Charles Batchelor's Recollections of Edison" (TAEB 4:866–67); Israel, *Edison*, 147–52.

CHAPTER 3. LIGHT

1. Quotation from *New York Daily Graphic*, August 28, 1878 (TAEM 94:338). Also see *New York Sun*, August 29, 1878 (TAEM 94:338); Israel, *Edison*, 163.

2. W. Bernard Carlson, *Innovation as a Social Process: Elihu Thomson and the Rise of General Electric, 1870–1900* (Cambridge: Cambridge University Press, 1991), 88; John A. Jakle, *City Lights: Illuminating the American Night* (Baltimore: Johns Hopkins University Press, 2001), 39–40; King, "Development of Electrical Technology," 344–407; C. Mackechnie Jarvis, "The Generation of Electricity," in *A History of Technology*, 5:179–83; Brian Bowers, "Electricity," in *An Encyclopaedia of the History of Technology*, 354–87.

3. Quotation from *New York Mail*, September 10, 1878 (TAEM 94:349).

4. Robert Friedel and Paul Israel, *Edison's Electric Light: Biography of an Invention* (New Brunswick, N.J.: Rutgers University Press, 1986), 7; Jakle, *City Lights*, 35–36, 52.

5. Quotation from *New York Sun*, October 20, 1878 (TAEM 94:382).

6. Friedel and Israel, *Edison's Electric Light*, 7–8.

7. Quotation from Edison to William Wallace, September 13, 1878 (TAEM 17:925). Also see Friedel and Israel, *Edison's Electric Light*, 8–13; Israel, *Edison*, 169–70.

8. *New York Sun*, September 16, 1878 (TAEB 4:503–6).

9. Edison to Theodore Puskas, October 5, 1878 (TAEM 28:828); Friedel and Israel, *Edison's Electric Light*, 22–23; Jean Strouse, *Morgan: American Financier* (New York: Random House, 1999), 181–83.

10. Israel, *Edison*, 171; Josephson, *Edison*, 187.

11. Quotation from Josephson, *Edison*, 163.

12. Israel, *Edison*, 119–21; Friedel and Israel, *Edison's Electric Light*, 33.

13. Friedel and Israel, *Edison's Electric Light*, 27–35; Israel, *Edison*, 86, 145, 179.

14. *New York Herald*, January 17, 1879 (TAEM 94:450); Friedel and Israel, *Edison's Electric Light*, 37–38; Alfred O. Tate, *Edison's Open Door: The Life Story of Thomas A. Edison, a Great Individualist* (New York: Dutton, 1938), 127.

15. *New York Herald*, January 10, 1880 (TAEM 94:562); Francis Jehl, *Menlo Park Reminiscences*, 3 vols. (Dearborn, Mich.: Edison Institute, 1937–41), 1:285; 2:515–16, 858; Conot, *Streak of Luck*, 184.

16. Quotation from *New York Herald*, January 10, 1880 (TAEM 94:562). Also see Jehl, *Menlo Park Reminiscences*, 3:1145; Conot, *Streak of Luck*, 87.

17. Friedel and Israel, *Edison's Electric Light*, 49–53; Israel, *Edison*, 180–84.

18. Quotation from Francis Upton to Elijah Wood Upton, July 6, 1879 (TAEM 95:543). Also see Friedel and Israel, *Edison's Electric Light*, 69–72; Israel, *Edison*, 175–76, 181–83.

19. *New York Herald*, March 27, 1879 (TAEM 94:464).

20. Quotations from *New York Daily Graphic*, February 6, April 11, 1879 (TAEM 94:456, 474). Also see *London Times*, March 22, 1879; *New York Sun*, July 7, August 3, 1879 (TAEM 94:464, 494, 504).

21. "American Electric Lights," *Telegraphic Journal and Electrical Review*, February 1, 1879 (TAEM 23:528).

22. Edison to Theodore Puskas, November 13, 1878 (TAEB 4:704–6).

23. Quotation from Friedel and Israel, *Edison's Electric Light*, 75. Also see TAEB 4:717–19; Friedel and Israel, *Edison's Electric Light*, 54–56, 116–17; Israel, *Edison*, 172; Thomas P. Hughes, *Networks of Power: Electrification in Western Society, 1880–1930* (Baltimore: Johns Hopkins University Press, 1983), 35–37.

24. Friedel and Israel, *Edison's Electric Light*, 78–79.

25. Quotation from *New York Herald*, December 21, 1879 (TAEM 94:537).

26. Quotation from Menlo Park Notebook #52, N-79-7-31, p. 115 (TAEM 33:594). Also see Friedel and Israel, *Edison's Electric Light*, 94–105; Israel, *Edison*, 187.

27. Quotations from Edison to George E. Gouraud, December 1, 1879 (TAEM 49:746); Israel, *Edison*, 187. Also see Friedel and Israel, *Edison's Electric Light*, 103–5.

28. Quotation from Josephson, *Edison*, 224–25.

29. For the New Year's Eve demonstration, see *New York Herald*, December 31, 1879, January 1, 1880 (TAEM 94:557–59); Friedel and Israel, *Edison's Electric Light*, 109–14; Israel, *Edison*, 187–88.

CHAPTER 4. ELECTRICITY AND LIFE

1. Quotation from Joseph Priestley, *The History and Present State of Electricity, with Original Experiments*, 4th ed. (London: C. Bathurst, 1775), 597–601. Also see Martin S. Pernick, "Back from the Grave: Recurring Controversies over Defining and Diagnosing Death in History," in *Death: Beyond Whole-Brain Criteria*, ed. Richard M. Zaner (Boston: Kluwer, 1988), 17–74; Jan Bondeson, *Buried Alive: The Terrifying History of Our Most Primal Fear* (New York: Norton, 2001).

2. Quotations from John [*sic*] Aldini, *An Account of the Galvanic Experiments Performed . . . on the Body of a Malefactor Executed at Newgate Jan. 17, 1803* (London: Cuthell and Marin, 1803), 8–10, 57. Also see Morus, *Frankenstein's Children*, 127.

3. Quotations from Morus, *Frankenstein's Children*, 128; *Authentic Confession of Jesse Strang* (New York: E. M. Murden and A. Ming, 1827), 20; *New York Times*, September 1, 1870. Also see Stuart Banner, *The Death Penalty: An American History* (Cambridge, Mass.: Harvard University Press, 2002), 175; Commission Report, 67–68; W. Mattieu Williams, "Electromania," *Popular Science Monthly* 21 (1882): 650–55.

4. Stanley Finger and Mark B. Law, "Karl August Weinhold and His 'Science' in the Era of Mary Shelley's Frankenstein: Experiments on Electricity and the Restoration of Life," *Journal of the History of Medicine* 53 (1998): 161–80.

5. Quotations from Mary Shelley, *Frankenstein*, ed. Johanna M. Smith (Boston: Bedford, 2000), 58–60; Edgar Allan Poe, *Poetry and Tales* (New York: Library of America, 1984), 670. Poe's "Some Words with a Mummy" addresses similar themes.

6. Quotation from advertisement in *Monthly Catalogue of the Eden Musée*, February 1887, p. 10, Billy Rose Theater Collection of the New York Public Library for the Performing Arts. Also see David E. Nye, *Electrifying America: Social Meanings of a New Technology* (Cambridge, Mass.: MIT Press, 1990), 154; Morus, *Frankenstein's Children*, 144–45, 234–54; Rowbottom and Susskind, *Electricity and Medicine*, 59–66, 113; M. Allen Starr, "Electricity in Relation to the Human Body," *Scribner's Magazine* 6 (November 1889): 589–99; Bonnie Ellen Blustein, *Preserve Your Love for Science: Life of William A. Hammond, American Neurologist* (New York: Cambridge University Press, 1991), 127–33.

7. Quotation from Oliver Wendell Holmes, *Medical Essays, 1842–1882* (Cambridge: Riverside Press, 1893), 203. Also see Charles Rosenberg, "The Therapeutic Revolution," in *Explaining Epidemics and Other Studies in The History of Disease* (Cambridge: Cambridge University Press, 1992), 9–31.

8. Quotations from George M. Beard and A. D. Rockwell, *A Practical Treatise on the Medical and Surgical Uses of Electricity*, 6th ed. (New York: William Wood, 1888), 217, 222. Also see Rowbottom and Susskind, *Electricity and Medicine*, 113–14; Lisa Rosner, "The Professional Context of Electrotherapeutics," *Journal of the History of Medicine* 43 (1988): 64–82; Rowbottom and Susskind, *Electricity and Medicine*, 103–14.

9. George Beard, "Neurasthenia, or Nervous Exhaustion," *Boston Medical and Surgical Journal* 80 (1869): 217–21; Charles Rosenberg, "George M. Beard and American Nervousness," in *No Other Gods: On Science and American Social Thought* (Baltimore: Johns Hopkins University Press, 1976).

10. Quotation from TAEB 2:204–9. On Beard's relationship with Edison, see Jarvis Edson to Edison, October 26, 1874 (TAEM 27:466); Israel, *Edison*, 110–15; TAEB 2:321, 754–55. On the induction coil, see Robert C. Post, "Stray Sparks from the Induction Coil: The Volta Prize and the Page Patent," *Proceedings of the IEEE* 64 (1976): 1279–86; Israel, *Edison*, 100. Inspired by his success in the medical market, Edison concocted "Edison's Polyform," a patent medicine for "neurologic pains" that contained morphine, alcohol, chloroform, ether, peppermint, and cloves. The inventor sold the name and formula for $5,000 to three promoters, who marketed Edison's Polyform well into the twentieth century. See TAEB 4:228–29; *New York Herald*, December 21, 1879 (TAEM 94:541–42); David Nye, *The Invented Self: An Anti-biography, from Documents of Thomas A. Edison* (Odense, Denmark: Odense University Press, 1983), 139–45.

11. Quotations from "Edisons Inductorum [*sic*]," draft essay, c. May 20, 1874 (TAEB 2:206); Israel, *Edison*, 22; TAEB 2:199.

12. Quotations from W. Mattieu Williams, "Electromania," *Popular Science Monthly* 21 (1882): 651; *Ward B. Snyder, Sportsmen's Goods* (New York, 1875), 152, Prints Division, Metropolitan Museum of Art; Kenneth Walter Cameron, ed., *The Massachusetts Lyceum During the American Renaissance: Materials for the Study of the Oral Tradition in American Letters* (Hartford, Conn.: Transcendental Books, 1969), 9; New York Museum of Anatomy, undated program, Billy Rose Theater Collection, New York Public Library for the Performing Arts. Also see *New York Herald*, December 21, 1879 (TAEM 94:539); "Prof. Anderson's Grand Drawing Room Entertainment of Natural Magic et Soirees Mysterieuse," undated handbill for Winter Garden Theater, Billy Rose Theater Collection. On Barnum and dime museums, see Neil Harris, *Humbug: The Art of P. T. Barnum* (Boston: Little, Brown, 1973).

13. Quotation from *New York Sun*, December 1, 1878 (TAEM 23:187). Also see *Annual Report of the Board of Managers of the Ohio Penitentiary . . . for the Fiscal Year 1895* (Columbus: Westbote, 1896), 28–29.

14. Quotations from B. C. Brodie, "Explaining the Mode in Which Death Is Produced by Lightning," *London Medical Gazette* 1 (1827–28): 80; Benjamin W. Richardson, "On Research with the Large Induction Coil of the Royal Polytechnic Institution, with Special Reference to the Cause and Phenomena of Death by Lightning," *Medical Times and Gazette* 1 (1869): 511–12, 595–99; 2 (1869): 183–86, 373–76; W. Bathurst Woodman and Charles Tidy, *A Handy-book of Forensic Medicine and Toxicology* (London: J. & A. Churchill, 1877), 969–70.

15. *New York Herald*, November 26, 30, December 14, 1879.

CHAPTER 5. "DOWN TO THE LAST PENNY"

1. Quotations from *Saturday Review*, January 10, 1880, as quoted in *Journal of Gas Lighting*, January 20, 1880 (TAEM 94:573); *Puck*, as quoted in *Telegraphic Journal and Electrical Review*, July 15, 1880 (TAEM 23:432). Also see unidentified clipping (TAEM 26:358); *Telegraphic Journal*, April 15, 1880 (TAEM 24:25)

2. *New York Sun*, September 16, 1878 (TAEM 94:354); Friedel and Israel, *Edison's Electric Light*, 177–78; Jakle, *City Lights*, 19–37.

3. Quotation from *New York Daily Tribune*, November 26, 1880 (TAEM 89:18). Also see *St. Louis Post-Dispatch*, May 1, 1882 (TAEM 95:191).

4. Quotation from Friedel and Israel, *Edison's Electric Light*, 198. Harold C. Passer, *The Electrical Manufacturers, 1875–1900: A Study in Competition, Entrepreneurship, Technical Change, and Economic Growth* (Cambridge, Mass.: Harvard University Press, 1953), 177–79; Frank Lewis Dyer and Thomas Commerford Martin, *Edison: His Life and Inventions*, 2 vols. (New York: Harper & Brothers, 1910), 2:910–17.

5. *New York Times*, August 9, 1880; *New York Truth*, March 6, 1881 (TAEM 94:609, 629); *Harper's Weekly* 26 (June 24, 1882), 394; Jehl, *Menlo Park Reminiscences*, 2:895–96; Friedel and Israel, *Edison's Electric Light*, 120–23.

6. Jehl, *Menlo Park Reminiscences*, 2: 558–63; Francis Upton to Elijah Wood Upton, May 9, 1880 (TAEM), 95:606; *Scientific American* 42 (May 22, 1880), 326; Friedel and Israel, *Edison's Electric Light*, 140–44; Conot, *Streak of Luck*, 170–71.

7. Quotations from John R. Segredor to Edison, October 8, 14, 1880 (TAEM 53:816, 893); Vesey Butler to Edison, October 27, 1880 (TAEM 53:912). Also see Friedel and Israel, *Edison's Electric Light*, 129–32, 154–56; Israel, *Edison* 196–97; Jehl, *Menlo Park Reminiscences*, 2:338, 563, 614.

8. Quotation from Francis Upton to Elijah Wood Upton, March 2, 1879 (TAEM 95:512). Also see Friedel and Israel, *Edison's Electric Light*, 146; David Trumbull Marshall, *Recollections of Edison* (Boston: Christopher Publishing, 1931), 38; Jehl, *Menlo Park Reminiscences*, 2:497, 852–54.

9. Jehl, *Menlo Park Reminiscences*, 2:516–17, 607.

10. *New York Evening Post*, December 21, 1880; *New York Herald*, November 18, 1880 (TAEM 94:617, 24).

11. Quotations from *Illustrated Scientific News*, December 21, 1880; *Hartford Daily Times*, undated clipping, c. December 10, 1880 (TAEM 89:14; 94:621).

12. Quotations from unidentified clipping, November 26, 1880; *New York Herald*, November 18, 1880 (TAEM 94:617, 620).

13. Friedel and Israel, *Edison's Electric Light*, 178–80; Jehl, *Menlo Park Reminiscences*, 2:721–24; Conot, *Streak of Luck*, 176–78; Marshall, *Recollections of Edison*, 29–30; Edison to Grosvenor Lowrey, July 20, 1880 (TAEM 54:65).

14. The account of the aldermen's visit is derived from the following New York newspapers for December 21, 1880: *Truth*, *Herald*, *Evening Post*, *World*, *Sun*, *Times*, and *Daily Tribune*.

15. Quotation from *New York Star*, December 21, 1880 (TAEM 94:623). Also see *New York Post*, December 21, 1880 (TAEM 94:624); Charles Brush, "The Arc-Light," *Century Magazine* 70 (1905): 110–18; Mel Gorman, "Charles F. Brush and the First Public Electric Street Lighting System in America," *Ohio Historical Quarterly* 70 (1961): 128–44; Carlson, *Innovation as a Social Process*, 80–81; Jakle, *City Lights*, 38–47. On electric light spectacles, see Carolyn Marvin, *When Old Technologies Were New: Thinking About Electric Communication in the Late Nineteenth Century* (New York: Oxford University Press, 1988), 158–62.

16. Passer, *Electrical Manufacturers*, 20.

17. This account of the night of July 13, 1881, is drawn from *Buffalo Daily Courier* and *Buffalo Morning Express*, July 14, 1881. On Buffalo in this time period, see *Atlas of the City of Buffalo, New York* (Philadelphia: G. M. Hopkins, 1884); Mark Goldman, *High Hopes: The Rise and Decline of Buffalo, New York* (Albany: State University of New York Press, 1983), 176–85; Brenda K. Shelton, *Reformers in Search of Yesterday: Buffalo in the 1890s* (Albany: State University of New York Press, 1976), 5–10.

18. This account of the night of August 7, 1881, is drawn from the *Buffalo Commercial Advertiser*, August 8, 1881; *Buffalo Morning Express*, August 8, 1881; *Buffalo Daily Courier*, August 8, 9, 1881; *Buffalo Evening News*, July 14 and August 8, 9, 11, 1881; testimony of Philip Fogerty, Wallace Harrington, and Charles Hayner in Kemmler Hearings.

19. Kemmler Hearings, 956–59; *Buffalo Evening News*, August 11, 1881; *Buffalo Morning Express*, August 9, 1881.

20. Quotation from *Buffalo Morning Express*, August 8, 1881.

CHAPTER 6. WIRING NEW YORK

1. Jehl, *Menlo Park Reminiscences*, 3:924–26, 967; *New York Times*, March 1, 1881; Dyer and Martin, *Edison*, 1:360–61.

2. Israel, *Edison*, 210–11; Friedel and Israel, *Edison's Electric Light*, 195; Jehl, *Menlo Park Reminiscences*, 2:503–5.

3. Quotation from *New York Sun*, September 16, 1878 (TAEM 94:354). Also see Friedel and Israel, *Edison's Electric Light*, 177–78; Jakle, *City Lights*, 19–37. For an incident of sabotage by line cutting, see *Frank Leslie's Illustrated Newspaper*, August 25, 1883.

4. Quotation from Dyer and Martin, *Edison*, 1:392–93.

5. *Harper's Weekly* 26 (June 24, 1882): 394; Friedel and Israel, *Edison's Electric Light*, 196, 207; Conot, *Streak of Luck*, 195–96.

6. *New York Times*, August 9, 1880; *New York Truth*, March 6, 1881 (TAEM 94:609, 629); *Harper's Weekly* 26 (June 24, 1882): 394; Jehl, *Menlo Park Reminiscences*, 2:895–96; *New York Evening Post*, December 1, 1881 (TAEM 95:4).

7. Quotation from Dyer and Martin, 1:380; Jehl, *Menlo Park Reminiscences*, 2:897, 3:959, 996.

8. *New York Times*, March 1, 1881.

9. *St. Louis Post-Dispatch*, May 1, 1882 (TAEM 95:191); Josephson, *Edison*, 261; Nye, *Electrifying America*, 32.

10. Jehl, *Menlo Park Reminiscences*, 2:741; Fred E. H. Schroeder, "More 'Small Things Forgotten': Domestic Electrical Plugs and Receptacles, 1881–1931," *Technology and Culture* 27 (1986): 529–30; Friedel and Israel, *Edison's Electric Light*, 169–71, 201–4, 219.

11. Friedel and Israel, *Edison's Electric Light*, 213–18; Israel, *Edison*, 205, 214–15; Conot, *Streak of Luck*, 189.

12. Quotations from Leslie Ward to Edison, January 18, 1882 (TAEM 50:698); Israel, *Edison*, 230, 163. Also see Conot, *Streak of Luck*, 122, 127, 19.

13. *New York Times*, March 9, 1882.

14. See, for example, Edison Electric Light Company, *Bulletin* 6 (March 27, 1882): 5–6; 7 (April 17, 1882): 9; *Morning Advertiser*, January 5, 10, 1882 (TAEM 95:111–12).

15. On Edison Electric's warnings on high-voltage cables, see Edward Johnson's letter to *New York Tribune*, February 7, 1881 (TAEM 94:633); *New York Truth*, March 6, 1881 (TAEM 94:629). On deaths from the current, see *New York Evening Post*, December 17, 1882, quoted in Payson Jones, *A Power History of the Consolidated Edison System, 1878–1900* (New York: Consolidated Edison, 1940), 113; R. H. Jaffé, "Electropathology," *Archives of Pathology* 5 (1928): 838; *New York Times*, February 23, 1882; *New York Herald*, October 5, 8, 10, 1882. Arc lamp wires also caused fires: see *New York Evening Post*, December 1, 1881 (TAEM 95:44); Edison Electric Light Company, *Bulletin* 4 (Feb. 24, 1882): 9.

16. Quotation from *New York Morning Advertiser*, April 12, 1882 (TAEM 95:182). Also see *New York Morning Post*, April 12, 1882 (TAEM 95:183); *Journal of Gas Lighting*, February 21, 1882 (TAEM 95:142); Edison Electric Light Company, *Bulletin* 7 (April 17, 1882): 3–4, 7–8; 8 (April 27, 1882): 11; 10 (June 5, 1882): 4–5.

17. Dyer and Martin, *Edison*, 1:380–81.

18. Friedel and Israel, *Edison's Electric Light*, 219–20; Josephson, *Edison*, 257; Jones, *Power History*, 133.

19. *Scientific American* 47 (August 26, 1882): 130.

20. *New York Herald*, *New York Tribune*, *New York Sun*, *New York Times*, September 5, 1882; Jones, *Power History*, 157–214, 293–94.

21. *New York Herald*, September 5, 1882 (TAEM 24:82).

22. *New York Sun*, September 5, 1882.

CHAPTER 7. THE HANGING RITUAL

1. Quotation from T. Commerford Martin, *Forty Years of Edison Service* (New York: Press of the New York Edison Company, 1922), 66. Also see *New York Times*, August 25, 1882; *Judge* 2 (September 9, 1882): 2, 16.

2. Quotation from Edison to C. F. Pond, August 22, 1882. Also see Pond to Edison, August 17, 1882 (TAEM 60:491, 81:816).

3. Quotation from *Manufacturer and Builder* 12 (February 1880): 39.

4. Pieter Spierenburg, *The Spectacle of Suffering: Executions and the Evolution of Repression* (Cambridge: Cambridge University Press, 1984); Peter Spierenburg, "The

Body and the State: Early Modern Europe," in *The Oxford History of the Prison: The Practice of Punishment in Western Society*, ed. Norval Morris and David J. Rothman (New York: Oxford University Press, 1995), 49–77; Michel Foucault, *Discipline and Punish: The Birth of the Prison*, trans. Alan Sheridan (New York: Vintage, 1979).

5. Spierenburg, "The Body and the State," 55–56.

6. Quotation from Karen Halttunen, *Murder Most Foul: The Killer and the American Gothic Imagination* (Cambridge, Mass.: Harvard University Press, 1998), 22–29.

7. This account of Strang's crime and execution draws on the following sources: *Albany Argus and City Gazette* and *Albany Daily Advertiser*, August 25, 27, 1827; *Authentic Confession of Jesse Strang* (New York: E. M. Murden and A. Ming, 1827); *Confession of Jesse Strang* (Albany: John B. Van Steenbergh, 1827); P. R. Hamblin, *United States Criminal History: Being a True Account of the Most Horrid Murders, Piracies, High-Way Robberies, &c., Together with the Lives, Trials, Confessions and Executions of the Criminals. Compiled from the Criminal Records of the Counties* (Fayetteville, N.Y.: Mason & De Puy, 1836), 258–66. Louis C. Jones, *Murder at Cherry Hill: The Strang-Whipple Case, 1827* (Albany: History Cherry Hill, 1982). White hanging shrouds were common as late as the 1820s but mostly disappeared by midcentury, replaced by common black suits: see Thomas M. McDade, *The Annals of Murder: A Bibliography of Books and Pamphlets on American Murders from Colonial Times to 1900* (Norman: University of Oklahoma Press, 1961), xxxi.

8. Quotations from *Albany Argus and City Gazette*, August 25, 1827; "The Order for the Burial of the Dead," *Book of Common Prayer, According to the Use of the Protestant Episcopal Church in the United States of America* (New York: Henry I. Megarey, 1818).

9. *Albany Daily Advertiser*, August 25, 1827.

10. Quotations from ibid. On public behavior at hangings, see Banner, *Death Penalty*, 146–51; Philip English Mackey, *Hanging in the Balance: The Anti-Capital Punishment Movement in New York State, 1776–1861* (New York: Garland, 1982), 108–11; V. A. C. Gatrell, *The Hanging Tree: Execution and the English People, 1770–1868* (New York: Oxford University Press, 1994), 90–105.

11. Paul E. Johnson, *A Shopkeeper's Millennium: Society and Revivals in Rochester, New York, 1815–1837* (New York: Hill and Wang, 1978), 37–61; Paul A. Gilje, *The Road to Mobocracy: Popular Disorder in New York City, 1763–1834* (Chapel Hill: University of North Carolina Press, 1987), 175–77; 235–64.

12. Quotation from Banner, *Death Penalty*, 150. Also see Gatrell, *Hanging Tree*, 32–38.

13. Quotations from Michael Madow, "Forbidden Spectacle: Executions, the Public and the Press in Nineteenth Century New York," *Buffalo Law Review* 43 (1995): 500.

14. Louis P. Masur, *Rites of Execution: Capital Punishment and the Transformation of American Culture, 1776–1865* (New York: Oxford University Press, 1989), 94; John D. Bessler, *Death in the Dark: Midnight Executions in America* (Boston: Northeastern University Press, 1997), 41–46.

15. Masur, *Rites of Execution*, 76–84; Foucault, *Discipline and Punish*; David J. Rothman, *Discovery of the Asylum: Social Order and Disorder in the New Republic* (Boston: Little, Brown, 1971); David J. Rothman, "Perfecting the Prison," in *Oxford History of the Prison*, 111–29; Lane, *Murder in America*, 79–80; Banner, *Death Penalty*, 88–111, 231–33; Gatrell, *Hanging Tree*, 7, 20–21, 201–3; David D. Cooper, *The Lesson of*

the Scaffold: The Public Execution Controversy in Victorian England (Athens: Ohio University Press, 1974), 27–33.

16. Quotation from Walt Whitman, "A Dialogue," *United States Magazine and Democratic Review* 17 (November 1845): 360–64. Also see Masur, *Rites of Execution*, 153–56; Reverend Charles Wiley, "Retributive Law and Capital Punishment," *American Presbyterian Review* 20 (1871): 414–31; A. Jacobi, William C. Wey, and B. F. Sherman, "Capital Punishment. Report of a Committee Appointed by the Medical Society of the State of New York at its Annual Meeting in 1891, and Presented Before the Society at the Session of 1892," *Sanitarian* 29 (1892): 47–57.

17. Quotations from Masur, *Rites of Execution*, 157; Banner, *Death Penalty*, 216–17; *New York Herald*, May 1, 1893. Also see E. S. Nadal, "The Rationale of the Opposition to Capital Punishment," *North American Review* 116 (1873): 138–50.

18. The Maine law stipulated that a prisoner could be executed no sooner than one year from the date of sentence, and then only by a specific order from the governor. After the law was passed, no executions took place in Maine for twenty-seven years. Philip English Mackey, "Introduction," in Mackey, ed., *Voices Against Death: American Opposition to Capital Punishment, 1787–1975* (New York: Burt Franklin, 1976), xxi–xxii; David Brion Davis, "The Movement to Abolish Capital Punishment in America, 1787–1861," in *From Homicide to Slavery: Studies in American Culture* (New York: Oxford University Press, 1986), 17–40; Banner, *Death Penalty*, 134–55.

19. Madow, "Forbidden Spectacle," 512–15; Banner, *Death Penalty*, 157–60.

20. Gatrell, *Hanging Tree*, 51–54; Banner, *Death Penalty*, 44–47, 171–74.

21. Madow, "Forbidden Spectacle," 530–31. For reports on bungled hangings, see *New York Times*, May 20, 1882; *New York World*, October 1, 1885; *Buffalo Courier*, July 11, 1889.

22. Quotations from Edmund Clarence Stedman, "The Gallows in America," *Putnam's Magazine* 13 (February 1869): 225–35 (reprinted in Mackey, *Voices Against Death*, 132–40, emphasis in original); *New York Times*, January 17, 1888. Also see Banner, *Death Penalty*, 220–21; Roger Lane, "Capital Punishment," in *Violence in America: An Encyclopedia*, ed. Ronald Gottesman (New York: Charles Scribner's Sons, 1999), 198–203. Maine restored the death penalty in 1883, then abolished it for good in 1887. See Mackey, *Voices Against Death*, xxx–xxxi.

CHAPTER 8. THE DEATH PENALTY COMMISSION

1. Lawrence Stone, "Interpersonal Violence in English Society, 1300–1800," *Past and Present* 101 (1983): 22–33; Halttunen, *Murder Most Foul*, 62; James Turner, *Reckoning with the Beast: Animals, Pain, and Humanity in the Victorian Mind* (Baltimore: Johns Hopkins University Press, 1980), 2–3.

2. Quotations from Turner, *Reckoning with the Beast*, 79, 81.

3. Halttunen, *Murder Most Foul*, 62–65; Keith Thomas, *Man and the Natural World: Changing Attitudes in England, 1500–1800* (London: Penguin, 1983), 143–91.

4. Turner, *Reckoning with the Beast*, 34–35.

5. Martin S. Pernick, *A Calculus of Suffering: Pain, Professionalism, and Anesthesia in Nineteenth-Century America* (New York: Columbia University Press, 1985); Turner, *Reckoning with the Beast*, 81–82; Halttunen, *Murder Most Foul*, 64–65.

6. Turner, *Reckoning with the Beast*, 50; Gerald Carson, *Men, Beasts, and Gods: A*

History of Cruelty and Kindness to Animals (New York: Charles Scribner's Sons, 1972), 95–96.

7. Quotations from Turner, *Reckoning with the Beast*, 70, 134. Also see Woman's Branch of the Pennsylvania Society for the Prevention of Cruelty to Animals, *Annual Report* (Philadelphia: M'Farland, 1870), 10.

8. Quotation from Edmund Clarence Stedman, "The Gallows in America," *Putnam's Magazine* 13 (February 1869): 225–35. Also see Jones, *Murder at Cherry Hill*, 114–15.

9. Quotation from G. W. Peck, "On the Use of Chloroform in Hanging," *American Whig Review* 8 (September 1848): 294. Also see Samuel Hand, "The Death Penalty," *North American Review* 133 (1881): 541–50; Pernick, *A Calculus of Suffering*, 111; Masur, *Rites of Execution*, 20–21; Foucault, *Discipline and Punish*, 91; Karl Shoemaker, "The Problem of Pain in Punishment: Historical Perspectives," in Austin Sarat, ed., *Pain, Death, and the Law* (Ann Arbor: University of Michigan Press, 2001), 15–41.

10. Benjamin Ward Richardson, "The Painless Extinction of Life," *Popular Science Monthly* 26 (1884–85): 641–52; Woman's Branch of the Pennsylvania Society for the Prevention of Cruelty to Animals, *Annual Report* (1874), p. 15, (1875), p. 12, (1878), pp. 13–4; Laurence Turnbull, *The Advantages and Accidents of Artificial Anaesthesia*, 2nd ed. (Philadelphia: P. Blakiston, 1885), 261–62; D. D. Slade, *How to Kill Animals Humanely* (Boston: Massachusetts SPCA, 1879); Diane L. Beers, "A History of Animal Advocacy in America: Social Change, Gender, and Cultural Values, 1865–1975" (dissertation, Temple University, 1998), 136.

11. Quotations from John H. Packard, "The Mode of Inflicting the Death Penalty," *Sanitarian* 6 (August 1878): 360–63; G. M. Hammond, "On the Proper Method of Executing the Sentence of Death by Hanging," *Sanitarian* 10 (November 1882): 664–68. Also see "What Hanging Is Like: The Enjoyable Experience of One Who Has Tried It," *New York World*, November 1, 1885; G. W. Peck, "On the Use of Chloroform in Hanging," *American Whig Review* 8 (September 1848): 294; F. H. Hamilton et al., "Committee on Substitutes for Hanging," *Physician and Bulletin of the Medico-Legal Society of New York* 13 (1880): 200–204; Benjamin Ward Richardson, "Modes of Death in the Execution of English Criminals," *Lancet* 2 (1883): 1006, 1066; N. E. Brill, "An Argument Against the Hangman's Bungling," *American Journal of Neurology* 3 (1884): 643–63; Wooster Beach, "The Death Penalty: Proper Mode of Its Infliction," *Medical Record* 30 (July 24, 1886): 89–90; J. B. Thornton, "Some Further Remarks on the Death Penalty and Method of Infliction," *Medical Record* 30 (August 21, 1886): 222–23; F. E. Maine, "A Few More Words Regarding the Death Penalty and the Mode of Infliction," *Medical Record* 30 (October 9, 1886): 417; Pernick, *Calculus of Suffering*, 111; Masur, *Rites of Execution*, 20–21.

12. Quotations from Edmund Clarence Stedman, "The Gallows in America," *Putnam's Magazine* 13 (February 1869): 225–35; *Scientific American* 28 (June 7, 1873): 352; *New York Herald*, November 26, 1879. Also see Benjamin Ward Richardson, "On Research with the Large Induction Coil of the Royal Polytechnic Institution," *Medical Times and Gazette* 1 (1869): 595. For other early tests on animals, see B. C. Brodie, "Explaining the Mode in Which Death Is Produced by Lightning," *London Medical Gazette* 1 (1827–28): 79. For other early references to electrical execution, see Alonzo Calkins, "Felonious Homicide: Its Penalty, and the Execution thereof Judicially," in *Papers Read Before the Medico-Legal Society of New York*, 2nd series, rev. ed. (New York: W. F. Vanden Houten, 1882), 273; *New York Daily Tribune*, February 25, 1885.

13. *New York Evening Post*, December 17, 1882, quoted in Jones, 113; Jaffé, "Electropathology," 838; *New York Times*, February 23, 1882; *New York Herald*, October 5, 8, 10, 1882; Kemmler Hearings, 956–59; *Buffalo Evening News*, August 11, 1881; *Buffalo Morning Express*, August 9, 1881.

14. Quotation from Alfred Southwick to Edison, December 5, 1887 (TAEM 119:321). Few details regarding Southwick's early experiments are known, but some information is available in *New York World*, February 7, 1888; *New York Times*, May 24, 1890; *Buffalo Express*, August 7, 1890; *Buffalo Courier*, June 12, 1898; Kemmler Hearings, 370–71.

15. A. P. Southwick, "Anatomy and Physiology of Cleft Palate," *Transactions of the Dental Society of the State of New York* 15 (1883): 50–55. For Southwick's biography, see *Men of New York*, vol. 1 (Buffalo: George E. Matthews, 1898), 229–30; Louis B. Lane, *Memorial and Family History of Erie County, New York* (New York: Genealogical Publishing, 1906–8), 272–74; *Dental Cosmos* 40 (1898): 597–98; *Buffalo Courier*, June 12, 1898; *Buffalo Express*, June 29, 1898; *Buffalo Commercial*, June 11, 1898.

16. Quotations from *New York Tribune*, January 13, 1885; *Laws of New York*, 1886, chap. 352. For McMillan's role, see *New York Times*, May 24, August 7, 1890; *New York Sun*, August 2, 1890; *Men of New York*, 1: 311–12.

17. On Gerry, see Allen Johnson and Dumas Malone, *Dictionary of American Biography* (New York: Charles Scribner's Sons), 4:227–28; *New York Sun*, February 18, 1927; *New York Times, New York World*, February 18, 1927. On the SPCC, see Stephen Murray Robertson, "Sexuality Through the Prism of Age: Modern Culture and Sexual Violence in New York City, 1880–1950" (dissertation, Rutgers University, 1998).

18. Commission Report, 30, 48–50, 81; Kemmler Hearings, 346–52, 366.

19. See the undated clippings from the *Buffalo Express* and other newspapers in scrapbook, box SB: 1886.001, Buffalo and Erie County SPCA.

20. Quotations from *Buffalo Morning Express*, July 17, 1887; "Report of Agent" and entry for meeting held July 29, 1887, Minutes Book, pp. 107–8, box SB: 1883: 002, Buffalo and Erie County SPCA. Also see *Annual Report, Buffalo Branch for Erie County of the American Society for the Prevention of Cruelty to Animals* (Buffalo, 1887), 8–9; Kemmler Hearings, 973–76.

21. Commission Report, 14–47.

22. Ibid., 19.

23. *Catalogue of the Eden Musée* (New York, 1884), Billy Rose Theater Collection of the New York Public Library for the Performing Arts; Louis Leonard Tucker, "'Ohio Show-Shop': The Western Museum of Cincinnati, 1820–1867," in *A Cabinet of Curiosities: Five Episodes in the Evolution of American Museums* (Charlottesville: University Press of Virginia, 1967), 73–105; Kathleen Kendrick, "'The Things Down Stairs': Containing Horror in the Nineteenth-Century Wax Museum," *Nineteenth-Century Studies* 12 (1998): 1–35; Halttunen, *Murder Most Foul*, 66, 78–82.

24. Commission Report, 81–82. During the Kemmler Hearings (p. 369), Gerry reported slightly different figures: 81 for hanging, 75 for electricity.

25. Quotation from Commission Report, 68–69.

26. Quotations from ibid., 49–50. Also see Daniel Arasse, *The Guillotine and the Terror*, trans. Christopher Miller (London: Allen Lane, 1989), 8–18. Quotation in footnote from *New York World*, December 28, 1886.

27. Kemmler Hearings, 368; Commission Report, 75.

28. Commission Report, 75–86.

29. On Southwick's advocacy of electricity and Gerry's initial reluctance, see Kemmler Hearings, 370–72; *New York World*, December 28, 1886; *New York Times*, January 24, December 17, 1887; *New York Herald*, January 27, 1888; *Electrical Review* 11 (December 24, 1887): 7.

30. Kemmler Hearings, 372, 391, 396.

CHAPTER 9. GEORGE WESTINGHOUSE AND THE RISE OF ALTERNATING CURRENT

1. Quotations from Mary Edison to Samuel Insull, April 30, 1884 (TAEM 71:615); Robert Lozier to John Tomlinson, August 9, 1884 (TAEM 71:627).

2. Quotations from Edison diary, July 12, 1885 (TAEM 90:9); Israel, *Edison*, 233; Notebook N-80-08-09, pp. 13, 57, 218 (TAEM 38:433, 454, 459). Also see Jehl, *Menlo Park Reminiscences*, 2:510–11; Baldwin, *Edison*, 87–88, 144; Conot, *Streak of Luck*, 225–26.

3. Edison Electric Light Company, *Bulletin* 22 (April 9, 1884): 22; Israel, *Edison*, 212–13.

4. Quotation from Josephson, *Edison*, 269.

5. Quotation from unidentified clipping, March 13, 1883 (TAEM 66:21).

6. Quotation from ibid. Also see Israel, *Edison*, 224–25; Passer, *Electrical Manufacturers*, 99.

7. Israel, *Edison*, 226–29; Passer, *Electrical Manufacturers*, 100–101.

8. *New York Sun*, undated clipping (TAEM 89:667).

9. John Hopkinson had patented the same design in Britain a few months before Edison.

10. See Israel, *Edison*, 219; Hughes, *Networks of Power*, 83–84; Passer, *Electrical Manufacturers*, 99, 112–23, 178; Jehl, *Menlo Park Reminiscences*, 3:1100.

11. Quotations from *Electrical Review* 16 (May 24, 1890): 3; *New York Herald*, March 16, 1890; Marvin, *When Old Technologies Were New*, 138. Also see Strouse, *Morgan*, 234–35; Ward McAllister, *Society as I Have Found It* (New York: Cassell, 1890), 353–54; Jehl, *Menlo Park Reminiscences*, 3:1000.

12. Passer, *Electrical Manufacturers*, 41–57.

13. Ibid., 21–31, 68–71.

13. Ibid., 65, 70; Joseph P. Sullivan, "From Municipal Ownership to Regulation: Municipal Utility Reform in New York City, 1880–1907" (dissertation, Rutgers University, 1995), 490.

15. Passer, *Electrical Manufacturers*, 148.

16. Frank Crane, *George Westinghouse: His Life and Achievements* (New York: William H. Wise, 1925), 5–8; Henry G. Prout, *A Life of George Westinghouse* (New York: American Society of Mechanical Engineers, 1921), 1–7; Curt Wohleber, "'St. George' Westinghouse," *American Heritage of Invention and Technology* 12 (Winter 1997): 30; Frank Wicks, "How George Westinghouse Changed the World," *Mechanical Engineering* 118 (October 1996): 74–79.

17. Quotation from Wohleber, "'St. George' Westinghouse," 31. Also see Prout, *Life of George Westinghouse*, 4, 8, 365; Crane, *George Westinghouse*, 8–9.

18. Quotation from Crane, *George Westinghouse*, 13.

19. Ibid., 13–14.

20. Steven W. Usselman, "Air Brakes for Freight Trains: Technological Innovation in the American Railroad Industry, 1869–1900," *Business History Review* 58 (1984): 30–50; Steven W. Usselman, "From Novelty to Utility: George Westinghouse and the Business of Innovation During the Age of Edison," *Business History Review* 66 (1992): 251–304; Prout, *Life of George Westinghouse*, 21–32; Crane, *George Westinghouse*, 5, 21–23.

21. Quotations from Francis G. Leupp, *George Westinghouse: His Life and Achievements* (Boston: Little, Brown, 1918), 287; Prout, *Life of George Westinghouse*, 5.

22. Quotation from Tate, *Edison's Open Door*, 150.

23. See Westinghouse to Edison, June 7, 1888 (TAEM 122:861); "The Westinghouse Electric Company's Dynamo and Incandescent Lamp," *Electrical World* 7 (April 3, 1886): 151–52; Prout, *Life of George Westinghouse*, 92–95; Passer, *Electrical Manufacturers*, 131; George Wise, "William Stanley's Search for Immortality," *American Heritage of Invention and Technology* 4 (Spring 1988): 42–49.

24. Hughes, *Networks of Power*, 85; Arthur A. Bright, *The Electric Lamp Industry: Technological Change and Economic Development from 1800 to 1947* (New York: Macmillan, 1949), 75–76; *New York Times*, December 27, 1882.

25. King, "The Development of Electrical Technology," 349–50; "The Distribution of Electricity by Secondary Generators," *Electrical World* 9 (March 26, 1887): 156–58.

24. The Jablochkoff arc lamp used alternating current. "The Distribution of Electricity by Secondary Generators," *Electrical World* 9 (March 26, 1887): 156–58; Hughes, *Networks of Power*, 86; Jarvis, "The Generation of Electricity," 212.

27. "The Distribution of Electricity by Secondary Generators," *Electrical World* 9 (March 26, 1887): 156–58; Hughes, *Networks of Power*, 87–102; Passer, *Electrical Manufacturers*, 131–32.

28. *Telegraphic Journal and Electrical Review* 12 (June 9, 1883): 467.

29. Hughes, *Networks of Power*, 98–104; Passer, *Electrical Manufacturers*, 131–32.

30. Stanley to Westinghouse, March 17, 1886, reprinted in *Electric Journal* 8 (June 1911): 493.

31. "The Gaulard and Gibbs System of Electric Distribution," *Electrical World* 7 (December 4, 1886): 271–72; Passer, *Electrical Manufacturers*, 137–38; Hughes, *Networks of Power*, 103–4.

32. "Copper," *Electrical Engineer* 7 (February 1888): 42; Reynolds and Bernstein, "Damnable Alternating Current," 1340. Also see William J. Hausman and John L. Neufeld, "Battle of the Systems Revisited: The Role of Copper," *IEEE Technology and Society Magazine* 11 (Fall 1992): 18–25.

33. F. L. Pope, "The Westinghouse Alternating System of Electric Lighting," *Electrician and Electric Lighting* 6 (September 1887), 332–42; Passer, *Electrical Manufacturers*, 137–38, 164–66.

34. "Notes on Distribution of Alternating Current," memo from Edison to Edward Johnson, 1886, pp. 7–10 (TAEM 148:3). Similar points appear in a longer report issued by Siemens & Halske, an Edison ally and major electrical manufacturer in Germany: "About the Use of Transformers in Electric Light Plants," November 1886 (TAEM 79:383).

35. Passer, *Electrical Manufacturers*, 166; Israel, *Edison*, 325.

36. "Notes on Distribution of Alternating Current," 7.

37. Quotation from W. S. Andrews to J. H. Vail, May 12, 1887 (TAEM 119:814). Also see "The Westinghouse Alternating Current System," *Electrical World* 10 (September 3, 1887): 125–27; Passer, *Electrical Manufacturers*, 121, 149.

38. Both agents were quoted in W. J. Jenks to J. H. Vail, November 12, 1887 (TAEM 119:849).

39. Edward Johnson to Edison, December 9, 1887 (TAEM 119:857).

40. Quotations from Alfred Southwick to Edison, November 8, 1887. Edison's response to Southwick's first letter does not survive, but its contents can be surmised from Alfred Southwick to Edison, December 5, 1887 (TAEM 119:321).

41. Quotation from Edison to Alfred Southwick, December 19, 1887 (TAEM 138:355).

CHAPTER 10. THE ELECTRICAL EXECUTION LAW

1. *New York Times,* March 22, 1890.

2. Quotation from Commission Report, 94.

3. Quotations from ibid., 88, 90. Also see Elbridge Gerry, "Capital Punishment by Electricity," *North American Review* 149 (1889): 321–25.

4. See Ruth Richardson, *Death, Dissection and the Destitute* (London: Routledge & Kegan Paul, 1987), 32–37, 51–53; Gatrell, *Hanging Tree*, 255–58; Banner, *Death Penalty*, 76–82. The bodies of executed criminals were also believed to possess magical healing qualities: see Mabel Peacock, "Executed Criminals and Folk-Medicine," *Folk-Lore* 7 (1896): 268–83.

5. Quotations from *New York Museum of Anatomy*, undated catalog, Billy Rose Theater Collection of the New York Public Library for the Performing Arts; *New York Herald*, July 9, 1891. Also see Banner, *Death Penalty*, 80; Luc Sante, *Low Life: Lures and Snares of Old New York* (New York: Vintage, 1992), 96–101.

6. Quotation from Commission Report, 89.

7. Quotations from Paul Avrich, *The Haymarket Tragedy* (Princeton, N.J.: Princeton University Press, 1984), 217, 393, 396. Also see *Additional Report of the Commissioners on Capital Punishment of the State of New York* (Albany: James B. Lyon, 1892), 23.

8. Quotations from Commission Report, 88–89; *New York World*, March 22, 1888; *New York World*, June 6, 1888. Also see *New York Herald*, January 27, 1888.

9. Quotations from *Additional Report of the Commissioners on Capital Punishment*, 8, 17.

10. *Additional Report of the Commissioners on Capital Punishment.* For further comments by Gerry, see *New York Herald*, January 27, 1888.

11. Quotation from *New York Herald*, April 18, 1888. Also see *Laws of New York* (1888), 778–81; *New York World*, January 18, March 22, May 11, June 5, 1888; *New York Times*, January 17, May 12, 1888; *New York Daily Tribune*, March 22, 1888; *New York Herald*, January 27, April 18, 26, 1888.

12. Quotations from Kemmler Hearings, 367; Elbridge Gerry, "Capital Punishment by Electricity," *North American Review* 149 (1889): 322; Park Benjamin, "The Infliction of the Death Penalty," *Forum* 3 (July 1887): 512. Also see Thomas D. Lockwood, "Electrical Killing," *Electrical Engineer* 7 (March 1888): 89–90; *New York Herald*, January 27, 1888.

13. Quotations from *Scientific American* 58 (June 30, 1888): 407; *New York Times*, December 17, 1887; *New York Daily Tribune*, June 11, 1888. Also see *New York Daily Tribune*, December 23, 1887; "Humanity in the Death Sentence," *JAMA* 10 (January 28, 1888): 114–15; "Report of the Committee on Best Methods of Executing Criminals," *Medico-Legal Journal* 5 (1888): 442–44.

14. Quotation from *Buffalo Express*, June 5, 1888. Also see *Buffalo Express*, December 24, 1886.

15. George Fell, who conducted the experiments with Southwick, later admitted the shortcomings: "The report has been criticized as not covering electrical measurements. I was limited in apparatus and accomplished as much as was possible with the means at my command." George Fell, "The Influence of Electricity on Protoplasm, with Some Remarks on the Kemmler Execution," *Physician and Surgeon* 10 (October 1890): 441.

16. Quotation from *Laws of New York* (1888), 780.

17. Quotations from "The Death Penalty by Electricity," *Medical Record* 33 (January 21, 1888): 34; *New York Herald*, March 9, 1888; *New York Herald*, June 24, 1888. Also see *Medical Record* 33 (February 11, 1888): 158.

18. *New York World*, June 24, 1888.

19. Israel, *Edison*, 244.

20. Quotations from Edison diary, July 12, 1885 (TAEM 90:9).

21. Quotations from ibid., July 15, 17, 20, 1885 (TAEM 90:9).

22. Quotations from ibid., July 17, 1885 (TAEM 90:9); Israel, *Edison*, 253.

23. See Larry Moore, *Selling God: American Religion in the Marketplace of Culture* (New York: Oxford University Press, 1994), 149–55.

24. Quotations from Edison diary, July 12, 1885 (TAEM 90:9); Israel, *Edison*, 247.

25. Edison to John and Theodore Miller, December 24, 1885 (TAEM 77:480).

26. *The Diary and Sundry Observations of Thomas Alva Edison*, ed. Dagobert D. Runes (New York: Greenwood Press, 1968), 54–55; Israel, *Edison*, 247.

27. Fort Myers Notebook N-86-03-18, March 20, 1886 (TAEM 42:843); Israel, *Edison*, 250–51.

28. Israel, *Edison*, 248–49.

29. Tate, *Edison's Open Door*, 140.

30. Passer, *Electrical Manufacturers*, 118–23.

31. Quotations from Edison to Wright, draft letter, c. August 1887 (TAEM 98:671–72); Millard, *Edison and the Business of Innovation*, 15. Also see Horace Townsend, "Edison: His Work and His Work-Shop," *Cosmopolitan Magazine* (April 1889): 598–607; "Edison's New Laboratory," *Scientific American* 57 (September 18, 1887): 184; David Trumbull Marshall, *Recollections of Edison* (Boston: Christopher Publishing, 1931), 90; Reginald A. Fessenden, "The Inventions of Reginald A. Fessenden," *Radio News* 7 (August 1925): 156; Andre Millard, Duncan Hay, and Mary Grassick, *Historic Furnishings Report. Edison Laboratory. Edison National Historic Site. West Orange, New Jersey*, 2 vols. (Harpers Ferry Center: National Park Service, 1995), 1:26–28; Millard, *Edison and the Business of Innovation*, 10–15; Israel, *Edison*, 260–66.

32. Quotations from Marshall, *Recollections of Edison*, 60; Edison to Wright, draft letter, c. August 1887 (TAEM 98:671–72). Also see Millard, *Edison and the Business of Innovation*, 16.

33. Arthur Kennelly to Dr. Muirhead, December 26, 1888 (TAEM 109:227). On

Kennelly's career, see Vannevar Bush, "Biographical Memoir of Arthur Edwin Kennelly, 1861–1939," *National Academy of Sciences Biographical Memoir* 22 (1943): 83–119; C. M. Worthington, "Interview with Dr. A.E. Kennelly on May 19th, at his home," Kennelly folder, Edison Pioneers collection, ENHS; "Autobiographical Sketch of the Life and Work of Arthur Edwin Kennelly," unpublished manuscript, 1936, Harvard University Archives; Israel, *Edison*, 272, 305–20; Millard, *Edison and the Business of Innovation*, 88–110.

34. The account of this experiment, including all quotations, is taken from the *New York World*, June 24, 1888. For another account, see the *New York Morning Sun*, June 24, 1888 (TAEM 146:353; misdated as November 4, 1888 in TAEM). Kennelly mentions the events of this day in Kemmler Hearings, 748–52. This is most likely the experiment described in W. K. L. Dickson and Antonia Dickson, *The Life and Inventions of Thomas Alva Edison* (New York: Thomas Y. Crowell, 1894), 326–30.

CHAPTER II. "A DESPERATE FIGHT"

1. Quotations from F. R. Chinook to Edison, April 17, 1888 (TAEM 123:612); Francis Upton to Edison, May 28, 1888 (TAEM 123:36). Also see Reynolds and Bernstein, "Damnable Alternating Current," 1340.

2. Nikola Tesla, "A New System of Alternate Current Motors and Transformers," *Electrical Engineer* 7 (June 1888): 252–57; Carlson, *Innovation as a Social Process*, 249–59; Ronald Kline, "Science and Engineering Theory in the Invention and Development of the Induction Motor, 1880–1900," *Technology and Culture* 28 (1887): 283–313; Reynolds and Bernstein, "Damnable Alternating Current," 1341–42; Curt Wohleber, "The Work of the World," *American Heritage of Invention and Technology* 7 (Winter 1992): 52.

3. Quotations from *A Warning from the Edison Electric Light Co.* (n.p., n.d. [February 1888]). On Johnson's plans for the circular, see Johnson to Edison, December 20, 1887 (TAEM 119:871).

4. Quotation from *Electrician* 21 (August 3, 1888): 398. The Chicago Electric Club had some of the most heated debates, reports of which appeared in the April, May, and June editions of *Western Electrician* 2 (1888). Also see *Electrical Engineer* 7 (April, May 1888): 154–55, 166–68, 220–24.

5. Quotation from George Westinghouse to Edison, June 7, 1888 (TAEM 122:861).

6. Quotation from Edison to Westinghouse, June 12, 1888 (TAEM 138:438).

7. On disputes over efficiency, see Westinghouse circular, July 3, 1888 (TAEM 122:877). On accidental deaths, see *New York Times*, November 7, 1886, January 21, October 7, December 6, 1887; *New York Tribune*, May 6, 1887.

8. Quotation from C. C. Haskins, "For High Insulation," *Electrician and Electrical Engineer* 6 (March 1887): 93–96. Also see Gilbert Wilkes and Cary T. Hutchinson, "Tests of American Insulated Wires," *Electrical World* 13 (January 26, 1889); Thomas Edison, "Insulation," *Electrical Engineer* 14 (July 13, 1892): 34–35; Sullivan, "From Municipal Ownership to Regulation," 132–42; Joseph P. Sullivan, "Fearing Electricity: Overhead Wire Panic in New York City," *IEEE Technology and Society*

Magazine 14 (Fall 1995): 8–16; *Electrical Engineer* 7 (August 1888): 330; Israel, *Edison*, 311–13; Marshall, *Recollections of Edison*, 67–69.

9. Quotation from "Overhead Wires in New York," *Electrical Engineer* 7 (August 1888): 330. On wire regulation in other cities, see William Thomson, "Electric Lighting and Public Safety," *North American Review* 149 (February 1890): 189–96; S. S. Wheeler, "Overhead and Underground Wires in New York," *Telegraphic Journal and Electrical Review* 23 (September 14, 1888): 291–94; *New York World*, January 14, 1890; George Cutter to Elihu Thomson, November 27, 1887, series II, 1887 folder, Thomson Papers; Harold L. Platt, *The Electric City: Energy and the Growth of the Chicago Area, 1880–1930* (Chicago: University of Chicago Press, 1991), 43–45. On efforts to regulate overhead wires, see Sullivan, "From Municipal Ownership to Regulation," 132–46; *New York Advertiser*, October 30, 1885 (TAEM 89:652); *New York Tribune*, January 6, 1888; *New York World*, July 17, 1888; *Electrical Engineer* 7 (August 1888): 330; *New York World*, *New York Tribune*, August 30, 1888; S. S. Wheeler, "Overhead and Underground Wires in New York," *Telegraphic Journal and Electrical Review* 23 (September 14, 1888): 291–94; Schuyler S. Wheeler, "Electric Lighting in New York," *Harper's Weekly* 33 (July 27, 1889): 593–96, 601–3.

10. "Overhead Wires in New York," *Electrical Engineer* 7 (August 1888): 330; Sullivan, "From Municipal Ownership to Regulation," 146.

11. *New York World*, April 16, 17, August 8, 1888; *New York Tribune*, April 17, 1888; *New York Times*, April 28, May 12, 1888.

12. *New York Morning Sun*, June 24, 1888 (TAEM 146:353; misdated as November 4, 1888, in TAEM).

13. Kemmler Hearings, 3–6; Harold Brown to Edison, December 29, 1879 (TAEM 50:483); Harold P. Brown folder, Edison Pioneers collection, ENHS; George Bliss to Edison, May 12, 1888; Edison to George Bliss, May 21, 1888 (TAEM 121:332, 122:313).

14. *New York Evening Post*, June 5, 1888; reprinted in "The Admission of Alternating Currents into New York City," *Electrical World* 12 (July 28, 1888): 40.

15. Quotations in this and the next paragraph from "The Admission of Alternating Currents into New York City," *Electrical World* 12 (July 28, 1888): 40–46. The same meeting transcript was also reprinted as "High Potential Systems Before the Board of Electrical Control of New York City," *Electrical Engineer* 7 (August 1888): 360–69.

16. This may have been an unwitting description of cardiac defibrillation, but its effectiveness in 1888 would have been entirely accidental, since the principle of defibrillation was not discovered until the middle of the twentieth century. See Theodore Bernstein, "Theories of the Causes of Death from Electricity in the Late Nineteenth Century," *Medical Instrumentation* 9 (1975): 267–73.

17. Quotation from *New York Evening Post*, June 5, 1888. On theories of the greater dangers of alternating current, see *New York Times*, June 5, 1882; Max Deri, "Alternating Currents and Their Practical Uses," *Electrician and Electrical Engineer* 5 (September 1886): 346–47; Edison, "Notes on Distribution of Alternating Current," memo to Edward Johnson, 1886, 7–10 (TAEM 148:3).

18. Quotation from Harold P. Brown, *The Comparative Danger to Life of the Alternating and Continuous Currents* (New York: n.p., 1889), 10.

19. See Peterson's obituary, *Journal of Nervous and Mental Disease* 88 (1938): 558–61.

20. Notebook N-88-06-06, pp. 74, 78–80, July 12, 15, 1888 (TAEM 102:529–35).

21. Ibid., pp. 78–80, July 17, 1888 (TAEM 102:540–3).

22. Alfred O. Tate to Henry Bergh, May 2, 1888 (TAEM 122:288). Bergh's letter to Edison does not survive, but its contents can be inferred from the letter from Tate, Edison's secretary, who wrote to Bergh on the inventor's behalf. Also see Kennelly's testimony in Kemmler Hearings, 748–52.

23. Edison to Henry Bergh, July 13, 1888 (TAEM 138:441).

24. Henry Bergh to Edison, July 14, 1888 (TAEM 122:882); Edison to Bergh, July 21, 1888 (TAEM 81:195).

25. Notebook N-88-06-10, pp. 31–32, 36–37, July 21 and 24, 1888 (TAEM 102:580–81, 584–87); Marshall, *Recollections of Edison*, 66; Harold P. Brown, "Experiments with Electric Currents on Dogs," *Electrical World* 12 (August 11, 1888): 72–73.

26. Quotations from *New York World*, July 31, 1888. Also see *New York Times*, August 4, 1888; "Mr. Brown's Rejoinder," *Electrical Engineer* 7 (August 1888): 330, 369–70; "Physiological Tests with the Electric Currents," *Electrical World* 12 (August 11, 1888): 69–72; Notebook N-88-06-10, p. 47 (TAEM 102:596).

27. *New York Times*, August 4, 1888; *New York Morning Sun*, August 4, 1888 (TAEM 146:292).

28. *Electrician* 22 (August 17, 1888): 478.

29. Harold Brown to Arthur Kennelly, August 4, 1888 (TAEM 122:924); Kennelly to Frank Hastings, August 8, 1888 (TAEM 109:121–22). Also see Hastings to Kennelly, August 6, 8, 1888 (TAEM 123:67–69); Kennelly to Brown, August 9, 1888 (TAEM 109:123).

CHAPTER 12. "CRIMINAL ECONOMY"

1. "Mr. Brown and the Dog—a Ballad," *Electrical Engineer* 7 (August 1888): 375.

2. Sullivan, "From Municipal Ownership to Regulation," 122–25, 146, 490; "The Subway Fight of the United States Company," *Electrical World* 13 (January 12, 1889): 21–22; "Meeting of the Board of Electrical Control," *Electrical World* 13 (January 26, 1889): 48; "New York Board of Electrical Control," *Electrical World* 13 (March 9, 1889): 149–50; "Meeting of the Board of Electrical Control," *Electrical World* 13 (March 16, 1889): 164.

3. Quotation from *Electrician* 21 (October 5, 1888): 691. Also see Arthur Kennelly to *Electrical Review*, September 7, 1888 (TAEM 109:141); Kennelly to W. H. Snell, September 24, 1888 (TAEM 109:174).

4. Quotations from *New York World*, August 19, 1888; Arthur Kennelly to *New York World*, August 20, 1888 (TAEM 109:124); *New York World*, August 27, 1888 (which reprints most of Kennelly's letter). Also see P. H. Van Der Weyde, "The Comparative Danger of Alternate vs. Direct Currents," *Electrical Engineer* 7 (September 1888): 451–54.

5. *Brooklyn Citizen*, November 4, 1888 (TAEM 25:580).

6. Quotations from Kennelly Notebook #2, p. 113, December 18, 1889 (TAEM 104:552); *Brooklyn Citizen*, November 4, 1888 (TAEM 25:580). On the relative velocities of nerve sensation and electricity, see Park Benjamin, "The Infliction of the

Death Penalty," *Forum* 3 (July 1887): 509–10; *New York Herald*, April 25, 1890; Kemmler Hearings, 243–46; A. D. Rockwell, *Rambling Recollections: An Autobiography* (New York: Paul B. Hoeber, 1920), 232. On research into the velocity of the nerve impulse, see Rowbottom and Susskind, *Electricity and Medicine*, 98–99.

7. Quotations from Commission Report, 77. Fell's experiment helped persuade Elbridge Gerry to support electrical execution: *New York Herald*, January 27, 1888. Footnote quotation from Thomas D. Lockwood, "Electrical Killing," *Electrical Engineer* 7 (March 1888): 89–90.

8. Quotations from Kemmler Hearings, 366; *Medical Record* 37 (May 17, 1890). Also see Commission Report, 75. On the Medico-Legal Society, see James C. Mohr, *Doctors and the Law: Medical Jurisprudence in Nineteenth-Century America* (New York: Oxford University Press, 1993), 219–24; Blustein, *Preserve Your Love for Science*, 112. For physician contributions to the search for a new killing method, see J. H. Packard, "The Mode of Inflicting the Death Penalty," *Sanitarian* 6 (1878): 360–63 (reprinted in *Bulletin of the Medico-Legal Society of New-York* 1 [1878–9]: 135–40); "Of the Death Penalty," *Bulletin of the Medico-Legal Society of New-York* 1 (1878–9): 141–56; G. M. Hammond, "On the Proper Method of Executing the Sentence of Death by Hanging," *Sanitarian* 10 (1882): 664–68; Alonzo Calkins, "Felonious Homicide: Its Penalty, and the Execution Thereof Judicially," in *Papers Read Before the Medico-Legal Society of New York*, 2nd series, rev. ed. (New York: W. F. Vanden Houten, 1882), 254–76; "The Method of Inflicting Capital Punishment," *Boston Medical and Surgical Journal* 110 (May 29, 1884): 508; Frederick Henry Gerrish, "The Hypodermic Administration of Morphine as a Substitute for Hanging in the Execution of Criminals," *Boston Medical and Surgical Journal* 113 (September 17, 1885): 270–71; W. Lindley, "The Methods of Capital Punishment," *Southern California Practitioner* 1 (1886): 73–81; J. Mount Bleyer, "Scientific Methods of Capital Punishment," *Humboldt Library of Popular Science* 9 (1887): 1–16. The American Medical Association now forbids physician participation in executions. See "Physician Participation in Capital Punishment," *JAMA* 270 (July 21, 1993): 365–68.

9. Quotation from "A Report on Execution by Electricity," *Electrical World* 12 (November 24, 1888): 275–76. Also see *New York World*, November 15, 1888 (TAEM 146:356); "Execution by Electricity," *Medical Record* 34 (November 17, 24, 1888): 597, 623; *New York Times*, March 9, 15, 1888; Clark Bell, "Electricity and the Death Penalty," *Journal of the American Medical Association* 12 (March 9, 1889): 32 (reprinted in *Medico-Legal Journal* 7 [1889–90]: 201–9).

10. Quotation from Arthur Kennelly to Frank Hastings, December 3, 1888 (TAEM 109:207).

11. The account of this experiment, including quotations, in this and the next two paragraphs is drawn from Kennelly Notebook #1, pp. 47–50, December 5, 1888 (misdated December 4 by Kennelly) (TAEM 104:302); Harold P. Brown, "Death-Current Experiments at the Edison Laboratory," *Electrical World* 12 (December 15, 1888). Also see *Scientific American* 59 (December 22, 1888): 393; *New York Times*, December 6, 1888; *Medical Record* 34 (December 8, 1888): 678.

12. S. Eaton to Edison, December 12, 1888 (ENHS: D-88-22-4; emphasis in original; my thanks to Matt Abramovitz for bringing this letter to my attention); *New York Tribune*, February 4, 1889; *New York World*, December 10, 14, 18, 1888. A Jer-

sey City man offered his sick spaniel to Edison in the hope that the dog might be killed "comparatively painlessly" and at the same time further "the interests of science." John C. Dewey to Edison, February 3, 1889 (ENHS: D-89-33).

13. "Report of the Committee of the Medico-Legal Society on the Best Method of Execution of Criminals by Electricity," *Medico-Legal Journal* 6 (1888–89): 276–81; Clark Bell, "Electricity and the Death Penalty," *Medico-Legal Journal* 7 (1889–90): 201–19.

14. *New York Daily Tribune*, December 10, 1888, as quoted in Brown, *Comparative Danger*, 27.

15. *New York Evening Post*, December 12, 1888.

16. Ibid.

17. See W. P. Hancock, "Report on Westinghouse Plant of Colorado Electric Company" (TAEM 122:996–1001); "Consolidation of Electric Light Companies," *Electrical Engineer* 8 (February 1889): 74–75; Passer, *Electrical Manufacturers*, 121, 149.

18. Quotations from Harold Brown, "An Electrical Duel," *New York Times*, December 18, 1888. Also see *Electrical World* 12 (December 22, 1888), 323; *Electrical World* 13 (March 9, 1889): 149.

19. Quotation from Harold Brown, circular, December 1888 (TAEM 122:193). Also see Frank Hastings to Edison, January 21, 1889 (TAEM 126:11); Alfred Tate to Hastings, January 24, 1889 (TAEM 138:765); Harold Brown to Edison, February 14, March 17, 1889 (TAEM 126:19, 39).

20. Quotation from Brown, *Comparative Danger*, iv. The third edition of this pamphlet was reprinted as Relator's Exhibit A in Kemmler Hearings.

21. For Rockwell's obituary, see *Journal of Nervous and Mental Disease* 79 (1934): 120–22. On the arrangements for these experiments, see Charles K. Baker to Harold Brown, February 20, 1889; Frank Hastings to Arthur Kennelly, February 25, 1889; Kennelly to Hastings, February 25, 1889; Brown to Kennelly, March 12, 1889 (TAEM 126:32, 33, 35; 109:285).

22. Kennelly Notebook #1, pp. 90–91, March 12, 1889 (TAEM 104:346); *New York Times*, March 9, 13, 1889; *New York Star*, March 18, 1889 (TAEM 146:431); *New York Tribune*, March 13, 1889.

23. *Scientific American* 60 (March 23, 1889): 181.

24. Carlos F. MacDonald to Harold Brown, March 19, 1889; Brown to Edison, March 27, 1889 (both reprinted in *New York Sun*, August 25, 1889. For a discussion of these letters, see chapter 15). For the terms of the contract, see *New York Times*, December 29, 1889, and Kemmler Hearings, 10, 1011–14.

25. Quotations from Brown, *Comparative Danger*, preface to the third edition; *New York Evening Post*, July 5, 1889. Also see Brown, *Electrical Distribution of Light, Heat, and Power, with a Partial List of Deaths from Electric Lighting Circuits* (n.p., n.d.), 21.

26. Edison Electric circular, May 1889 (TAEM 147:237).

27. Quotation from *New York Times*, May 8, 1889. Also see *Pittsburgh Post*, May 23, 1889 (TAEM 89:235); *New York Star*, May 9, 1889; *New York Tribune*, May 23, 1889.

28. Quotations from *New York World*, February 24, 1889; *Electrical World* 14 (September 28, 1889): 214.

29. *American Notes and Queries* 3 (June 8, July 13, 1889): 66, 131.

30. Quotations from *American Notes and Queries* 3 (May 25, June 1, 1889): 45, 57. For other coinages, see *Electrician* 23 (May 24, 1889): 73; *Electrical Review* 14 (August

17, 1889): 20; *Medical Record* 36 (August 24, 1889): 223; *Scientific American* 63 (August 16, 1890): 101.

31. Quotations from Alfred Tate to Sherburne Eaton, May 20, 1889 (TAEM 139:184); Eugene Lewis to Sherburne Eaton, June 1, 1889 (TAEM 126:46, emphasis in original); Sherburne Eaton to Alfred Tate, June 6, 1889 (TAEM 126:49).

32. On June 10 the *World* used the word apologetically—"'Electrocution' Is Not Certain," a headline read—but two weeks later it dispensed with the quotation marks to note that Bourke Cockran's "argument against electrocution was able and telling." *New York World*, June 10, 26, 1889.

33. Quotations from *New York Times*, July 11, 1889; *Buffalo Courier*, July 15, 1889; *Law Quarterly Review* 7 (1891): 111. Also see *New York Tribune*, July 11, 1891; *The American* 20 (August 16, 1890): 357. Controversy over the word dissipated fairly quickly, and by the late 1890s even the newspapers that protested most strongly against it had succumbed to popular usage. The term soon came to refer to accidental electrical deaths as well—an even less appropriate usage considering that such deaths were not executions. Only purists continued to complain: "This barbarism jars the unhappy latinist's nerves much more cruelly than the operation denoted jars those of its victim," H. W. Fowler wrote in his classic *Dictionary of Modern English Usage* in 1930. Three decades later the editor of the volume's second edition conceded defeat in the matter of electrocution: "as it is established, protest is idle." H. W. Fowler, *A Dictionary of Modern English Usage* (Oxford: Clarendon Press, 1930), 130; H. W. Fowler, *A Dictionary of Modern English Usage*, 2nd ed., revised by Ernest Gowers (Oxford: Clarendon Press, 1965), 148. My thanks to Bill Gordy for bringing the Fowler references to my attention.

CHAPTER 13. CONDEMNED

1. *Buffalo Express*, May 8, 1889.

2. Quotation from *Buffalo Times*, April 1, 1889.

3. Quotations from *Buffalo Express*, May 7, March 30, 1889. Also see *Buffalo Times*, April 1, 1889.

4. Quotations from *Buffalo Times*, May 8, 1889; *Buffalo Evening News*, May 7, 1889. Also see *Buffalo Evening News*, April 2, May 8, 1889.

5. Quotations from *Buffalo Courier*, May 9, 1889; *Buffalo Times*, May 9, 1889; *Buffalo Express*, May 8, 1889.

6. Quotation from *Buffalo Evening News*, May 10, 1889.

7. Quotations from ibid., May 7, 15, 1889.

8. Cockran later served three more terms in Congress; he also became a mentor to the young Winston Churchill: James H. Andrews, "Winston Churchill's Tammany Hall Mentor," *New York History* 71 (1990): 133–71. On Cockran's work for Cravath, see Robert T. Swaine, *The Cravath Firm and Its Predecessors, 1819–1947* (New York: Ad Press, 1946), 588–89. Also see James McGurrin, *Bourke Cockran: A Free Lance in American Politics* (New York: Charles Scribner's Sons, 1948), 94–95; *New York Herald*, May 27, 1889; *New York World*, June 6, 1888.

9. Quotations from *New York World*, June 11, 1889; *New York Times*, July 12, 1889. Also see *Buffalo Express*, June 12, 1889.

10. Quotations from James E. Browne to Edison, July 23, 1888 (TAEM 122:909); Elihu Thomson to Charles Coffin, May 16, 1889, 1889 folder, series II, Thomson Papers. Also see Carlson, *Innovation as a Social Process*, 283.

11. Quotations from Elihu Thomson to F. P. Fish, May 6, 1890, Letterbook vol. 41, p. 510, series IIA, Thomson Papers; Carlson, *Innovation as a Social Process*, 298.

12. Quotation from unidentified clipping, 1889 (TAEM 95:255). On patent disputes, see "Schedule of Suits Brought Against Light Company and Licensees," October 1, 1888 (TAEM 123:86); Edison Lamp Co. to Edward Johnson, April 3, 1889 (TAEM 95:848); "Patent Litigation Committee. Mr. Eaton's Mem. of Business for Meeting," October 14, 1889 (TAEM 126:292).

13. "Consolidation of Electric Light Companies," *Electrical Engineer* 8 (February 1889): 74–75; Carlson, *Innovation as a Social Process*, 281–83; Bright, *Electric Lamp Industry*, 87–92; Passer, *Electrical Manufacturers*, 142–54; "Annual Convention of Edison Illuminating Companies," *Electrical World* 14 (August 24, 1889): 135–36; *New York Tribune*, February 17, 1889.

14. Quotation from Edison to E. D. Adams, February 2, 1889, as quoted in Passer, *Electrical Manufacturers*, 174.

15. "Annual Convention of Edison Illuminating Companies," *Electrical World* 14 (August 24, 1889): 135–36; Edison to Henry Villard, December 11, 1888 (TAEM 123:281); Israel, *Edison*, 321–23.

16. Quotation from *New York Star*, May 9, 1889. Also see *New York Daily Graphic*, May 28, 1889; *Frank Leslie's Illustrated Newspaper*, June 8, 1889, 306. Harold Brown mentions the *Leslie's* article in a letter to Charles Coffin, June 3, 1889, reprinted in *New York Sun*, August 25, 1889.

17. Quotations from *Evening Post*, May 14, 1889, as quoted in *Electrical Engineering* 8 (June, 1889): 247.

18. Quotations from *Medical Record* 34 (November 24, 1888): 623; *New York World*, June 7, 1889; *New York Tribune*, June 20, 1889; *New York Sun*, July 17, 1889. The views of several newspapers were quoted in the *Buffalo Express*, June 11, 1889. Although news articles in the *World* criticized the execution law (see June 6–11, 19), the editorial page supported it (see May 25). Also see *Buffalo Express*, August 2, 1889; *New York World*, June 8, 1889; *Electrician* 23 (June 28, 1889): 204–5; *Medical Record* 36 (July 20, 1889): 69; Ludwig Gutmann, "A Review of Mr. Harold P. Brown's Experiments," *Electrical World* 14 (July 13, 1889): 25–26; "Mr. Harold P. Brown's Reply to Mr. Gutmann," *Electrical World* 14 (July 27, 1889): 57; Ludwig Gutmann, "A Reply to Mr. Harold P. Brown," *Electrical World* 14 (September 8, 1888): 134–35.

19. Quotation from *New York World*, June 26, 1889.

CHAPTER 14. SHOWDOWN

1. The presidents of the neurological association were Frederick Peterson and Bernard Sachs; of the engineering society, Franklin Pope, Arthur Kennelly, and Schuyler S. Wheeler. See *Dictionary of American Medical Biography*, ed. Martin Kaufman et al. (Westport, Conn.: Greenwood Press, 1984), 659–60; "Frederick Peterson, M.D.," *Archives of Neurology and Psychiatry* 4 (1938): 1021–22; Hughes, "Harold P. Brown and the Executioner's Current," 158. On Cockran, see Andrews, "Winston Churchill's Tammany Hall Mentor," 133–71; *Buffalo Times*, August 1, 1889.

2. For West's testimony, see Kemmler Hearings, 466–76; *New York Evening Post*, July 19, 1889; *New York Sun*, July 20, 1889.

3. See Kemmler Hearings, 116–17.

4. For Smith's testimony, see ibid., 433–35, 476–521.

5. *New York World*, July 16, 1889.

6. *New York Sun*, July 16, 1889. For Tupper's testimony, see Kemmler Hearings, 208–23.

7. Ricky Jay, *Learned Pigs and Fireproof Women* (New York: Farrar, Straus and Giroux, 1986), 156–62; Jan Bondeson, *Buried Alive: The Terrifying History of Our Most Primal Fear* (New York: Norton, 2001), 180–81.

8. Martin S. Pernick, "Back from the Grave: Recurring Controversies over Defining and Diagnosing Death in History," in *Death: Beyond Whole-Brain Criteria*, ed. Richard M. Zaner (Boston: Kluwer, 1988), 17–74.

9. See *Electrician* 21 (May 11, 1888): 2.

10. Kemmler Hearings, 250–52.

11. Ibid., 278.

12. Ibid., 60–66.

13. Ibid., 382.

14. Quotations from ibid., 978, 581. On Rockwell's testimony, also see the clippings in the A. D. Rockwell Notebook-Scrapbook, Bakken Library. For further comment about the cause of electrical death, see *Electrician* 21 (June 8, 1888): 155; Edwin J. Houston, "On Death by the Electric Current," *Proceedings of the American Philosophical Association* 25 (1888): 127–29; Philip E. Donlin, "The Pathology of Death by Electricity," *Medico-Legal Journal* 7 (1888–89): 470–85; Harold P. Brown, "The New Instrument of Execution," *North American Review* 149 (November 1889): 586–93.

15. Kemmler Hearings, 882–84.

16. Ibid., 24–27.

17. Ibid., 47–48.

18. Ibid., lxxxiii.

19. Ibid., 125. On Pope's death, see Hughes, *Networks of Power*, 106.

20. *New York Times*, July 12, 1889.

21. *New York Sun*, July 12, 1889.

22. Kemmler Hearings, 1016–17; *New York Evening Post*, July 12, 16, 1889; *New York Sun*, July 13, 1889; A. E. Kennelly, "The Law of Probability of Error as Applied to the Observed Electrical Resistance of the Human Body," *Electrical World* 14 (August 3, 1889): 73–74.

23. Quotation from *New York World*, July 16, 1889. Apparently Hale's only further participation in the matter was as author of an article, "The Kemmler Case," *Albany Law Journal* 41 (May 10, 1890): 364–67. Also see *New York Evening Post*, July 9, 17, 1889; *New York World*, July 10, 1889.

24. Kemmler Hearings, 372, 391–96.

25. Quotations from *New York Sun*, July 24, 1889; Harold Brown to Samuel Insull, July 17, 1889 (TAEM 126:50).

26. *New York Times*, July 24, 1889; *New York Daily Graphic*, July 23, 1889.

27. Kemmler Hearings, 623–64.

28. Ibid., 628–36.

29. *New York Sun*, July 24, 1889.

30. On Cockran's sarcasm, see *New York Times*, July 24, 1889.

31. Kemmler Hearings, 644–45.

32. *Electrical Review,* 14 (Aug. 3, 1889): 2.

33. Kemmler Hearings, 652–53.

34. *New York World*, July 24, 1889.

35. Quotations from *New York World*, July 25, 1889; *Albany Journal*, July 23, 1889 (TAEM 146:464).

CHAPTER 15. THE UNMASKING OF HAROLD BROWN

1. Quotations from Kemmler Hearings, 25; *New York Sun*, August 25, 1889, letter #2.

2. Quotation from *New York Sun*, August 25, 1889, letter #5.

3. Brown to Edison, March 17, 1889; Edison to Brown, March 22, 1889, *New York Sun*, August 25, 1889, letters #8, 9.

4. *New York Sun*, August 25, 1889, letter #11. On the formation of Edison General, see Israel, *Edison*, 321–33. My interpretation of this "consolidation" differs from that of Reynolds and Bernstein ("Edison and 'The Chair,'" 28, n. 37), who suggest that it refers to merger talks between Thomson-Houston and Edison General. Although such discussions had taken place by this time, Edison was strongly opposed to such a move, and it therefore seems unlikely that Brown would have referred to it as "approaching." The merger with Thomson-Houston did not take place until 1892. On Edison's attitudes toward the possible merger with Thomson-Houston, see Carlson, *Innovation as a Social Process*, 292–93.

5. Harold Brown to Edison, May 13, 1889 (TAEM 126:45).

6. Quotation from *Pittsburgh Post*, May 23, 1889 (TAEM 89:235). Also see Harold Brown to Charles Coffin, April 23, 1889, *New York Sun*, August 25, 1889, letter #13.

7. Harold Brown to Thomson-Houston, March 14, 1889; Charles Coffin to Brown, May 4, 1889, *New York Sun*, August 25, 1889, letters #7, #18.

8. Harold Brown to Charles Coffin, May 13, 29, 1889, *New York Sun*, August 25, 1889, letters #20, 25.

9. Harold Brown to Charles Coffin, May 13, 1889, *New York Sun*, August 25, 1889, letter #20.

10. Arthur Kennelly to Harold Brown, June 29, 1889, *New York Sun*, August 25, 1889, letter #40. Brown thanked Kennelly for the "kind letter" and said he was prepared "to make the test suggested, or to offer to do so before the referee." Brown to Kennelly, July 9, 1889, *New York Sun*, August 25, 1889, letter #45.

11. See "Some Inside History," *Electrical Engineer* 8 (August 1889): 335. An anonymous satiric pamphlet, almost certainly published by the Edison interests, hints that the letter thief was John H. Noble, a Westinghouse employee: *Dangers of Electric Lighting. A Reply to Mr. Edison*, 5, 18 (ENHS: D-89-33). On the theft of the letters, see *New York Journal*, September 4, 1889; *New York Tribune*, September 5, 1889.

12. Arthur Kennelly to Harold Brown, June 29, 1889, *New York Sun*, August 25, 1889, letter #40; original in TAEM 109:323. The originals of *Sun* letters #8 and #9 (Brown to Edison, March 17, 1889; Edison to Brown, March 22, 1889) can be found in TAEM 126:39; 138:1089. Another letter in the Edison archives (Brown to Edison,

March 23, 1889, ENHS: D-89-33) is Brown's response to Edison's *Sun* letter #9. Previous scholars who have examined this episode also consider the letters to be authentic: Hughes, "Harold P. Brown and the Executioner's Current," 156 n. 55; Reynolds and Bernstein, "Edison and 'The Chair,'" 27–28, n. 36.

13. Elihu Thomson to Charles Coffin, May 16, 1889, 1889 folder, Series II, Thomson Papers.

14. Quotations from *New York Sun*, August 25, 28, 1889; *Electrical World* 14 (August 31, 1889): 143; *Electrician* 23 (September 27, 1889): 534.

15. *New York Sun*, August 25, 1889; *Electrical World* 14 (August 31, 1889): 143.

16. *New York Journal*, September 4, 1889 (TAEM 146:452).

17. Notebooks N-88-06-06, pp. 69, 74, 78–80, July 6, 12, 17, 1888 (TAEM 102:516, 529, 540–43); N-88-06-10, pp. 58–59, 64, 68, 77, 89 (TAEM 102:607–8, 613, 619, 629, 642–43); Kemmler Hearings, 28, 668–70.

18. Quotations from Arthur Kennelly to *Electrical Review*, September 7, 1888 (TAEM 109:141); Kennelly to Frank Hastings, September 6, 1888 (TAEM 109:138–39); Edison to Henry Bergh, July 13, 1888 (TAEM 138:441). Also see Kennelly to W. H. Snell, September 24, 1888 (TAEM 109:174). Kennelly explained that he conducted the experiments "because Mr. Edison told [him] to." Kemmler Hearings, 752.

As Kennelly experimented on dogs in the fall of 1888, Brown tested leakage from light wires in Manhattan using equipment that Frank Hastings arranged for him to borrow from the Edison laboratory. See Frank Hastings to Arthur Kennelly, September 18, 1888 (TAEM 122:934); Hastings to Kennelly, October 8, 12, 1888 (TAEM 109:1039; 124:1041); Kennelly to Harold Brown, October 12, 16, 1888 (TAEM 109:186, 189); Brown to Kennelly, October 14, 1888 (TAEM 124:1042).

My interpretation of the physiological experiments differs from those of Hughes ("Harold P. Brown and the Executioner's Current") and Reynolds and Bernstein ("Edison and 'The Chair'"), who consider Brown the lead investigator. My reading of the laboratory notebooks, combined with Kennelly's statements in letters from August and September that he conducted the experiments himself on orders from Edison, lead me to conclude that Brown's role was secondary.

19. Quotation from Frank Hastings to Charles Batchelor, July 20, 1888 (TAEM 123:59–60). Also see Hastings to Alfred Tate, July 26, 1888; Harold Brown to Arthur Kennelly, August 4, 1888; Kennelly to Hastings, August 8, 1888; Hastings to Kennelly, August 6, 8, 1888; Kennelly to Brown, August 9, 1888 (TAEM 123:62; 122:924; 109:121–22; 123:67–69; 109:123).

Brown may have been working for the Edison interests even in early June 1888, when he mounted his first public attack on alternating current, but there is no clear evidence to support this charge. Just as likely, Brown criticized alternating current out of personal conviction and thereby brought himself to the attention of the Edison interests, who realized that they could put him to use.

20. Quotation from Edward Johnson to Edison, September 18, 1888 (TAEM 122:933). Also see Alfred Tate to Johnson, September 18, 1888 (TAEM 122:565). Brown may have been the first to suggest the use of alternating current at the New York pound: see *New York Herald*, August 19, 1888. The city ultimately decided to use lethal gas instead of electricity: see Henry Bergh to Edison, July 28, 1888 (TAEM 122:918); American Society for the Prevention of Cruelty to Animals, *Twenty-Third Annual Report for 1888* (New York, 1889), 12.

21. Quotation from Harold Brown to Arthur Kennelly, December 6, 1888 (TAEM 122:976). On arrangements for the December tests, see Frank Hastings to Kennelly, November 20, 24, 26, 28, 1888 (TAEM 122:955, 959, 960; 123:110); Kennelly to Hastings, November 21, 1888 (TAEM 109:196); Kennelly to Professor Marks, December 4, 1888 (TAEM 109:196, 206). Edison also encouraged *Scientific American* to send an illustrator to the laboratory to make sketches of the equipment used in the tests: Alfred O. Tate to *Scientific American*, December 18, 1888 (TAEM 138:628).

Edison wanted his lobbying to remain secret. After the tests, Arthur Kennelly made some technical recommendations to the Medico-Legal Society that he believed were off the record. When he learned that the society was planning to include his statement in its official report, Kennelly begged Frederick Peterson to delete the comments: "Mr. Edison is desirous that no expression of opinion on this point should publicly emanate from the laboratory." Kennelly to Peterson, December 6, 10, 1888. Also see Kennelly to Brown, December 6, 1888; Kennelly to Peterson, December 6, 1888; Peterson to Kennelly, December 10, 1888; Peterson to Hastings, December 13, 1888 (TAEM 109:209, 210, 215, 218; 122:979). For the aftermath of the December experiments, see Peterson to Edison, December 10, 11, 1888; Peterson to Kennelly, December 10, 26, 1888; Kennelly to Peterson, December 19, 1888; Peterson to Kennelly, December 26, 1888; Clark Bell to Edison, December 26, 1888; Clark Bell to Peterson, December 26, 1888 (TAEM 122:978, 980, 988, 989; 109: 225).

In the spring and summer of 1889, both Edison and the officers of Edison Electric continued to work closely with Brown, arranging for him to borrow equipment and to carry out the March tests for the state electrocution commission. See Kennelly to Brown, May 16, 30, June 6, 1889 (TAEM 109:306, 309, 313); Hastings to Edison, March 8, 1889 (TAEM 126:34); Brown to Edison, March 12, 1889 (TAEM 126:35); Hastings to Kennelly, February 25, 1889 (TAEM 126:32); Kennelly to Hastings, February 25, 1889 (TAEM 109:285); Charles K. Baker to Brown, February 20, 1889 (TAEM 126:33); Hastings to Edison, January 21, 1889; Alfred Tate to Hastings, January 24, 1889 (TAEM 126:11; 138:765).

22. Quotation from Kennelly to Charles Wirt, December 7, 1888 (TAEM 109:214). Also see Kennelly to Brown, December 6, 1888 (TAEM 109:210); Billbook #3, pp. 390–94, 426. For the expenses of later tests, see Billbook #6, pp. 63, 82, ENHS; "Edison Electric Light Company Memoranda," April 25, 1890 (TAEM 140:1088).

CHAPTER 16. PRIDE AND REPUTATION

1. Quotation from Israel, *Edison*, 370. Also see Tate, *Edison's Open Door*, 234–36; Israel, *Edison*, 369–70.

2. *JAMA* 13 (Sept. 21, 1889): 431.

3. *Electrical Review* 14 (June 8, 1889): 5; Israel, *Edison*, 371.

4. *New York Times*, February 15, 1890; A. D. Rockwell, "Discussion of Electrical Execution," *JAMA* 19 (September 24, 1892): 363.

5. Quotations from Tate, *Edison's Open Door*, 164.

6. Quotations from Dickson and Dickson, *Life and Inventions of Thomas Alva Edison*, 330; *New York Morning Sun*, June 24, 1888 (TAEM 146:353; misdated as November 4, 1888 in TAEM). Also see James S. Evans, "Edison Regrets Electric Chair Was Ever Invented," *New York American*, February 10, 1905 (TAEM 221:289).

7. Quotations from Arthur Kennelly to C. F. MacDonald, June 5, 1889 (TAEM 109:311); *New York Morning Sun*, June 24, 1888 (TAEM 146:353; misdated as November 4, 1888 in TAEM). Also see *Brooklyn Citizen*, November 4, 1888 (TAEM 25:580); *New York Herald*, September 29, 1889 (TAEM 146:534); Kennelly to W. H. Snell, September 24, 1888; Kennelly to Frederick Peterson, December 6, 1888 (TAEM 109:311, 174).

8. *New York Evening Post*, December 12, 1888; Kemmler Hearings, 652–53.

9. Carlson, *Innovation as a Social Process*, 284, n. 20; Israel, *Edison*, 326; Passer, *Electrical Manufacturers*, 171–73; Hughes, *Networks of Power*, 95–97; *Electrical World* 9 (January 29, 1887): 51; *Electrician* 25 (May 23, 1890): 72.

10. Quotation from "Annual Convention of Edison Illuminating Companies," *Electrical World* 14 (August 24, 1889): 135–36.

11. Quotation from Edison to J. H. Herrick, October 30, 1889 (TAEM 139:825). Also see Association of Edison Illuminating Companies, *Minutes of Semi-Annual Meeting* (New York, 1888), 120–21; "Explanation of Experiments Covered by Bills Rendered Edison General Electric Co.," May 14, 1890 (TAEM 130:679); W. J. Jenks to Edison, April 8, 22, 1889 (TAEM 126:259, 265); Francis Upton to Edison, May 28, 1888 (TAEM 123:36); Israel, *Edison*, 326–27.

12. Quotation from "Annual Convention of Edison Illuminating Companies," *Electrical World* 14 (August 24, 1889): 135–36. Also see Association of Edison Illuminating Companies, *Minutes of Semi-Annual Meeting*, 120–21.

13. Quotation from Edison to Mrs. C. F. Pond, August 22, 1882. Also see Mrs. C. F. Pond to Edison, August 17, 1882 (TAEM 60:491, 81:816). On Edison's fear of alternating current, see his testimony in Kemmler Hearings, 633–34; "Notes on Distribution of Alternating Current," 7–10; *New York Times*, June 5, 1882; Max Deri, "Alternating Currents and Their Practical Uses," *Electrician and Electrical Engineer* 5 (September 1886): 346–47.

14. *New York World*, October 20, 1889. Edison was probably quoting from Jacques Arsène d'Arsonval, "Electro-Physiology," *Electrician* 23 (August 30, 1889): 431–42. On d'Arsonval, see Rowbottom and Susskind, *Electricity and Medicine*, 120–40.

15. *New York Times*, December 27, 1882.

16. Quotation from "Notes on Distribution of Alternating Current," 9. Also see Millard, *Edison and the Business of Innovation*, 102.

17. Quotations from Millard, *Edison and the Business of Innovation*, 65; Josephson, *Edison*, 260.

18. See W. Bernard Carlson and A. J. Millard, "Defining Risk Within a Business Context: Thomas A. Edison, Elihu Thomson, and the a.c.–d.c. Controversy, 1885–1900," in *The Social and Cultural Construction of Risk: Essays on Risk Selection and Perception*, ed. Branden B. Johnson and Vincent T. Covello (Boston: Reidel, 1987).

19. Westinghouse appealed the case to the U.S. Supreme Court, which in 1895 affirmed the decision of the lower court. Bright, *Electric Lamp Industry*, 92.

20. Quotation from S. B. Eaton to Edison, October 7, 1889 (TAEM 127:308).

21. Reynolds and Bernstein, "Edison and 'The Chair,'" 20–21.

22. *New York Times*, September 3, 14, October 1, 9, 10, 1889; Sullivan, "From Municipal Ownership to Regulation," 149–54.

23. *New York Evening Sun*, October 14, 1889 (TAEM 146:545).

24. "Safety Devices with Transformers," Thomson to *Electrician*, February 27, 1888, series IV, Thomson Papers; Thomson to editor of *Electrical Review*, September

26, 1888; Thomson to Coffin, December 11, 1888, Letterbook vol. 23, pp. 314–15, 547–50, series IIA, Thomson Papers; Carlson, *Innovation as a Social Process*, 250–60.

25. Quotation from Thomson to A. C. Bernheim, October 16, 1889, Letterbook vol. 30, pp. 559–66, series IIA, Thomson Papers.

26. Quotations from Thomson to J. R. Lovejoy, November 6, 1889; Thomson to *Electric Age*, December 27, 1889; Thomson to A. C. Bernheim, October 16, 1889; Thomson to Charles Coffin, November 6, 1889, Letterbook vol. 30, pp. 662–67, 889, 559–66, 658–61, series IIA, Thomson Papers. Also see Elihu Thomson, "Insulation and Installation of Wires and Construction of Plant," *Electrical Engineer* 7 (March 1888): 90–91.

27. Quotation from Thomson to Coffin, December 24, 1889, Letterbook vol. 30, pp. 903–8, series IIA, Thomson Papers.

28. Elihu Thomson, "Safety and Safety Devices in Electric Installations," *Electrical World* 14 (February 22, 1890): 145–46.

39. Quotations from *New York Evening Post*, December 12, 1888; George Westinghouse, "A Reply to Mr. Edison," *North American Review* 149 (December 1889): 658.

30. Reynolds and Bernstein, "Edison and 'The Chair,'" 20.

31. 7 N.Y.S. 145 (1889); *New York Tribune*, September 18, 1889; *New York Times*, October 10, 1889; *Buffalo Evening News*, October 9, 1889; *New York World*, October 10, 1889; *Electrical World* 14 (October 19, 1889): 268; Deborah W. Denno, "Is Electrocution an Unconstitutional Method of Execution? The Engineering of Death over the Century," *William and Mary Law Review* 35 (1994): 578–85.

32. Quotations from 7 N.Y.S. 145 (1889).

33. *New York World*, October 10, 1889.

CHAPTER 17. THE ELECTRIC WIRE PANIC

1. The following account of the events of October 11 is derived from *New York World*, *Tribune*, *Times*, and *Sun*, October 12, 1889. I borrow the title of this chapter from Sullivan, "From Municipal Ownership to Regulation," chap. 4.

2. *New York World*, October 12, 1889.

3. Quotations from *New York Tribune*, October 13, 1889; *New York Sun*, October 17, 1889.

4. *New York World*, October 12, 1889.

5. Quotation from ibid., January 19, 1890. Also see *New York Sun*, October 17, 1889.

6. *New York Sun*, October 13, 1889; *New York Sun*, October 17, 1889.

7. Quotation from *New York World*, December 14, 1889. Also see Sullivan, "From Municipal Ownership to Regulation," 149; Swaine, *Cravath Firm*, 588–89.

8. *New York Tribune*, October 12, 1889.

9. Quotation from *New York Tribune*, December 3, 1889. Also see Mark Aldrich, *Safety First: Technology, Labor, and Business in the Building of American Work Safety* (Baltimore: Johns Hopkins University Press, 1997); John Fabian Witt, "Toward a New History of American Accident Law," *Harvard Law Review* 114 (2001): 690–837.

10. "Causes of Death from Accident in New York City," *Electrical World* 14 (July 13, 1889): 26; John Trowbridge, "Dangers from Electricity," *Atlantic Monthly* 65 (1890): 413–18; Sullivan, "From Municipal Ownership to Regulation," 166–69; Neil

Harris, "Utopian Fiction and Its Discontents," in *Cultural Excursions: Marketing Appetites and Cultural Tastes in Modern America* (Chicago: University of Chicago Press, 1990), 150–73.

11. Quotations from *New York World*, October 12, 1889; *Harper's Weekly* 33 (December 14, 1889): 990.

12. Quotations from *New York World*, October 12, 1889; *New York Times*, December 12, 1889.

13. Quotations from *New York Times*, December 2, 1889; *Telegraphic Journal and Electrical Review* 25 (October 25, 1889): 461. Also see *New York World*, October 15, 1889; Harold P. Brown, "The New Instrument of Execution," *North American Review* 149 (November 1889): 586–93.

14. Quotation from Eugene Crowell to Edison, October 14, 1889 (TAEM 126:86). Also see Harold Brown to Edison, October 22, November 7, December 30, 1889; Alfred Tate to Brown, November 11, 1889; Bergmann and Co. to Kennelly, November 30, 1889 (TAEM 126:94, 129, 690; 139:910; 125:58).

15. Quotations from *New York World*, October 12, 1889; *News* (Wilmington, Del.), October 14, 1889 (TAEM 146:542); *New York Evening Sun*, October 14, 1889 (TAEM 146:545); *New York Evening Sun*, October 14, 1889 (TAEM 146:545).

16. Thomas A. Edison, "The Dangers of Electric Lighting," *North American Review* 149 (November 1889): 625–34. Also see Edison to *North American Review* and Lloyd Stephens Bryce, October 14, 1889 (TAEM 139:747). Commenting on this article, the *Electrical Engineer* wrote, "Mr. Edison appears to be wholly unable to rise above a very low plane of self-interest." *Electrical Engineer* 8 (December 1889): 505–6.

17. *New York Times*, December 2, 11, 12, 15, 28, 1889; *New York World*, December 2, 11, 15, 16, 1889. Edison was asked to testify at the coroner's inquests regarding these deaths, but he did not do so. See J. B. Messemer to Edison, December 5, 10, 1889 (TAEM 126:139–42, 140:4, 29).

18. Quotations from *New York Tribune*, December 3, 1889; *Post* as quoted in *Electrical World* 14 (December 28, 1889): 411; *World* (Charleston, S.C.), October 14, 1889 (TAEM 146:542); unidentified clipping (TAEM 95:252). Also see *Electrician* 24 (November 8, 1889): 18; *New York Evening Sun*, October 16, 1889 (TAEM 146:542); *Star* (Worcester, Mass.), October 15, 1889 (TAEM 146:546).

19. Quotations from Cyrus Edson and Edward Martin, report to the Health Department of the City of New York, October 15, 1889 (ENHS: D-89-33); *New York World*, October 16, 1889; *New York Commercial Advertiser*, October 16, 1889. Also see *New York World*, October 13, 15, 1889; *New York Evening Post*, October 15, 1889.

20. Quotation from *New York World*, December 11, 1889. Also see George Westinghouse, "A Reply to Mr. Edison," *North American Review* 149 (December 1889): 653–64; *New York Tribune*, October 13, 1889; Sullivan, "From Municipal Ownership to Regulation," 157.

21. *New York Times*, December 14, 1889.

22. *New York World*, December 16, 1889.

23. Quotations from *New York Times*, December 16, 25, 1889. Also see *New York Times*, December 15–22, 24–25, 27–31, 1889; *New York World*, 16, 19, 21–22, 29–30, 1889; Sullivan, "From Municipal Ownership to Regulation," 159.

24. *New York World*, December 17, 19, 1889; *New York Tribune*, December 28, 1889; *New York Times*, December 15, 1889; "The Electric Light in New York, January 1,

1890," *Electrical Engineer* 9 (January 1890): 1; Sullivan, "From Municipal Ownership to Regulation," 490.

CHAPTER 18. DESIGNING THE ELECTRIC CHAIR

1. Quotations from 7 N.Y.S. 813 (Sup. Ct. 1889). Also see Denno, "Is Electrocution an Unconstitutional Method of Execution?" 585–88.

2. *New York Times*, August 6, December 29, 1889, January 19, 1890; *Buffalo Express*, August 7, 1889; *New York World*, July 17, October 10, December 27, 1889.

3. David J. Rothman, "Perfecting the Prison," in *Oxford History of the Prison*, 117–19; John N. Miskell, *Executions in Auburn Prison, Auburn, New York, 1890–1916* (Auburn: n.p., 1996).

4. George Fell, "The Influence of Electricity on Protoplasm, with Some Remarks on the Kemmler Execution," *Physician and Surgeon* 10 (October 1890): 441–42; *New York Times*, January 1, February 12, 1890; *New York World*, February 12, 1890; *New York Herald*, February 12, 1890; *Medical Record* 37 (January 11, 1890): 47; "Electric Executions," unidentified clipping, A. D. Rockwell Notebook-Scrapbook, Bakken Library; A. D. Rockwell, *Rambling Recollections: An Autobiography* (New York: Paul B. Hoeber, 1920), 229–30.

5. Quotations from J. Mount Bleyer, "Best Method of Executing Criminals," *Medico-Legal Journal* 5 (1887–88): 429–32; *New York Herald*, December 17, 1887. Also see Park Benjamin, "The Infliction of the Death Penalty," *Forum* 3 (July 1887): 503–12.

6. Quotations from *New York World*, December 28, 1886; Commission Report, 80. Also see *Scientific American* 28 (June 7, 1873): 352; *Manufacturer and Builder* 12 (February 1880): 39; *Electrical Review* 9 (December 18, 1886): 7; R. D. Blackwood, "Electricity as a Means of Inflicting the Death Penalty Through Process of Law," *Medical and Surgical Reporter* 57 (December 10, 1887): 761–64.

7. Quotation from *New York Advertiser*, August 3, 1888 (TAEM 146:291). Also see *New York World*, August 19, 1888. On the question of resistance in electrocution, see A. E. Kennelly, "The Law of Probability of Error as Applied to the Observed Electrical Resistance of the Human Body," *Electrical World* 14 (August 3, 1889): 73–74; *New York Evening Post*, July 12, 16, 1889; *New York Sun*, July 13, 1889.

8. Quotations from Kemmler Hearings, 52; *New York World*, June 22, 1888; "Report of the Committee of the Medico-Legal Society on the Best Method of Execution of Criminals by Electricity," *Medico-Legal Journal* 6 (1888–89): 278. Edison's manacle proposal was endorsed by *Electrical Review* 12 (June 30, 1888): 4; and the *Electrician* 21 (July 13, 1888): 319.

9. "Report of the Committee of the Medico-Legal Society on the Best Method of Execution of Criminals by Electricity," *Medico-Legal Journal* 6 (1888–89): 278. Also see Banner, *Death Penalty*, 181.

10. "Report of the Committee of the Medico-Legal Society on the Best Method of Execution of Criminals by Electricity," 278.

11. Kennelly Notebook #1, pp. 90–91, March 12, 1889 (TAEM 104:346); *New York Times*, March 9, 13, 1889; *New York Star*, March 18, 1889 (TAEM 146:431); *New York Tribune*, March 13, 1889.

12. *New York Star*, May 9, 1889; *New York Daily Graphic*, May 28, 1889; *Frank Leslie's Illustrated Newspaper*, June 8, 1889, p. 306; Kemmler Hearings, 1015, lxxxiii.

13. Quotation from Fell, "The Influence of Electricity on Protoplasm," 442–43. Also see *Buffalo Daily Times*, April 30, 1890. Works suggesting, incorrectly, that Stickley did design the first chair include Barry Sanders, *A Complex Fate: Gustav Stickley and the Craftsman Movement* (New York: Wiley, 1996), 8; Mary Ann Smith, *Gustav Stickley: The Craftsman* (Syracuse: Syracuse University Press, 1983), 4; Rita Reif, "The Master of Mission," *New York Times*, July 9, 1978, sec. 2, p. 27. On the 1893 electrocution, see *New York Herald*, July 28, 1893.

14. Quotations from Edison to Henry Villard, February 8, 1890 (TAEM 140:510); *Richmond Times*, February 12, 1890.

While attacking Westinghouse on the question of safety, the Edison forces continued to pursue the matter of alternating current's efficiency. Harold Brown's *Sun* letters revealed that he wanted to send a Westinghouse dynamo to Johns Hopkins University for efficiency tests. Despite the bad publicity, the tests went ahead as planned, and they showed that the Westinghouse system was far less efficient than claimed. Brown, Kennelly, and Edison worked together to publicize the inefficiency of the Westinghouse system. See Harold P. Brown, *A Test of the Efficiency of a Westinghouse Alternating Current Electric Lighting Plant* (New York: J. W. Pratt, 1890); *Electrician* 24 (April 4, 1890): 554; Kennelly to Edison, February 15, 1890; Kennelly to Brown, February 15, March 30, 1890; Alfred Tate to Edison, March 3, 1890; Harvey Ward Leonard to Tate, March 10, 1890 (TAEM 109:417, 413; 130:666; 140:633; 129:449).

15. Quotations from *Richmond Daily Times*, February 12, 13, 1890. Also see *Richmond Dispatch*, February 12, 13, 1890; *Electrical World* 15 (February 22, 1890): 156; Lewis B. Stillwell, "Alternating Versus Direct Current," *Electrical Engineering* 53 (1934): 708–10; Hughes, "Harold P. Brown and the Executioner's Current," 155.

16. Quotation from Kennelly to Edison, March 8, 1890 (TAEM 109:433). Also see Kennelly Notebook #3, March 8, 1890; Jacob Herrick to Charles Batchelor, March 3, 1890; Charles Batchelor to Edison, March 3, 1890 (TAEM 104:621, 128: 1098–1100).

17. Quotations from *New York Herald*, January 23, 1890; *New York Times*, January 23, 1890. Also see *New York Herald*, January 24, February 16, 1890; *New York Times*, February 16, 1890.

18. *New York Times*, April 4, 1890.

19. *New York Herald*, February 21, March 28, 1890; *New York Times*, February 21, 25, March 1, 2, 4, 12, 16, 18, 19, 21, 27, 29, April 4, 19, 1890; *New York World*, March 1, 2, 15, 1890; *New York Tribune*, March 1, 2, 4, 15, 28, April 4, 1890.

20. Quotations from *New York World*, December 15, 1888, which printed the opinions of many newspaper reporters on this topic.

21. Quotations from *New York Herald*, February 13, 1890; *Additional Report of the Commissioners on Capital Punishment of the State of New York*, 19–20. Also see *New York Herald*, February 15, 1890; *New York Times*, March 12, 1890.

22. 119 N.Y. Reports 569–79 (1890); Denno, "Is Electrocution an Unconstitutional Method of Execution?" 588–89; *New York Times*, *New York World*, *New York Tribune*, February 26, March 22, 1890.

CHAPTER 19. THE CONVERSION OF WILLIAM KEMMLER

1. Quotations from *New York World*, April 1, 1890.
2. Quotation from ibid. Also see *New York World*, December 31, 1889.
3. *New York World*, April 29, 1890.
4. Quotations from *New York Herald*, April 7, 1890.
5. Quotations from ibid., April 7, 1890.
6. Quotation from ibid., April 20, 23, 1890. Also see *New York World*, April 27, 1890.
7. Quotation from *Buffalo Evening News*, April 29, 1890. Also see *New York Herald*, April 29, 1890.
8. *New York World*, April 30, 1890.
9. Quotations from *New York World*, April 27, 1890; *Buffalo Evening News*, April 29, 1890. Also see *Auburn Bulletin*, April 24, 28, 1890; *New York World*, April 28, 1890.
10. Quotations from *New York Herald*, April 7, 1890; *New York World*, April 28, 1890.
11. *New York World*, April 26, 1890.
12. Quotation from *New York Herald*, April 20, 1890. At the time, an ordinary barber's chair meant a simple wooden chair with a footrest. See, for instance, the "American Shaving Chair" in R. Hovenden and Sons, *Revised and Illustrated Catalogue of Perfumery, Combs, Brushes . . .* (London, 1867), 119, Department of Drawings and Prints, Metropolitan Museum of Art.
13. *New York Herald*, April 20, 27, 29, 1890; *New York World*, April 26, 29, 1890.
14. *New York Times*, April 29, 30, 1890.
15. Reprinted in *Electrical Engineer* 9 (May 14, 1890): 350.
16. *New York World, New York Herald*, April 30, 1890.
17. *New York Sun*, April 30, 1890; *New York World*, April 30, 1890; *New York Times*, April 30, 1890. Also see Fell, "Influence of Electricity," 444–45. Brown's association with the Edison interests continued for some years. He worked with Thomas Edison on insulation, metal alloys, and other products past the turn of the century. He later became a member of the "Edison Pioneers," an association of those who worked with the inventor. In his application to join the Pioneers, he cited his work at the laboratory on "Determination of current best suited for electrical execution" and claimed to have "appeared as his [Edison's] personal representative before the Ohio Legislature" in the matter of restricting alternating current. See Brown folder, Edison Pioneers collection, ENHS. Also see Brown's correspondence with the Edison laboratory, 1893–1905, in TAEM. For Brown's later career as an inventor, see *National Cyclopedia of American Biography*, volume B (New York: James T. White, 1927), 329–30.
18. Quotations from *New York World*, April 30, 1890; *New York Herald*, May 4, 1890.
19. Quotations from *New York Herald*, May 4, 1890; *New York World*, May 2, 1890; *New York Times*, May 24, 1890. Also see Matthew Hale, "The Kemmler Case," *Albany Law Journal* 41 (May 10, 1890): 364–47.
20. *New York Herald*, May 4, 1890.
21. Ibid., May 1, 1890. For Westinghouse's letter to Sherman, see *Buffalo Express*, August 9, 1890. For estimates of Westinghouse expenses, see *New York Tribune*, August 7, 1890; *Scientific American* 63 (August 16, 1890): 96.

22. 136 U.S. 436 (1890); Denno, "Is Electrocution an Unconstitutional Method of Execution?" 590–92.

23. *New York Herald*, June 15, 25, 1890; *New York Times*, June 25, 1890; *Utica Saturday Globe*, June 21, 1890, typescript reproduction, New-York Historical Society.

24. *New York Herald*, June 25, 1890.

CHAPTER 20. THE FIRST EXPERIMENT

1. *New York World*, August 7, 1890. This chapter is based primarily on coverage in the following newspapers for August 7, 1890: *New York World, New York Times, New York Herald, New York Tribune, New York Sun, Buffalo Evening News, Buffalo Express, Buffalo Courier, Auburn Bulletin, Auburn Daily Advertiser*. Most of the articles were based on accounts provided by the reporters for the Associated Press and the United Press, the only newspaper representatives to witness the execution. The *World* is a particularly good source because it printed, in addition to its own story, the full A.P. and U.P. reports. Unless otherwise noted, all quotations are from the *World*. Also see Fell, "Influence of Electricity"; Charles R. Huntley, "The Execution as Seen by an Electrician," *Electrical World* 16 (August 16, 1890): 100; Carlos F. MacDonald, "The Infliction of the Death Penalty by Means of Electricity," *New York Medical Journal* 55 (May 7, 1892): 505–9, 535–42.

2. *New York Times*, August 2, 4, 6, 1890; *New York Herald*, August 3, 6, 1890; Fell, "Influence of Electricity," 447.

3. On Gerry's whereabouts, see *New York Sun*, August 2, 1890.

4. On Spitzka, see Charles E. Rosenberg, *The Trial of the Assassin Guiteau: Psychiatry and Law in the Gilded Age* (Chicago: University of Chicago Press, 1968); *Dictionary of American Medical Biography*, ed. Martin Kaufman et al. (Westport, Conn.: Greenwood Press, 1984), 709–10.

5. For details of this discussion, see *New York Herald*, August 7, 1890; MacDonald, "Infliction of the Death Penalty," 506.

6. Fell, "Influence of Electricity," 449.

7. Ibid., 435.

CHAPTER 21. AFTER KEMMLER

1. The headlines appeared in the August 7, 1890, editions of these newspapers.

2. Quotation from *Buffalo Express*, August 5, 1890. Also see *New York Times*, August 7, 1890; *New York Sun*, August 6, 1890.

3. Quotation from "The First Execution by Electricity," *Medical Record* 38 (August 9, 1890): 154–56. Also see *New York Times, New York Herald*, August 8, 1890.

4. Quotations from *New York Herald*, August 7, 8, 1890.

5. *Utica Saturday Globe*, August 16, 1890, typescript reproduction, New-York Historical Society; *Buffalo Courier*, August 8, 1890; *New York Tribune*, August 7, 1890.

6. Quotations from *Buffalo Evening News*, August 7, 1890; *Electrical Review* 16 (August 16, 1890). Also see George F. Shrady, "The Death Penalty," *Arena* 2 (1890): 513–23; "Some Remarks on the Kemmler Vivisection," *Electrical Engineer* 10 (August

13, 1890): 169–70; *New York Herald*, August 7, 1890; *Auburn Daily Advertiser*, August 7, 1890.

7. Quotations from *Public Opinion* 9 (1890): 432–35; *New York Times*, August 7, 1890. Also see *Electrical World* 16 (August 16, 1890): 105.

8. *New York Times*, August 7, 1890.

9. Quotations from *Buffalo Evening News*, August 6, 1890; *New York World*, August 7, 1890; *New York Sun*, August 7, 1890. Also see *Public Opinion* 9 (1890): 432–35.

10. Quotation from *New York Tribune*, October 9, 1890. Also see "The Official Report of the Execution by Electricity," *Medical Record* 38 (October 18, 1890): 438; MacDonald, "Infliction of the Death Penalty," 505–9; 535–44.

11. Quotations from E. C. Spitzka, "Preliminary Report Concerning the Post Mortem Changes in the First Person Executed by Electricity," *Atlanta Medical and Surgical Journal* n.s. 7 (1890–91): 460–61; Fell, "Influence of Electricity," 449–50. Also see *New York Herald*, August 7, 1890.

12. Quotation from *New York Tribune*, November 22, 1890. Also see *New York Times*, November 25, 1890; *Jugiro v. Brush*, 140 U.S. 686.

13. *New York Tribune*, November 22, 1890.

14. Quotations from *New York Evening Sun*, July 7, 1891, Rockwell Scrapbooks, Bakken Library; *New York Herald*, July 8, 1891. Also see *New York Times*, July 8, 1891. A few months later Kennelly did similar work on the ammeter for the Clinton Prison execution plant. Kennelly Notebook #4, pp. 27, 37, 112, December 13, 1890, March 21, 1891 (TAEM 105: 41, 54, 165). On the appeals of the other prisoners, see *New York Times*, November 26, 28, 30, December 2, 1890, January 8, March 13, 27, April 11, 19, May 12, June 5, 12, 13, 1891; *New York Herald*, January 9, February 18, 1891.

15. Quotations from Carlos MacDonald and S. B. Ward to W. R. Brown, July 30, 1891, printed in *Medico-Legal Journal* 9 (1891–92): 167–73.

16. See Marlin Shipman, "'Killing Me Softly'? The Newspaper Press and the Reporting on the Search for a More Humane Execution Technology," *American Journalism* 13 (1996): 176–205.

17. See *New York Herald*, July 9, 1891; clippings in Rockwell Scrapbooks, Bakken Library.

18. Quotation from Kennelly Notebook #5, p. 88, September 15, 1891 (TAEM 105:403). Also see *New York Herald*, July 8, 1891; Kemmler Hearings, 630; *New York Times*, August 7, 1890. Harold Brown, most likely at Edison's suggestion, also proposed liquid electrodes at one point: Harold P. Brown, "The New Instrument of Execution," *North American Review* 149 (November 1889): 586–93.

19. Quotations from Arthur Kennelly to Alfred Southwick, September 15, 1891 (the microfilmed version of this letter [TAEM 109:874] is largely indecipherable; for the somewhat more legible original, see Kennelly Letterbook #5, p. 263, ENHS); Southwick to Kennelly, September 21, 1891 (TAEM 142:802). Edison's secretary provided the inventor with copies of the Kennelly-Southwick correspondence. See Alfred Tate to Edison, September 23, 1891 (TAEM 142:801).

20. *New York Herald*, December 8, 1891. Also see the *New York World*, *New York Sun*, December 8, 1891; MacDonald, "Infliction of the Death Penalty."

21. *New York Herald*, August 5, 1891, February 5, 1892; *New York World*, February 8, 1892; *New York Times*, January 6, February 9, 1892.

22. Quotation from *New York Sun*, February 9, 1892. Also see *New York Herald*, *New York World*, *New York Times*, February 9, 1892; Kennelly to Carlos MacDonald, February 3, 1892 (TAEM 109:968); Reynolds and Bernstein, "Edison and 'The Chair,'" 25–26.

23. Quotations from *New York Herald*, February 9, 1892.

24. *New York Times*, February 9, 1892. Edison had no comment, but Kennelly admitted the error in a brief paper about the execution: A. E. Kennelly, "Physiological Observations at the McIlvaine [*sic*] Electrocution," *Electrical Engineer* 13 (February 17, 1892): 157–58. Also see W. J. Jenks, "Electrical Execution," *New York Medical Journal* 55 (May 14, 1892): 542–44; Kennelly Notebook #6, p. 32, February 8, 1891 (TAEM 105:580).

25. On electrocution techniques, see A. E. Kennelly and Augustin H. Goelet, "Does Execution by Electricity, as Practiced in New York State, Produce Instantaneous, Painless and Absolute Death?–Observations Made at the Execution of David Hampton, at Sing Sing, Jan. 28, 1895," *Electrical World* 25 (February 16, 1895): 197; for the next Auburn execution, see *New York Herald*, May 19, 1892. In August 1892 the electric chair at Clinton Prison in Dannemora was pressed into service for the first time, and proved reasonably successful: see *New York Herald*, August 3, 1892; "Electrical Execution," *JAMA* 19 (August 1892): 236. For coverage of later electrocutions, see, for example, *New York Herald*, January 21, 1897. For a list of executions, see Daniel Allen Hearn, *Legal Executions in New York State: A Comprehensive Reference, 1639–1963* (Jefferson, N.C.: McFarland, 1997): 81–91.

26. *New York Herald*, July 28, 29, 1893; J. W. Brown, "The Latest Electrocution," *Medical Record* 44 (August 12, 1893): 222–23. The controversy over this case revived two years later: *New York Herald*, July 23, August 2, 1895. For a similar controversy, see "A Ghastly View of Electrical Execution," *New York Medical Journal* 68 (October 1, 1898): 487; W. M. Hutchinson, "Electrical Execution," *New York Medical Journal* 68 (December 10, 1898): 861.

27. Quotations from *New York Herald*, February 11, 1892; "Electrical Execution," 236. For other responses to electrocution in medical journals, see "Electrocution So-Called and Its Lesson," *JAMA* 15 (August 23, 1890): 290; MacDonald, "Infliction of the Death Penalty"; MacDonald and S. B. Ward to W. R. Brown, July 30, 1891, in *Medico-Legal Journal* 9 (1891–92): 167–73. On the attempt to return to hanging, see *New York Sun*, *New York World*, February 9, 1892; *New York Herald*, March 17, 1892; *New York Times*, March 31, 1892. On poison and other scientific killing methods, see S. W. Abbott, "Carbonic Oxide vs. Electricity," *Boston Medical and Surgical Journal* 121 (August 15, 1889): 171–72; M. S. Macy, "What Is the Proper Mode of Executing Criminals," *Transactions of the Illinois State Medical Society* 44 (1894): 526–32; Leonard S. Taylor, "The Humane Execution of Condemned Criminals," *Chicago Medical Times* 28 (1895): 323–24; "A Plea for Humane Executions," *Medical Review* (St. Louis) 33 (1896): 148–50; F. O. Marsh, "Some Medical Aspects of Capital Punishment," *Transactions of the Ohio State Medical Society* (1898): 416–21.

CHAPTER 22. THE END OF THE BATTLE OF THE CURRENTS

1. Quotation from Samuel Insull to Edison, July 16, 1890 (TAEM 129:228). Also see "Annual Convention of Edison Illuminating Companies," *Electrical World* 14 (August 24, 1889): 135–36.

2. For the work on alternating current, see John Kruesi to Samuel Insull, July 31, 1890 (TAEM 129:236); Samuel Dana Greene to Edison, October 10, 1890 (TAEM 129:263); Alfred Tate to Greene, October 14, 1890 (TAEM 141:638); Tate to Insull, November 1, 1890 (TAEM 141:700); Insull to J. C. Henderson, November 5, 1890 (TAEM 129:277); Insull to Tate, November 6, 1890 (TAEM 129:276); Kennelly Notebook #3, August 22, 1890, pp. 101, 128, 136–39, 148, 150 (TAEM 104: 730, 740, 768, 782, 787, 800, 803); "Memorandum of Mr. Edison's Work During the Year 1891," January 6, 1892 (TAEM 143:16); Israel, *Edison*, 333.

3. Quotations from Carlson, *Innovation as a Social Process*, 284, n. 20; *New York Herald*, November 16, 1890. When the state needed a new execution generator at the end of 1890, the Edison company considered adapting one of its own new alternating generators for the purpose. It is unclear whether the machine was actually sold to the state. See Kennelly to J. C. Henderson, December 14, 1890 (TAEM 109:648). Five years later, Edison said that if 2,000-volt alternating-current wires were properly maintained, "the danger to human life, by reason of electric shocks, is insignificant." Edison to G. W. Grant, July 1895 (TAEM 135:793; month conjectured by TAEM).

4. Carlson, *Innovation as a Social Process*, 287–91; Israel, *Edison*, 335–36.

5. Quotation from Edison to Henry Villard, April 1, 1889 (TAEM 144:216).

6. Carlson, *Innovation as a Social Process*, 281–83, 294; Bright, *Electric Lamp Industry*, 87–88.

7. Carlson, *Innovation as a Social Process*, 294–301; Israel, *Edison*, 336; Passer, *Electrical Manufacturers*, 150, 321–27.

8. Quotations from *New York World*, February 21, 1892.

9. Quotations from Israel, *Edison*, 336; Edison to Henry Villard, April 1, 1889 (TAEM 144:216). Also see Tate, *Edison's Open Door*, 260–61; *New York Tribune*, February 20, 1892; *Newark Call*, February 28, 1892 (TAEM 146:752); Carlson, *Innovation as a Social Process*, 281–82.

10. Edison to Henry Villard, February 8, 1890 (TAEM 140:510).

11. Quotations from Tate, *Edison's Open Door*, 278. Also see Israel, *Edison*, 335–37.

12. Quotation from Israel, *Edison*, 292. Also see Charles Musser, *Thomas A. Edison and His Kinetographic Motion Pictures* (New Brunswick: Rutgers University Press, 1995).

13. Israel, *Edison*, 338–62.

14. Ibid., 335–37.

15. Kline, "Science and Engineering Theory," 288–92; Hughes, *Networks of Power*, 129–35.

16. Quotation from Edward Dean Adams, *Niagara Power: History of the Niagara Falls Power Company, 1886–1918*, 2 vols. (Niagara Falls, N.Y.: privately printed, 1927), 1:144. Also see Steven Lubar, "Transmitting the Power of Niagara: Scientific, Technological, and Cultural Contexts of an Engineering Decision," *IEEE Technology and Society Magazine* 8 (March 1989): 11–13; Robert Belfield, "The Niagara System: The Evolution of an Electric Power Complex at Niagara Falls, 1883–1896," *Proceedings of the IEEE* 64 (1976): 1344–45.

17. Passer, *Electrical Manufacturers*, 142–43; Daniel H. Burnham, *The Final Official Report of the Director of Works of the World's Columbian Exposition* (New York: Garland, 1989), 5–6.

18. Quotations from J. P. Barret, *Electricity at the World's Columbian Exposition* (Chicago: R. R. Donnelley & Sons, 1894), xi, 451; Marian Shaw, *World's Fair Notes: A Woman Journalist Views Chicago's 1893 Columbian Exposition* (St. Paul, Minn.: Pogo Press, 1992), 4.

19. Hughes, *Networks of Power*, 122–25; Barret, *Electricity at the World's Columbian Exposition*, 166.

20. Belfield, "The Niagara System," 1349–50; Hughes, *Networks of Power*, 137–39; Passer, *Electrical Manufacturers*, 292–93.

21. Andre Millard, "Thomas Edison, the Battle of the Systems and the Persistence of Direct Current," *Material History Review* 36 (Fall 1992): 18–28; Hughes, *Networks of Power*, 81, 120–26; Passer, *Electrical Manufacturers*, 175; Prout, *Life of George Westinghouse*, 115–16.

CHAPTER 23. THE AGE OF THE ELECTRIC CHAIR

1. On the Massachusetts law, see *Commonwealth v. Storti*, 178 Mass. 549 (1901); Hiller B. Zobel, "The Death of Luigi Storti," *Massachusetts Legal History* 7 (2001): 81–93; *New York Tribune*, April 15, 1898. On Ohio, see *Annual Report of the Board of Managers of the Ohio Penitentiary . . . for the Fiscal Year 1896* (Norwalk, Ohio: Laning, 1897), 10, 43–44; *Annual Report of the Board of Managers of the Ohio Penitentiary . . . for the Fiscal Year 1897* (Norwalk, Ohio: Laning, 1898), 17, 57; *New York Tribune*, November 27, 1896; *New York Herald*, November 20, 1897; F. O. Marsh, "Some Medical Aspects of Capital Punishment," *Transactions of the Ohio State Medical Society* (1898): 416–21. Harold P. Brown, on behalf of Edison, wrote to German officials explaining electrocution: see Richard J. Evans, *Rituals of Retribution: Capital Punishment in Germany, 1600–1987* (New York: Oxford University Press, 1996), 426.

2. Ronald Kline, "Science and Engineering Theory"; Theodore Bernstein, "Theories of the Causes of Death from Electricity"; Reynolds and Bernstein, "Edison and 'The Chair,'" 26; A. J. Jex-Blake, "Death by Electric Currents and by Lightning," *British Medical Journal* 1 (1913): 425–30, 492–98, 548–52; Jaffé, "Electropathology," 837–70.

3. Quotation from Edwin J. Houston and A. E. Kennelly, "Death by the Alternating Current," *JAMA* 25 (August 1895): 283–85. Also see Kennelly and Goelet, "Does Execution by Electricity," 197; Arthur E. Kennelly and E. R. W. Alexanderson, "The Physiological Tolerance of Alternating-Current Strengths up to Frequencies of 100,000 Cycles per Second," *Electrical World* 56 (July 21, 1910): 154–56; Arthur E. Kennelly, "The Danger of Electric Shock from the Electrical Engineering Standpoint," *Physical Therapeutics* 45 (1927): 16–23.

4. Quotation from *Cleveland Plain Dealer*, August 7, 1895. Also see *Buffalo Times*, August 8, 1895 (TAEM 146:970). As late as 1938 *Popular Science Monthly* reported that research at Harvard Medical School indicated that the electric chair's victim "may only be shocked into a semblance of death and that the final spark of life is extinguished unwittingly in the autopsy room." Robert E. Martin, "Electric Shocks . . . Do They Really Kill?" *Popular Science Monthly* 133 (July 1938): 44–45, 101. Also see Louise G. Robinovitch, "Electrocution: An Experimental Study

with an Electric Current of Low Tension," *Journal of Mental Pathology* 7 (1905): 75–85; Banner, *Death Penalty*, 191–92.

5. Quotation from *Encyclopedia Britannica*, 11th ed. (1910), s.v. "electrocution." Also see *Newark Evening News*, *Trenton Evening Times*, December 11, 1907. For an abstract of a Spitzka article, see James W. Garner, "Infliction of the Death Penalty by Electricity," *Journal of Criminal Law and Criminology* 1 (1910): 626. For other medical journal response, see A. D. Rockwell, "Discussion of Electrical Execution," *JAMA* 19 (September 24, 1892): 363–65; George E. Fell, "Electrodes, and Their Application in Electrocution," *JAMA* 19 (September 24, 1892): 365–67; Edmund W. Holmes, "Anatomy of a Hanging," *Pennsylvania Medical Journal* 4 (1901): 737–45; S. R. Klein, "Capital Punishment in the Electric Chair," *New York Medical Journal* 99 (May 30, 1914): 1089–90; A. H. Werner, "Death by Electricity," *New York Medical Journal* 118 (October 1923): 498–500. For criticism of the electric chair, see "Failure of Electrocution," *Medical Record* 65 (June 25, 1904): 1050; Frederic Rowland Marvin, "Execution by Electricity," *Medical Record* 66 (July 23, 1904): 145–46; S. R. Klein, "Capital Punishment in the Electric Chair," *New York Medical Journal* 99 (May 30, 1914): 1089–90.

6. *Execution of Czolgosz, with Panorama of Auburn Prison* (Thomas A. Edison, 1901). The film was first advertised in the *New York Clipper*, November 16, 1901, p. 832. For a brief discussion of the film, see Charles Musser, *Thomas A. Edison and His Kinetographic Motion Pictures* (New Brunswick: Rutgers University Press, 1995), 32–3. On the actual Czolgosz case, see Carlos F. MacDonald, "The Trial, Execution, Autopsy, and Mental Status of Leon F. Czolgosz," and Edward Anthony Spitzka, "A Report of the Post-Mortem Examination," *New York Medical Journal* 75 (January 4, 1902): 12–23.

7. *New York World*, *New York Herald*, *New York Times*, January 5, 1903. *Electrocuting an Elephant*, (Thomas A. Edison, 1903). The film was first advertised in the *New York Clipper*, January 17, 1903, p. 1052, as *Electrocution of the Baby Elephant, 'Topsy'* (Topsy was in fact more than thirty years old). A clip from this Edison film appears in Errol Morris's documentary, *Mr. Death* (Lion's Gate Films, 1999). In 1901 officials of the Pan-American Exposition in Buffalo tried to kill the rogue elephant Jumbo II. Electricians gave the elephant six shocks at 2,200 volts, but the "shocks had simply tickled him." *New York Herald*, November 9, 10, 1901.

8. For New Jersey's first electrocution, see *Newark Evening News*, December 11, 1907; *Trenton Evening Times*, December 11, 1907. On Davis, see *New York Times*, April 29, 1890; *New York Sun*, April 30, 1890, July 8, 1891; *New York Times*, August 6, 1890; Thomas P. Dimitroff and Lois S. Janes, *History of the Corning-Painted Post Area* (Corning, N.Y.: Corning Area Bicentennial Committee, 1977), 147; Robert G. Elliott, *Agent of Death: The Memoirs of an Executioner* (New York: Dutton, 1940), 51.

9. Quotations from Adams Electric Company to E. F. Morgan, March 30, 1908; Adams Electric to E. F. Morgan, September 25, 1908, Virginia folder, Miscellaneous Correspondence folder, Carl F. Adams Papers, Special Collections and University Archives, Rutgers University Libraries (hereinafter cited as "Adams Papers"). Also see North Carolina and South Carolina folders, Adams Papers; "Trentonian Built Death Chair at Prison Here; Consulted with Edison, then Went to Work," unidentified, undated clipping (probably 1936), Adams Papers; Charles V. Carrington, "The History of Electrocution in the State of Virginia," *Virginia Medical Semi-*

Monthly (November 11, 1910): 353–54; *Daily State Gazette* (Trenton), August 20, 1910. When a state official in Canton, China, wanted to adopt electrocution, his representatives wrote to Montgomery Ward of Chicago, apparently under the impression that electric chairs might be available by mail order, alongside washing machines and dining room tables. Montgomery Ward referred the Chinese official to the Virginia penitentiary, which referred the matter to Adams, who offered an electrocution plant for $3,000 plus all expenses for his trip to China to install it. J. A. Cheony to Virginia State Penitentiary, January 14, 1914; Adams Electric Company to Cheony, February 28, 1913, China, Kansas, Kentucky folder, Adams Papers.

10. James W. Clarke, "Without Fear or Shame: Lynching, Capital Punishment and the Subculture of Violence in the American South," *British Journal of Political Science* 28 (1998): 269–89; Philip Dray, *At the Hands of Persons Unknown: The Lynching of Black America* (New York: Random House, 2002). For the phonograph recording of the lynching, see Edward L. Ayers, *The Promise of the New South: Life After Reconstruction* (New York: Oxford University Press, 1992), 159.

11. For articles explicitly promoting capital punishment as an alternative to lynching, see "Lynch Law and Its Remedy," *Yale Law Journal* 8 (October 1898): 335–40; J. E. Cutler, "Capital Punishment and Lynching," *Annals of the American Academy of Political and Social Science* 29 (1907): 622–25. Virginia's prison surgeon preferred the electric chair because, he said, it prevented its victims from finding glory on the gallows: "A negro likes nothing better than to be the central figure, be it a cake-walk or a hanging." Carrington, "The History of Electrocution in the State of Virginia," 353. For death penalty statistics, see William J. Bowers, *Legal Homicide: Death as Punishment in America, 1864–1982* (Boston: Northeastern University Press, 1984), appendix A.

12. Banner, *Death Penalty*, 196–202, 349, n. 36; Bowers, *Legal Homicide*, 9–15; Roger Lane, "Capital Punishment," in *Violence in America: An Encyclopedia*, ed. Ronald Gottesman (New York: Charles Scribner's Sons, 1999), 201; Stephen Trombley, *The Execution Protocol* (London: Century, 1993), 12–13, 96–99.

13. Bungled electrocutions may have increased in frequency as the century progressed. Professionals of the stature of Alfred Kennelly or Alfred A. Spitzka became less willing to offer their services, and inept prison handymen sometimes found themselves in charge of electric chair maintenance, with disastrous results. In the 1980s America's foremost engineer in matters of execution was a man named Fred Leuchter, who designed and sold equipment to several state prisons. After Leuchter became involved with Holocaust denial groups, reporters looked into his background and discovered that his only college degree was a B.A. in history. He was convicted on criminal charges of practicing engineering without a license. Denno, "Is Electrocution an Unconstitutional Method of Execution?" 37–43; Errol Morris, *Mr. Death* (Lion's Gate Films, 1999); Harold Hillman, "The Possible Pain Experienced During Execution by Different Methods," *Perception* 22 (1993): 745–53; Harold Hillman, "An Unnatural Way to Die," *New Scientist* (October 27, 1983): 276–78; A. Sances et al., "Electrical Injuries," *Surgery, Gynecology, and Obstetrics* 149 (July 1979): 97; Mark Costanzo, *Just Revenge: Costs and Consequences of the Death Penalty* (New York: St. Martin's, 1997), 43–45; Jacob Weisburg, "This Is Your Death," *New Republic* 205 (July 1, 1991): 23–27; Austin Sarat, "Killing Me Softly: Capital Punishment and the Technologies for Taking Life," in *Pain, Death,*

and the Law (Ann Arbor: University of Michigan Press, 2001), 43–70; Robert Jay Lifton and Greg Mitchell, *Who Owns Death? Capital Punishment, the American Conscience, and the End of Executions* (New York: William Morrow, 2000), 56; Susan Lehman, "A Matter of Engineering: Capital Punishment as a Technical Problem," *Atlantic Monthly* 265 (February 1990): 26–29.

14. Banner, *Death Penalty*, 220–23; John F. Galliher, Gregory Ray, and Brent Cook, "Abolition and Reinstatement of Capital Punishment During the Progressive Era and Early Twentieth Century," *Journal of Criminal Law and Criminology* 83 (1992): 538–78; Norman S. Hayner and John R. Cranor, "The Death Penalty in Washington State," *Annals of the American Academy of Political and Social Science* 284 (1952): 101–4; Ellen Elizabeth Guillot, "Abolition and Restoration of the Death Penalty in Missouri," *Annals of the American Academy of Political and Social Science* 284 (1952): 105–14.

15. Banner, *Death Penalty*, 223–27, 240; Bowers, *Legal Homicide*, 21–24, 67–102; Hugo Adam Bedau, ed., *The Death Penalty in America: Current Controversies* (New York: Oxford University Press, 1997), 11–17.

16. Quotation from *Furman v. Georgia*, 408 U.S. 238 (1972). Also see Banner, *Death Penalty*, 228–30, 247–66; Bedau, *Death Penalty in America*, 183–85.

17. Banner, *Death Penalty*, 257–78, 284–90; Bedau, *Death Penalty in America*, 16–7, 185–87.

18. Banner, *Death Penalty*, 267, 275–78, 284–90; Bedau, *Death Penalty in America*, 17–8; Phoebe C. Ellsworth and Samuel R. Gross, "Hardening of the Attitudes: Americans' Views of the Death Penalty," in Bedau, *Death Penalty in America*, 90–115.

19. Banner, *Death Penalty*, 300; Amnesty International Web site, http://www.web.amnesty.org/rmp/dplibrary.nsf/ff6dd728f6268d0480256aab003d14a 8/ 46e4de9db9087e358025688100 50f05f!OpenDocument, July 10, 2002.

20. Quotation from Evans, *Rituals of Retribution*, 427. Also see *Report of the Royal Commission on Capital Punishment, 1949–1953* (London: Her Majesty's Stationery Office, 1953), 256; Evans, *Rituals of Retribution*, 427. Political differences also played a role. In Britain and France, the decision to abolish the death penalty was extremely unpopular at first. But the national governments refused to reinstate it, and eventually the people came to support the ban. After the *Furman* decision, politicians in many American states proved more responsive to public opinion and immediately brought back the death penalty. Banner, *Death Penalty*, 281–82, 300–301; Roger Hood, *The Death Penalty: A World-Wide Perspective*, 2nd ed. (Oxford: Clarendon Press, 1996); Bedau, *Death Penalty in America*, 10; Evans, *Rituals of Retribution*, 537.

21. Quotation from Trombley, *Execution Protocol*, 277. Also see Denno, "Is Electrocution an Unconstitutional Method of Execution?"

22. Quotations from *New York Times*, March 7, 2001. The "offending standards of decency" phrase is drawn from a 1958 Supreme Court decision, *Trop v. Dulles*, that had become central to arguments concerning the Eighth Amendment. See Banner, *Death Penalty*, 237.

23. Statistics courtesy Death Penalty Information Center, personal correspondence with author, May 15, 2002; Carol Robinson, "Block First Woman Executed Since '57," *Birmingham News*, May 10, 2002; Maria Glod, "Family's Killer Dies in Va. Electric Chair," *Washington Post*, April 10, 2003, p. B1.

EPILOGUE. THE NEW SPECTACLE OF DEATH

1. The anecdote is recounted in Wachhorst, *Thomas Alva Edison*, 3.

2. See "Chronology of Thomas Edison's Life," Thomas A. Edison Papers Web site, http://edison.rutgers.edu/chron2.htm, August 4, 2002; Wachhorst, *Thomas Alva Edison*, 21–22; Baldwin, *Edison*, 370.

3. James S. Evans, "Edison Regrets Electric Chair Was Ever Invented," *New York American*, February 10, 1905 (TAEM 221:289). In 1909 Edison again endorsed electrical execution in a private letter written by his secretary: "Mr. Edison directs me to write you that his belief, based on experiments with animals in this Laboratory is that a person electrocuted is mentally dead in less than 1/1000 of a second, and physically dead in two seconds; any motions made after that time is reflect [*sic*] action." Harry Frederick Miller to William H. Roberts, May 29, 1909 (TAEM 198:108). The official biography is Frank Lewis Dyer and Thomas Commerford Martin, *Edison: His Life and Inventions*, 2 vols. (New York: Harper & Brothers, 1910).

4. See Reynolds and Bernstein, "Edison and 'The Chair.'"

5. *New York World*, December 21, 1889. As late as 1907, Edison was still throwing his support behind efforts to limit high-voltage transmission within cities. See "Edison Condemns High Voltages," *Newark Advertiser*, December 26, 1907, clippings file, ENHS.

6. Quotation from *Electrician* 24 (November 15, 1889): 41. Also see *New York World*, December 17, 1889.

7. Quotation from *New York World*, October 20, 1889.

8. On low-voltage deaths, see A. E. Pain, "A Case of Death from the Electric Current While Handling the Telephone and an Electric Light Fixture," *Boston Medical and Surgical Journal* 155 (1906): 741; F. E. Jones, "A Case of Death from the Electric Current as a Result of Turning on an Electric Light," *Boston Medical and Surgical Journal* 160 (1909): 239; C. Van Zwaluwenburg, "A Case of Accidental Electrocution from Ordinary One Hundred Volt Alternating Lighting Current," *JAMA* 41 (1903): 967. Most accidental deaths occurred in industry. See "Fatal Accidents from Electric Shock in Recent Years in the United States and Canada, in England and Wales, and in Canada: A Report by the Engineering Committee of the Conference on Electric Shock," *Journal of Industrial Hygiene* 10 (1928): 111–16; Mark Aldrich, *Safety First: Technology, Labor, and Business in the Building of American Work Safety* (Baltimore: Johns Hopkins University Press, 1997), 84. Also see Carlson and Millard, "Defining Risk Within a Business Context."

9. Suspicions of Westinghouse interference persisted even after Kemmler's death. When Shibuya Jugiro filed an appeal claiming that electrocution was cruel, the *Herald* complained, "It is within the power of the Westinghouse or any other electric light concern to baffle the law of capital punishment." When a state senator moved to abolish electrocution in 1892, the move was dismissed as a ploy by "the interested electric machinery companies." *New York Herald*, January 9, 1891; March 17, 1892.

10. Quotation from *New York World*, June 22, 1888.

11. Quotation from Bill Sloat, "Ohio's Electric Chair Is History," *Cleveland Plain Dealer*, February 27, 2002. On requests to attend executions, see Edmund W.

Holmes, "Anatomy of a Hanging," *Pennsylvania Medical Journal* 4 (1901): 737–74. On electrocution exhibits at a dime museum, see *Monthly Catalogue, Eden Musée*, March 1893, November 1906, Billy Rose Theater Collection of the New York Public Library for the Performing Arts; Kathleen Kendrick, "'The Things Down Stairs': Containing Horror in the Nineteenth-Century Wax Museum," *Nineteenth-Century Studies* 12 (1998): 1–35. On the chair at the world's fair, see Moses P. Handy, *Official Directory of the World's Columbian Exposition* (Chicago: W. B. Conkey, 1893), 648; Shaw, *World's Fair Notes*, 44. On electric chair carnival sideshows, see Jay, *Learned Pigs and Fireproof Women*, 129–31.

12. Mark Twain, *A Connecticut Yankee in King Arthur's Court* (Berkeley: University of California Press, 1979), 467, 489, emphasis in original. Also see H. Bruce Franklin, *War Stars: The Superweapon and the American Imagination* (New York: Oxford University Press, 1988), 62–64; H. Bruce Franklin, "*Billy Budd* and Capital Punishment: A Tale of Three Centuries," *American Literature* 69 (1997): 337–59. Twain may have been inspired by a statement made by Edison in his first public statement on the electrocution law: "An electric light current will kill a regiment in the ten-thousandth part of a second." *New York World*, June 22, 1888.

13. William Dean Howells, "Execution by Electricity," *Harper's Weekly* 32 (1888): 23. Puzzled by Howells's irony, one electrician wrote, "So far as I can make out from his language, he favors the use of electricity." Thomas D. Lockwood, "Electrical Killing," *Electrical Engineer* 7 (March 1888): 89–90. Howells had a good reason for expressing his opposition to the new method in such an indirect style. In 1887 he courageously criticized the conviction of innocent men for the Haymarket bombing and as a result became the target of harsh public criticism. See Kenneth S. Lynn, *William Dean Howells: An American Life* (New York: Harcourt Brace Jovanovich, 1971), 288–97; Timothy L. Parrish, "Haymarket and *Hazard*: The Lonely Politics of William Dean Howells," *Journal of American Culture* 17 (Winter 1994): 23–32.

ART CREDITS

Images on the pages noted have been supplied by the following sources:

Jean Antoine Nollet, *Essai sur l'Electricité des Corps* (Paris, 1746), frontispiece. Courtesy Bakken Library: page 7

Illustrations by Martie Holmer: pages 13 and 111

Courtesy U.S. Department of the Interior, National Park Service, Edison National Historic Site: pages ii, 17, 21, 24, 33, 35, 38, 47, 63, 65, 127, 130, 144, 201, and 287

Giovanni Aldini, *Essai Théorique et Expérimentale su le Galvanisme* (Paris, 1804). Courtesy National Library of Medicine: page 41

Adapted from Frank Lewis Dyer and Thomas Commerford Martin, *Edison: His Life and Inventions,* vol. 2 (New York: Harper & Brothers, 1910), 917. Courtesy Cornell University Libraries: page 52

Scientific American 44 (April 2, 1881): 207. Courtesy Cornell University Libraries: page 59

Courtesy Consolidated Edison Company of New York: page 70

Scientific American 47 (August 26, 1882): 130. Courtesy Edison National Historic Site: page 72

Author collection: page 76

Louis B. Lane, *Memorial and Family History of Erie County, New York,* vol. 2 (N.Y.: N.Y. Genealogical Publishing Co., 1906–8), 106. Courtesy Cornell University Libraries: page 91

Sydney H. Coleman, *Humane Society Leaders in America* (Albany, 1924), facing page 65. Courtesy Cornell University Libraries: page 93

Scientific American 58 (June 30, 1888): 407. Courtesy Cornell University Libraries: page 94

Courtesy Historical Society of Western Pennsylvania: page 107

National Cyclopedia of American Biography, vol. 5 (New York: James T. White, 1945), 22. Courtesy Cornell University Libraries: page 132

Courtesy Museum of the City of New York: page 138

National Cyclopedia of American Biography, vol. B (New York: James T. White, 1927), 329. Courtesy Cornell University Libraries: page 140

Electrical World 15 (April 5, 1890): 237. Courtesy Cornell University Libraries: page 145

Harper's Weekly 33 (July 27, 1889): 601. Courtesy Cornell University Libraries: page 149

Scientific American 59 (December 22, 1888): 393. Courtesy Cornell University Libraries: page 154

Art Credits

National Police Gazette, August 23, 1890, p. 16. Courtesy Clements Library at the University of Michigan: page 165

The Bookman 21 (1905): 490. Courtesy Cornell University Libraries: page 168

Frank Leslie's Illustrated Newspaper, June 8, 1889, p. 306. Courtesy Library of Congress: page 172

People of the State of New York, ex rel. William Kemmler against Charles F. Durston, as Warden of the State Prison at Auburn, N.Y., vol. 2 (Buffalo: J. D. Warren's Sons, 1889), following lxxxv. Courtesy Buffalo and Erie County Public Library: page 182

Courtesy American Philosophical Society: page 208

New York World, October 12, 1889. Courtesy Library of Congress: page 214

Judge 17 (October 26, 1889): front cover. Courtesy Cornell University Libraries: page 217

Judge 17 (December 21, 1889): back cover. Courtesy Cornell University Libraries: page 222

Medico-Legal Journal 5 (1887–8): 430. Courtesy Cornell University Libraries: page 227

Clark Bell, Medico-Legal Studies, vol. 2 (New York: *Medico-Legal Journal,* 1889): 93. Courtesy Cornell University Libraries: page 229

Electrical World 16 (August 16, 1890): 99. Courtesy Cornell University Libraries: page 239

Courtesy Cayuga Museum: pages 246, 255, and 292

New York Herald, August 7, 1890. Courtesy Library of Congress: page 251

New York Medical Journal 55 (May 14, 1892): following 542. Courtesy Cornell University Libraries: pages 262 and 263

Courtesy Chicago Historical Society ICHi-02237: page 275

Harper's Weekly 33 (July 27, 1889): 601; Herbert Newton Casson, *The History of the Telephone* (Chicago, A. C. McClurg & Co., 1910), facing 132. Courtesy Cornell University Libraries: page 289

INDEX

Abolition, 87
Accidental deaths, 98, 124, 150, 205, 207, 219, 220, 224, 228–29, 259, 289, 290
 from alternating current, 135, 136, 139, 159, 209
 J. Feeks, 212–16, 214*f*
 New York City, 136, 216–17
Accidental shocks, 155
 survivors, 180
 victims reviving, 178
Adams, Carl F., 280, 281
Adams Electric Company, 280
African-Americans, 280–81, 283
Albany Republican Military, 77
Aldini, Giovanni, 41–42, 41*f*
Alternating current, 14, 110–11, 114, 134, 186, 270, 273
 advantages over direct, 272
 banning, 148, 212, 220, 232, 266
 became industry standard, 288
 Brown's attack on, 141
 danger from, 114, 135, 141–42, 146–47, 150, 152, 155, 156, 157, 159, 186, 198, 204–5, 207, 209, 210, 218, 243, 278, 290
 deaths from, 136, 139, 159, 209, 211
 defense of, 135
 Edison's and Kennelly's experiments on, 265–66
 Edison's battle against, 288–90
 efficiency, 329*n*14
 execution by, 117, 149–50, 152, 153, 154–55, 156–57, 198–99, 202
 government regulation, 203
 killing with, 158, 182
 limiting/restricting, 141, 147, 157, 191, 212, 219, 288, 289–90
 single-phase, 274
 symbolic victory over direct, 275
 Thomson opposed, 207–9
 transmitting, 112
 triumph of, 276
 voltage, 114
 voltage limits, 221, 230–31
 Westinghouse and, 112, 113, 141, 276, 290
Alternating-current plant, first, 113
Alternating-current system
 danger from, 136
 drawbacks to, 134–35
 Edison Electric's lack of, 171
 Edison patent rights, 186
 Edison testing, 131
 Edison's, 266
 first, 163
Alternating equipment, market for, 270–76
Alternating generators/dynamos, 117, 198, 265, 266, 275
 acquiring for state prisons, 159, 160
 Edison built, 203–4
 efficiency, 193
 in electrical execution, 133
 in executions, 193–94
 two-phase, 274
Amber, 5
America, as classless society, 120–21
American Electric Company, 104
American Institute of Electrical Engineers, 146, 174
American Nervousness (Beard), 45
American Revolution, 80–81
American Society for the Prevention of Cruelty to Animals (ASPCA), 87, 88, 93, 146, 279
American Telegraph, 15

Index

Amperage, 28, 29, 124, 142*n*
Amusement, electricity source of, 40, 46
Anarchists, 120-21
Andrews, George, 216, 221
Animal electricity, 10-11, 45
Animal-killing experiments, 2, 10, 48, 91, 93-94, 98, 124, 132-33, 142-47, 148, 149-50, 152, 153, 154*f*, 158, 177-78, 180, 219, 225-26, 228, 230, 231, 241, 278
  Brown and, 181, 196-99, 228
Animals
  and humane movement, 87-88
  killing, 89
Anti-capital punishment movement, 81-82
Anticruelty movement, 89, 96
Arc lamp(s), 26-28, 29, 59*f*, 223, 272
  current for, 273
  impact of, 104
  limitations of, 28
  New York City, 137
  number of, 105
Arc lighting/lights, 58-60, 74-75, 104-5, 110, 204
  accidents, 90-91
  safety issues, 69
Arc lighting manufacturers, 64, 104-5
Armature, 13-14, 13*f*, 27
Arsonval, Jacques Arsène d', 204-5, 278
Ashley, James, 19
Asphyxiation, 42, 69, 94, 205
Astor, John Jacob, Jr., 87
Atlantic & Pacific, 20
Auburn prison, 118, 158, 194, 230, 246*f*
  death chamber, 238-39, 241, 247
  electrocutions, 263-64
  Kemmler in, 167, 235-37, 240-41
  tests at, 225-26
"Auburn system" of prison discipline, 225
Auto da fe, 95
Automatic braking systems, 107-8
Autopsy(ies), 91, 278
  Kemmler, 254-55
Ball Electrical Illuminating, 136
Bamboo, 54-55
Barnes, Charles, 249, 256
Barnum, P. T., 46, 59, 67, 154
Batchelor, Charles, 31, 37, 103, 132, 198
Battery(ies), 8, 10, 12, 40, 90, 123
  current from, 14, 44
"Battle of the currents," 135, 140-42, 169, 288, 293
  Edison company's surrender in, 268-69
  end of, 265-76
Beard, George, 44-45, 48, 158
Becker, Tracy C., 174, 180-81, 183, 184, 185, 189, 190
Beheading, 95
Beating with clubs, 95
Bell, Alexander Graham, 21, 22
Benjamin, Park, 123
Bergh, Henry, 87, 143-45, 146
Bergmann, Sigmund, 63
Bergmann & Company, 63, 66, 67, 129, 170
  product development for, 131
Bishop, Washington Irving, 178
Bismarck, Otto von, 277
Bliss, George, 140
"Bloody Code," 80
Blowing from cannon, 95
Bly, Nellie, 214
Boehm, Ludwig, 55, 56
Boiling, execution by, 95
Boston, 58, 87, 103, 129
Breaking on the wheel, 95, 97
Bribery, 231, 232, 242
Brodie, B. C., 48
Brown, Harold Pitney, 140-42, 140*f*, 143, 146-47, 148, 149, 150, 152, 153, 156-57, 158, 211, 212, 221, 231, 288
  animal experiments, 181, 186, 196-99, 228
  Auburn prison tests, 225
  contract to acquire equipment, 158-59, 160, 168, 192-93, 199, 241, 243
  and Edison, 186, 188
  and Edison Electric, 148, 155-56, 159, 194, 197-98

and Edison laboratory, 172–73
electric chair, 171, 229–30, 239
Kemmler hearings, 185
letters stolen from, 190–91, 194–95
not invited to Kemmler execution, 247
unmasking of, 190–99, 200, 202
and wire panic, 218–19
witness, 179, 180–81, 183–84
Brush, Charles, 58–59, 109–10
Brush Electric Company, 58, 60, 104, 140, 205, 223
Brush Electric Illuminating, 136, 207, 216, 218, 223
Brush lighting plant, 91
Bryant, William Cullen, 81
Buffalo, New York, 59–60, 113, 163, 274
SPCA, 87, 98, 143, 145
Burglar alarm, 45–46
Burners, 33–35, 37
theory of, 36
Burning
death by, 95
in electrocutions, 227, 260, 262
in Kemmler execution, 253, 254–55
Burning alive, 81
Burning at the stake, 97
Burroughs, John, 286

Capital murder trial, 163–67
Capital punishment, 81, 84
arbitrariness of, 283–84
debates on, 91–92
deterrent effect of, 119, 123
Edison opposed, 3, 116–17, 202, 287
humane method of, 92
new method of, 95
New York considering abolishing, 242
support for, 282–84
see also Death penalty
Capital punishment bill, 118–19, 122–23
Carbon, 29, 36–37
Carbon burners, 34
Carbon monoxide, 89, 90, 97
Carbon rods, 27, 29, 37
Cataract Construction Company, 271
Cement, 286
Central stations, 58, 62, 67, 102, 103, 105, 109, 129
alternating-current, 113
model, 68
number of, 115
Chaffee, G. W., 60, 61
Chautauqua Institute, 126, 127
Chemical battery, 41–42
Chicago, 129, 288
Haymarket Square, 120–21
underground wires, 137–39
world's fair, 272–74, 275*f*, 291
Child labor laws, 86
China, 284
Circuit design, 51–52
Cities, 79, 102, 120, 134
Civil War, 16, 43, 216
Civilization, 86, 97, 123, 124, 257
Clinton prison, 118, 158, 241
Coal, 27, 28, 50
Cockran, W. Bourke, 167–68, 168*f*, 173, 174–76, 177, 179–83, 184–85, 186–89, 203, 235, 243–44
and Kemmler's death, 255
Coffin, Charles, 169, 192–94, 195, 267, 268
Columbia College, 146, 197, 198
Commercial lighting station (New York City), 50, 51–53, 56, 62, 64–66, 70–74
model of, 56–58, 65
Commutator, 14
Comparative Danger to Life of the Alternating and Continuous Electrical Currents, The (Brown), 157, 159, 191
Compassion, 86
Competition, 134, 135
Edison/Westinghouse, 115–116, 169, 203, 265, 266–67
General Electric/Westinghouse, 270–76
in telegraph industry, 15
Conductors, 6, 27, 28, 51, 62
copper, 113
human, 7*f*, 8–9
in New York lighting station, 57, 58
prime, 8, 10

Conductors (*continued*)
resistance of, 36
underground, 63, 64–65, 65*f*, 69, 71, 105
Coney Island, 59
Connecticut Yankee in King Arthur's Court, A (Twain), 293
Consolidated Electric Light Company, 170, 206
Consolidated Telegraph and Electrical Subway Company, 139
Constitution, 86
Bill of Rights, 81
Continuous current
see Direct current
Cooke, W. F., 14
Cooper, Peter, 87
Copper costs, 51, 52, 109, 113, 134, 141
Copper wire, 27, 29
Cornell, Ezra, 87
Corporal punishment, 80–81
Corpses, disposition of, 119–20, 122, 123
Corruption, 139, 231, 242
Cotton-picking machine, 128
Cravath, Paul, 168
Cream separator, 128
Criminal classes, 119, 120–22
Criminals
executing with electricity, 3, 48, 75
humane killing of, 89–90
punishment of, 80–83
Cruel and unusual punishment, 3
banned, 81, 86
defined, 243
electrical execution as, 173, 211
Cruelty, 86
contagiousness of, 88–89
Current, 12–13, 14, 27
and body systems' balance, 44
danger of, 142*n*
high-voltage, 140–42, 208
physiological effects of, 146
and resistance, 183
types of, 273–74
Curtis, N. M., 242
Czolgosz, Leon, 279
"Dangers of Electric Lighting, The" (Edison), 219–20
Davis, Edwin F., 249, 252, 280
Davy, Humphry, 12, 26
Day, S. Edwin, 173, 174, 210–11
Dead (the), reviving, 40–42
Deadly Parallel, The, 159
Dean, Charles, 63
Death
catalog of, 95–96
diagnosing, 40–41
electrical, 291
new spectacle of, 286–94
painless, instantaneous, 2, 61, 98, 180, 224, 278, 279
public fascination with, 96
uncertainties in definition of, 178–80, 258
Death(s) from electricity, 2–3, 9–10, 61, 69, 74–75, 140, 179–80, 215, 243, 290
alternating-current, 135, 139, 186, 211, 219
arc lighting accidents, 90–91
how and when occurred, 151
see also Accidental deaths; Killing with electricity
Death penalty, 292
abolished, 80–81, 284
constitutionality of, 283
debate over, 81–83, 84, 85, 89
doctors and, 152
humane, 89–90
morality of, 282
racial disparities, 283
see also Capital punishment
Death penalty commission, 92–93, 95–99, 116–17, 118, 120, 121, 123, 151, 152, 184, 285
report, 95, 96, 99, 118, 124–25, 151, 227, 294
Death row, 235, 259
Democracy in America (Tocqueville), 225
Deterrence, 82, 282
Detroit Young Men's Society, 17
Dickson, William, 270
Dime museums, 46, 96, 120, 291

Direct current, 14, 110, 111, 129, 186, 207, 268, 270, 273
advantages of, 113–14
alternating current's symbolic victory over, 275
animal experiments, 146
danger from, 150
in electrical execution, 152
rival to, 271
safety of, 141, 142
voltage, 114
Direct-current dynamos, 110, 113, 143
Direct-current generating plants, 113
Direct-current system(s), 203
Discharge, 8, 10, 11, 12
Dismemberment, 119
Dissection, 119, 122, 178
District of Columbia, 281, 282
Dog-killing cage, 94*f*
Dogs, recovering from shock, 176–78
see also Animal-killing experiments
Drawing and quartering, 81
Drexel, Morgan & Company, 30, 57, 72
Duplex telegraphy, 20
Durston, Charles, 225, 234, 238, 240–41, 247, 248–49, 250–51, 255–56
Durston, Gertrude, 236, 238, 240, 245
Dynamos, 27, 60, 69, 105, 123
alternating, 159, 160, 198, 203–4
building, 66
direct-current, 110, 113, 143
efficiency, 131, 193
for electrocution, 90, 91, 175, 257–58
Jumbo, 67–68, 70, 71, 72*f*
"long-legged Mary-Ann," 34, 35*f*, 38
see also Generators

Economic growth, toll in lives, 216
Edison, Charles, 128
Edison, Madeleine, 128
Edison, Marion (Dot), 26, 68, 125
Edison's relationship with, 101, 126
Edison, Mary Stilwell, 19, 26, 68, 127
illness, death, 100–101, 125
Edison, Mina Miller, 125–28, 127*f*, 200
Edison, Theodore, 128
Edison, Thomas Alva, 17–18, 17*f*, 23, 24*f*, 63*f*, 286–91, 287*f*
advice on electrocution, 260–61
advocacy of electrical execution, 3, 99
biography, 287
and Brown, 181
and Brown letters, 191–92, 194, 196
business dealings, 19–20, 101–4
business hurt by Westinghouse, 156
children, 26, 28, 128
deafness, 17, 186
death of, 286
and death of wife, 101
and dog-killing experiments, 197
education, 16–17
experiments on alternating current, 265–66
experiments in electrical killing, 125, 132–33, 142–47, 152–53, 158
as expert witness, 1–3, 124
flair for promotion, 115
and formation of General Electric, 268–70
headquarters in New York, 62
helping create better ways to kill, 202–3
as husband/father, 68
interview on electricity's dangers, 150–51
library, 129
loss of fortune, 270
meaning of business to, 205–6, 267
and name for electrical execution, 161
opposed capital punishment, 3, 116–17, 202, 287
in Paris, 200–202
personal characteristics, 19
physical appearance, 23
place in history, 287–88
and religion, 126–27
resentment of Westinghouse, 169–70
role in creation of electric chair, 287–88
role in lamp experiments, 55
second wife, 125–28
skills, 16
wealth, 104, 129

Edison, Thomas Alva (*continued*)
  and G. Westinghouse, 108–9, 114, 135–36, 188, 206–7, 288
  witness at Kemmler hearings, 174, 185–89
  working life, 16–20
Edison, Tom, Jr. (Dash), 26, 101
Edison, Will, 68, 101
Edison & Swan United Electric, 206
Edison Electric Light Company, 51, 53, 66, 72, 101, 103, 156, 170, 193
  alternating-current-for-executions experiments, 198–99
  Brown and, 148, 155–56, 159, 194, 197–98
  and Brown letters, 190, 191, 192, 195–96
  Bulletin, 69
  contest between Westinghouse Electric and, 181
  directors, 102
  Edison's policy involvement, 136
  Edison sold stock of, 63
  incorporated, 30
  investors, 51, 57
  losing ground to Westinghouse, 170–71
  officers, 103
  plot to discredit Westinghouse, 199
  product development for, 131
  rights to alternating-current system, 203–4
  suits with Westinghouse, 170, 206
  wires from, 249
Edison Electric Tube Company, 63, 129, 170
Edison General Electric, 170–71, 192, 265–66, 267
  merger with Thomson-Houston, 267–70
Edison Illuminating Company of New York, 57, 136–37, 223
"Edison Kinetoscopic Record of a Sneeze" (motion picture), 270
Edison Lamp Company, 63, 129, 131, 170
Edison Machine Works, 63, 65–66, 67, 70, 129, 131, 170, 171
Edison Manufacturing Company, 279
Edison Pioneers, 330*n*17
Edison system, 66, 67, 71–73, 74, 105, 134, 220
  copper conductors, 113
  efficiency, 135, 136
  electrical craze inspired by, 104
  expansion of, 171
  limited range of, 115
  promotion, 103
  safety issues, 69–70
Edison utility companies
  competition with Westinghouse, 265–66
Edison's Polyform, 302*n*10
Electric chair, 75, 171–72, 172*f*, 181, 182*f*, 257, 262, 263*n*, 278, 279, 280
  age of, 277–85
  designing, 226–30
  Edison's role in creating, 287–88
  first, 239*f*, 255*f*
  in Kemmler execution, 238, 239–40, 246, 247, 248–49, 250, 251, 252, 256
  number died in, 285
  in South, 280–81
  spectacle of death, 291–92
  switchboard, 249–50, 251, 252, 256, 262
  terror of, 294
Electric Chair, 292
Electric devices, 273
Electric Girl Illuminating Company, 104
Electric light, 25
  danger of, 69–70
  demand for, 115–16
  Edison's announcement of, 29–31, 34–35, 36, 49–50
  Edison's inventions, 40
  Edison's research on, 26–29
  Edison's work on, 31–39
Electric table (proposed), 228, 229*f*
Electric utilities, 205
Electric wire panic, 217–23, 254, 266, 288, 289, 290, 293
Electrical engineering, 277
Electrical execution, 139–40, 150, 202, 224
  alternating current for, 156–57, 198–99

combined with hanging, 153–54
Edison advocating, 3, 99
Kennelly's experiments in, 132–33
most humane, 116–17
in New York, 3, 288
opposition to, 173
plan for, 182*f*
proposals for machinery of, 226–30
questions about, 151–55
quick, clean, painless, 252
restraint in, 228
ruled not cruel, 243
technical issues, 183–84
term for, 160–62
Westinghouse's response to experiments in, 155–57
wire panic and, 218
see also Electrocution
Electrical execution law, 2–3, 118–25, 140, 152, 157–58, 160, 231, 243, 285
calls for repeal of, 261
challenge to, 210–11
constitutionality of, 167–68, 173, 233, 244
doubts about, 172–73
first man sentenced under, 163–67
gag law provision challenged, 232–33
opposition to, 124–25
Electrical experiments fad, 42–43
Electrical force, manipulating, 27–28
Electrical hut (proposed), 227*f*
Electrical industry, 205, 267
Electrical manufacturing market, 274–75
Electrical supply, universal system of, 274
Electricity
behavior of, 109
dangers of, 142, 143, 150
domestic dangers of, 290
Edison as expert witness on, 1–3
expansion of, 276
experience of, 59–60, 272–73
Franklin's experiments with, 9–10
method of execution, 90, 97, 98–99, 118
as movement of electrons in conductor, 27–28
mysterious force, 3, 123, 217, 272
regulation of, 148, 221, 232, 266, 290
study of, 5–12
stunning but not killing, 178–79, 226, 254, 278
traditions of, 40–43
see also Electrical execution; Killing with electricity
Electrochemistry, 12, 14, 110, 275
Electrocuting an Elephant (film), 279, 291
Electrocution(s), 257, 260–64, 277–79, 294
bungled, 261, 262, 263–64, 282, 337*n*13
constitutionality of, 174–89, 285
controversy surrounding, 282
cruelty of, 258–59
efficient, painless, 257
experimental nature of, 260
in films, 279
humane, 284, 287–88
legitimacy of, 291
in New York, 288
physiological effects of, 278
scientific foundations of, 180
spread of, 279–82
states adopting, 282
support for, 278–79
term, 161–62
see also Electrical execution
Electrocution commission, 225–26, 229, 230
Electrocution machinery, tests of, 225–26
Electrodes
attaching to body, 124, 133
liquid, 260–61
placement of, 153, 158, 171, 226–30, 239, 259, 261–62, 263, 263*f*
placement of: Kemmler execution, 250
Electroliers, 67, 72
Electromagnetism, 40
Electromagnets, 21–22, 27
Electrophysiology, 277–78
Electroplating, 14, 67, 110, 114
Elevators, 273
England
"Bloody Code," 80

English Bill of Rights, 81
English justice system, 75–76
Europe, 58, 95
  capital punishment, 81
  justice in, 75
  underground wires, 137
Ex parte Kemmler, 243, 258
Execution(s)
  by alternating current, 149–50
  audience for, 83
  as deterrence, 82
  effect on spectators, 88
  humane methods of, 284
  with hypodermic injection of drugs, 284–85
  legal machinery of, ground to halt, 283
  number of, 282
  painless and instantaneous, 278, 279
  privatized, 80, 122
  public ceremonies, 75–79, 80
  scripted, 75–76
  in South, 281
  see also Electrical execution
Execution methods, 75, 91, 97, 116–17, 118, 281, 285
  humane, 92, 116–17, 123
  information about, 93
  painless, 89
Execution of Czolgosz (film), 279, 291
Execution protocol, 291
Execution rituals, Kemmler, 237
Execution techniques, refining, 277–78
Experimentation, Edison's interest in, 18–19

Fabbri, Egisto, 30
Faraday, Michael, 12–13, 18, 27, 40, 110
Farmer, Moses, 21, 29
Feeder-and-main circuit, 52–53, 52*f*, 64, 73, 109
Feeks, John, 212–26, 214*f*, 217, 218, 219, 221, 254
Fell, George, 94, 98, 124, 151, 180, 225–26, 230, 241, 288
  electric chair, 239–40
  at Kemmler execution, 247, 250, 253, 254, 258
Fell Motor, 241, 254
Female Trouble (film), 291–92
Fibrillation, 151, 278
Fielding, Sarah, 86
Filament, 37, 53, 54, 55, 67, 131
  high-resistance, 51
  paper, 170, 206
Films, electrocution in, 291–92
Fire alarm systems, 15, 19, 137
Fires, 28, 66, 69
Firestone, Harvey, 286
Five-wire circuit, 203
Ford, Henry, 286
Fort Wayne Electric, 104
Fowler, Joseph, 91
Frankenstein (Shelley), 43
Frankfurt, Germany, electrical exhibition, 271–72
Franklin, Benjamin, 3, 9–10, 46, 90, 272
French Academy of Sciences, 202
Freud, Sigmund, 45
Furman v. Georgia, 283, 338*n*20

Gag law, 122, 123, 259, 260
  challenged, 232–33, 259
  press opposition to, 261
Gallows, 83, 97, 228, 291
Galvani, Luigi, 11, 41, 45
Galvinism, 42
Garfield, James, 57, 120
Garrote, 90, 97, 98, 119
Gas chamber, 281–82, 284
Gas lighting, 28, 50–51, 58
  dangers of, 66, 69, 205
Gas lighting companies, 50–51, 69
Gaulard, Lucien, 111–12
General Electric, 268–70
  competition with Westinghouse, 270–76
General faradization, 44
Generators, 11, 26, 27, 28, 34, 50, 62, 63, 66, 117
  direct-current, 14
  earliest, 110

Edison's work on, 65
efficient, 51
in electrical executions, 124
see also Dynamos
George III, King, 81
Georgia Supreme Court, 285
Germany, 42–43, 277
Gerry, Elbridge T., 92–93, 93*f*, 99, 118, 122, 123, 152, 160, 198, 288
fighting rescinding of gag law, 233
not at Kemmler execution, 247
witness at Kemmler hearings, 174, 179, 184–85, 189
Gibbeting, 80, 119
Gibbs, John, 112
Gilbert, William, 5
Glenmont, 128
Gold & Stock Telegraph Company, 19–20
Gould, Jay, 20
Government regulation, 203, 216, 221
Gramme machine, 204
Grant, Hugh, 215–16, 221
Gray, Elisha, 21
Gray, Stephen, 5–6
Great Depression, 282
Greeley, Horace, 81, 87
Green, Norvin, 30
Guillotin, Joseph Ignace, 160
Guillotine, 90, 97–98, 119, 202, 284
name, 160
Guiteau, Charles, 120

Hale, Matthew, 92–93, 99, 118, 184
Hammond, William A., 124
Hanging(s), 92, 97, 98, 179, 264, 285
abolished, 279
abolished in New York State, 2, 210
bungled, 84, 97, 264
combined with electrical execution, 153–54
in Great Britain, 284
humane, 83–84, 89–90
opposition to, 88–89
privatized, 83–84, 122
resuscitation after, 42
Hanging rituals, 75–80, 83, 84, 122
Hard rock mining, 270
Harrison, Benjamin, 237
Hastings, Frank, 191, 197–98, 199
Hawthorne, Julian, 161
Hayes, Rutherford B., 23
Healing, electricity in, 40–41
Hearing aid, 128
Helmholtz, Hermann von, 31–32
Henry, Joseph, 110
High-voltage current, 208
debates over, 140–42
High-voltage systems, 109–10, 115
Higginson, Henry, 268
Hill, David B., 92, 123, 140, 261, 285, 288
Hobbes, Thomas, 86
Holmes, Oliver Wendell, 44
Homicide rate, 284
Horror writing, 96
Houghton, Oscar A., 236, 240, 245
Houston, Edwin J., 104
Howells, William Dean, 294
Human body
effects of electricity on, 189
electrical properties of, 6–8
flow of electricity through, in electrocution, 263*f*
Human feeling, revolution in, 86–87
Human nature, 81–82, 86
Humane movement, 3, 87–88
Humane societies, 40
Humanitarian sensibility, 86–87, 88, 89, 280, 281
dark side of, 96
Huntley, Charles, 257
Hutcheson, Francis, 86

"Illuminated body," 95
Incandescence, 28, 29
Incandescent lamp, 29, 37–39, 38*f*, 48, 55–56, 272
burners, 36, 37
current for, 273
first, 1
first commercial installation of, 53–56
patent infringement suit, 267
patents, 170, 203

Incandescent light/lighting, 30–31, 34, 105, 109, 110
competition in, 134
Edison's demonstration of, 37–39, 49
Edison's monopoly in, 105
Edison's premature announcement of success of, 64
New York City, 136–37
obstacles to, 51
for *SS Columbia*, 53–56
Induction (principle), 110–12
Induction coils, 18, 45–46, 48, 60, 90, 111, 112, 127
Inductorium, 45, 46, 47*f*
Insulation, 6, 11, 36, 57, 64, 131, 137, 139, 176–77, 209, 290
Insull, Samuel, 170–71, 185, 265, 266, 269
Intermittent current
see Alternating current
Invention(s) (Edison), 2, 19–20, 21–23, 35–36, 40, 102, 267, 287
appropriated by Westinghouse, 169–70
control of, 269
Edison's lack of faith in, 290
electric light, 30–31
first patented, 19
Orange laboratory, 131
Investors, 50, 57, 63, 102
arc lighting system, 105
Edison Electric, 51, 57
Iron ore industry, 270
Isolated lighting plants, 66–67, 102, 104, 108, 109

Jablochkoff, Paul, 26, 27
James, William, 86
Johns Hopkins University, 193, 329*n*14
Johnson, Edward H., 23, 69–70, 103, 114, 115, 116, 135, 141, 169, 185, 192, 197–98, 199
opposition to alternating current, 204
resigned, 171
Jugiro, Shibuya, 258, 259, 260, 339*n*9
Jumbo dynamo, 67–68, 70, 71, 72*f*
Justice, new ideas about, 81
Justice system, 75–76

Kelvin, Lord, 52
Kemmler, William, 164–67, 165*f*, 171, 224, 225
appeals, 167–68, 173, 244, 290
appeals rejected, 210–11
conversion of, 235–37, 240
execution, 237–40, 244, 245–53, 251*f*, 265, 278
execution: possible sabotage in, 255–56
execution: problems in, 252–53, 254–56, 258, 277
execution: witnesses' assessment of, 256–58
execution delayed, 240–44
execution report, 257–58
sentence, 233, 234–235
Kemmler hearings on constitutionality of electrocution, 174–89, 190, 194, 196, 200, 203, 260, 291
Kennelly, Arthur E., 131–32, 132*f*, 142–43, 146, 147, 152, 153, 157, 158, 183, 194, 195, 198, 231, 259, 277, 288, 320*n*1, 337*n*13
dog experiments, 196–97, 199
and execution, 148–49, 260, 261
experiments on alternating current, 265, 266
experiments in electrical execution, 132–33, 186
not at Kemmler execution, 247
page from notebook of, 144*f*
shock, 150
support for electrical execution, 202, 278
Killing, humane, 92, 122, 292
see also Killing with electricity; Scientific killing
Killing experiments, 46–48, 90
see also Animal-killing experiments
Killing tests, Auburn prison, 225–26
Killing with electricity, 2–3, 40, 75, 90–91, 92, 94, 113, 122, 123, 124, 202
disagreement about, 189

experiments in, 125, 142–47, 152–53
technical requirements of, 152–55
Kinetograph, 270
Kinetoscope, 270
Kruesi, John, 31, 63

Labor violence, 121
Laboratory, Menlo Park, 2, 19, 20–22, 21*f*, 31–33, 37–39, 56
exhibiting electric light at, 37–39, 49
model of New York lighting station, 56–58, 65
staff, 31–32, 33*f*, 55
staff of, heading new companies, 63
Westinghouse at, 108–9
Laboratory, Orange River Valley, 128–32, 130*f*, 183–84, 188
animal-killing experiments, 152, 158, 178
archives, 195, 196, 197, 199
Brown and, 172–73
dynamo room, 129, 132, 145*f*
machine shop, 129
precision machine shop, 130
private experimental room, 130
staff, 131
stockroom, 130–31
Lacey, Reverend, 77, 78
Lathrop, Austin, 157–58, 230
Laudy, Louis, 225
Lethal injection, 284–85
Leuchter, Fred, 337*n*13
Leyden jar(s), 8, 10, 11, 41, 48, 90
charging, 11
experiment on Parisian monks, 60
Life
and death, 40
electricity and, 43, 147, 178, 217
Lightbulb, 1, 30–31, 34, 270
lawsuit defending, 170
Lighting business, 129
Lighting companies, 137–39, 148, 212
safety issue, 221–22
Lightning, 9, 10, 48, 123, 175
physiological effects of, 48
victims of, 41
Llewellyn Park, 128
London Inventions Exhibition, 112
Loppy, Martin, 261
Lynchings, 121, 280–81

MacDonald, Carlos F., 158, 225, 248, 257–58, 261–62, 288
report on executions, 259–60
McElvaine, Charles, 261–62, 262*f*
McKinley, William, 279
McMillan, Daniel, 92
McNaughton, Daniel, 234, 235, 236, 238, 244, 245
Machine, civilization of, 292
Magnetic field, 27
Magnetism, 6, 13
Magneto-electric generators (magnetos), 13–14, 13*f*, 42, 43, 44
Manhattan, 213, 288
central stations in, 129
electric light utility companies, 136–37
overhead wires, 148
underground wires, 289*f*
Manhattan, Lower, 52–53, 100
Manhattan Electric, 136
Manhattan Melodrama (film), 291
Manufacturing companies, 66, 101–2, 129, 170–71
financing, 62–63
merging with Edison Electric, 192
Massachusetts, 80, 86, 280
Maxim, Hiram, 55–56
Medical anesthesia, 87
Medical machinery, 45–46
Medical uses of electricity, 10, 40–41, 43–45
Menlo Park, New Jersey, 25, 26, 28, 100
see also Laboratory, Menlo Park
Mercantile Safe Deposit Company, 55
Meter(s), 63, 67, 114
alternating-current, 169
lack of, 114, 134
Meter patent(s), 169
Middle class, 86, 88
Miller, Lewis, 125, 126–27
Mob violence, 79, 121
Moleyns, Frederick De, 29

Morgan, J. P., 30, 66, 69, 72, 104, 268, 271
Morphine, 98, 99, 118, 123
Morse, Samuel F. B., 14–15, 40
Motion picture business, 270, 279
Motors, 114, 131
  alternating-current, 135, 270–71, 273
  industrial, 275
  lack of, 114, 134
Mt. Morris Electric, 136
Murder, 76–77, 85
  degrees of, 81
Musschenbroek, Pieter van, 8
Mutilation, 97, 119
  possibility of, in electrocution, 187, 194

National Association for the Advancement of Colored People, 281
  Legal Defense Fund, 283
Nebraska, 281, 285
Neurasthenia, 45
New Jersey, 80, 279–80
New Orleans, 115, 129
New York City, 209, 231
  accidental deaths, 136, 216–17
  arc street lights, 58, 105
  Board of Aldermen, 57–58
  Board of Electrical Control, 139, 141, 146, 215, 218
  Board of Health, 221
  central stations in, 58, 62, 102, 103
  Department of Public Works, 222
  Edison moved to, 62, 100
  hanging rituals, 79–80
  incandescent lighting, 136–37
  overhead wires, 138*f*, 139, 141, 148, 215–16
  Pearl Street district, 273–74
  plan to build commercial lighting station in, 50, 51–53, 56, 62, 63, 64–66, 70–74
New York Medico-Legal Society, 151–52, 154–55, 156–57, 158, 198, 228, 229
  Committee on Substitutes for Hanging, 90
New York State, 151
  abolished hanging, 167, 210
  abolished public executions, 80
  electrical execution, 150, 280, 288
  law on criminal execution with electricity, 2–3, 122–24, 210–11, 233, 261 (*see also* Electrical execution law)
  law requiring burial of wires, 212
  official executioner, 280
  Senate Committee on Laws, 231–32
  state prisons, 118, 120, 123, 158–59
New York State Assembly, 122–24
New York State Court of Appeals, 233, 235, 241–42, 243
New York State Legislature, 118, 137–39, 243
New York Supreme Court, 224, 225, 233
Newspapers, 69, 258
  gag law, 232–33, 259
  and hangings, 83, 84
  on Kemmler execution, 257
  reporting on executions, 122, 254
  represented at Kemmler execution, 247
  and telegraphy, 15
  and wire panic, 218
  see also Press
Niagara Falls, 271–72, 274–75

Ohio, 277, 280
Oklahoma, 281, 284
One-, two-, three-phase current, 273, 274
Orange River Valley, 128–32, 130*f*, 202
  see also Laboratory, Orange River Valley
Oregon Railway and Navigation Company, 53
Overhead wires, 64, 105, 137, 209
  danger from, 207, 210
  New York City, 138*f*, 139, 141, 148, 215–16
  removal of, 217–18, 221–23, 289

Pain, 80, 85, 86
  in electrocution, 282
  revulsion at, 87, 96
  from shocks, 8, 150–51, 176

Paris, 200–2, 288
Paris International Electrical Exposition, 67
Parisian monks, Leyden jar experiment on, 60
Park, Roswell, 166
Patent infringement, 203, 267
Patent suits/litigation, 2, 103, 170, 206, 207, 266, 267
Patents, 19, 22, 30, 101, 181, 270
  alternating-current motor, 135
  alternating system, 186
  electrocution-chair, 280
  incandescent lamp, 170
  licensing, 62
  paper-filament lamp, 206
  safety devices, 208
  stolen, 210
  Tesla, 270–71
  violations of, 135
  Westinghouse, 169
Pearl Street Station, 70–74, 70*f*, 100, 102
Pennsylvania, 80, 81, 281
Pennsylvania SPCA, 88, 89
Peterson, Frederick, 142–43, 146, 152, 178–79, 228, 320*n*1
Phase-convertors, 274
Philadelphia, 58, 129, 272
  SPCA, 87
Phonograph, 23, 24*f*, 25, 31, 36, 270
Physicians
  and death penalty, 152
  and electrocutions, 260, 264
  at Kemmler execution, 248, 258
Pittsburgh Reduction Company, 274
Platinum, 29–30, 31, 33–34, 36, 37
Poe, Edgar Allan, 43, 96
Poison, 90, 97, 98, 151, 264
Politics, and death penalty, 338*n*20
Poor (the), 88, 120
Pope, Franklin, 19, 35, 181–83, 195, 320*n*1
Poste, William, 2, 180, 185–86
Potassium, 12
Practical Treatise on the Medical and Surgical Uses of Electricity, A (Beard and Rockwell), 44
Premature burial, 178
"Premature Burial, The" (Poe), 43
Press, 50
  access to death chamber, 261
  and battle of currents, 149
  and Brown letters, 195–96
  on electrical execution law, 172–73
  gag clause, 122, 123, 260
  and Kemmler hearings, 174
  restrictions on, at executions, 119
  see also Newspapers
Priestley, Joseph, 41
Prime conductors, 8, 10
Prison discipline, 225
Product development, 131
Pulitzer, Joseph, 218
Punishment, 80–83, 237
  corporal, 80–81
  electricity in, 46
  see also Cruel and unusual punishment

Quadruplex telegraph, 20, 30
Quicklime, 119, 123, 255
Quinby, George, 238, 253

Railroad brake industry, 107–8, 207
Railways, 270, 274
  and telegraphy, 15
Reid, Mary, 163–64, 166
Researches in Electricity (Faraday), 18
Resistance, 28–29, 36, 37, 51, 109, 227–28, 278
  of human beings, 183–84, 260
  in Kemmler execution, 256
Resuscitation, 40–42, 254, 278
Richardson, Benjamin Ward, 48, 89, 90
Richardson, Samuel, 86
Richmann, Georg Wilhelm, 9
Richmond, Virginia, 230–31
Rockwell, A. D., 44, 158, 180, 225, 259
Rotary convertor, 274
Royal Society, 5, 12, 42
Royal SPCA, 87

Sachs, Bernard, 320*n*1
Safety devices, 53, 64, 67, 208, 209
Safety Electric, 136
Safety issues, 69–70, 75, 209, 221, 231–32, 290
  Edison's concern with, 203, 205, 219, 230–31, 232
  New York City, 217, 217*f*
  Westinghouse's intransigence on, 209–10
Safety regulations, 216
Scientific killing, 3, 89–94, 123, 294
  methods of, 264, 284
Screw socket, 67
Segredor, John, 54–55
Shelley, Mary, 43
Sherman, Roger M., 240, 241–42, 243, 258–59
Shocks, 8, 46, 59, 74, 142, 209
  accidental, 155, 178, 180
  effects of, 44
  in electrocution, 259
  killing with, 91
  nonfatal, 154
  pain from, 150–51
  survivors of, 176
Siemens, Werner, 112
Sing Sing prison, 118, 158, 241, 259
  executions, 260, 261, 278
Skaggs, John, 42
Slocum, James, 259
Smiler, Harris, 259–60
Smith, Carpenter, 176
Smith, Lemuel, 60–61, 69, 91, 163, 248
Social reforms, 86–87
Society for the Prevention of Cruelty to Animals (SPCAs), 87–88, 94, 97, 98
Society for the Prevention of Cruelty to Children, 93
Sockets, 63, 66–67
Sodium, 12
Sounders, 15
Southwick, Alfred Porter, 91–92, 91*f*, 93–94, 95, 97, 98, 99, 113, 116–17, 118, 124, 163, 184, 288
  and electric chair, 226–27
  and electrocution, 260–61
  at Kemmler execution, 238, 247, 252, 254, 256–57
Spanish Inquisition, 95
Spitzka, Edward A., 278–79, 280, 337*n*13
Spitzka, Edward C., 248, 250, 252, 253, 255, 257, 258, 278
SS *Columbia*, 53–54, 56, 66
Stanley, Henry M., 202
Stanley, William, 109, 110, 112–13
Starr, J. W., 29
States
  abolished public executions, 80
  adopting electric chair, 279–82
  capital punishment, 282
  capital punishment statutes invalidated, 283
  ending death penalty, 82–83
Static, 12
Stedman, Edmund Clarence, 84
Stickley, Gustav, 230*n*
Stock and gold quotation systems, 15, 19, 137
Storage battery electric lighting system, 110, 205, 287
Storti, Luigi, 277
Strang, Jesse, 77–78, 88, 237
Streetcars, 273, 275
Suffering, 87, 96, 98, 280
  in death penalty debates, 84, 85, 89
  eliminated with electrical execution, 123
  social consequences of, 89
  spectacle of, 92, 98
Swan, Joseph, 205
Switches, 63, 67

Tammany Hall, 57, 64, 70–71
Tate, Alfred, 269
Taylor, William, 263–64
Telegraph, 17–18, 30, 40, 128
Telegraph printer, 19
Telegraph services, 19, 64, 137
Telegraphy, 14–15, 18–20, 45, 50
Telephone, 21–22, 36, 128, 137
Tesla, Nikola, 135, 270–71, 274

Tesla Polyphase System, 274
Texas, 281, 284
Thermal regulator, 29–30, 31
Thomas A. Edison Construction Department, 102–3
Thomson, Elihu, 104, 169, 195, 207–9, 208*f*
Thomson-Houston Electric Company, 104, 134, 159
- alternating generators, 193–94
- alternating system, 207, 208
- and Brown letters, 190, 195–96
- charter revision, 169
- competition, 266
- merger with Edison General, 267–70
- patent violations, 135
- Westinghouse dynamos from, 192–93, 195

Three-wire distribution system, 109, 203
Tocqueville, Alexis de, 225
Tombs, 83
Transformer, 111*f*, 112, 113, 114, 265–66, 276
Transmission, 111–12, 113, 271, 276
Tree circuit, 51, 52
Tubes, 63
Tupper, Charles, 176–77
Twain, Mark, 293

Underground conductors, 63, 64–65, 65*f*, 69, 71, 105
Underground wires, 56–57, 69, 137–39, 149*f*, 209, 223, 288
- Manhattan, 289*f*

United States, 79
- capital punishment in, 284
- physical punishment in, 80–81

United States Electric Lighting Company, 55–56, 104, 105, 170
United States Illuminating Company, 136, 139, 148, 207, 216, 218, 223
Universal Exposition (Paris), 26, 200–2
- Edison exhibit, 201–2, 201*f*

Universal stock printer, 19–20
Upton, Francis, 31–32, 55, 63, 103, 134
U.S. Circuit Court, 206
U.S. Patent Office, 135, 169
U.S. Supreme Court, 121, 206, 234, 243, 258–59, 283
U.S. Telegraph, 15

Vacuum pumps, 34
Vanderbilt, Cornelius, 66, 69
Vanderbilt, William H., 30
Velling, Joseph, 245–46, 250
Villard, Henry, 53, 170, 230, 267, 269
Violence, 79, 85, 121
- public fascination with, 96

Volta, Alessandro, 11–12, 40
Voltage, 28, 51, 52, 74
- in alternating current, 114, 276
- boosting, 109–10
- deadly, 74–75
- Edison's call for limits to, 220–21
- in execution, 124, 152, 153, 155, 183, 257
- high, 137, 205
- high-to-low/low-to-high, 111, 112, 113
- physiological effects of, 278

Voltaic pile, 11–12
Vote recorder, 19

Wallace, William, 28, 29, 34
Wanamaker, John, 59
Warhol, Andy, 292
Warning from the Edison Electric Light Company, A, 135, 141
Waters, John, 291
Weinhold, Karl August, 42–43
Wesley, John, 10
West, Alfred, 175
Western Electric Manufacturing Company, 21, 140
Western Union Telegraph Company, 15, 18, 19, 20, 30, 212, 215, 247
Westinghouse, George, 106–9, 107*f*, 110, 114, 117, 134, 140,157, 163
- and alternating current, 112, 113, 141, 276, 290
- Brown's participation in plot against, 196–99

Westinghouse, George (*continued*)
competition from, 131
and Edison, 135–36, 188, 206–7, 288
and Edison's help in electrical execution, 204
and electrocution, 291
and execution law, 168–69
on Kemmler execution, 257
and Kemmler execution delay, 241, 242–43
marketing, 115–16
patents, 135
response to electric execution experiments, 155–57
safety issue, 209–10, 216, 221, 232
and theft of Brown's letters, 195
universal system of electrical supply, 274
unscrupulous businessman, 169, 170
Westinghouse Air Brake Corporation, 108
Westinghouse dynamo, 186, 198, 204, 329*n*14
in executions, 159, 168, 182–83, 192–94, 195, 199, 203, 241, 243
in Kemmler execution, 249–50, 251, 253, 256
Westinghouse Electric, 113, 135–36, 156, 207, 243
alternating system, 208
competition, 266
competition with General Electric, 270–76
contest between Edison Electric and, 181
and danger of alternating current, 218
objections to use of their machines in executions, 182–83, 194
patent infringement suit against, 267
reorganization, 267
suit against Edison, 206
underground wires, 212, 288
Wheatstone, Charles, 14, 183
Wheatstone bridge, 183
Wheeler, Schuyler S., 320*n*1
Whitman, Walt, 82
Witnesses to executions, 118–19, 123, 259, 261
Kemmler execution, 238, 246, 247–49, 253, 256–57
Women, execution of, 97
Wood, Joseph, 259, 260
Working class, 79, 120–21
World's Columbian Exposition, 272–74, 275*f*
Wyoming Territory, 24–25

Yates, Horatio, 236

Ziegler, Tillie, 164–65, 165*f*, 166, 234, 235